Sex and Subjection

Sex and Subjection

Attitudes to Women in Early-Modern Society

Margaret R. Sommerville

A member of the Hodder Headline Group
LONDON • NEW YORK • SYDNEY • AUCKLAND

First published in Great Britain in 1995 by
Arnold, a division of Hodder Headline PLC
338 Euston Road, London NW1 3BH
175 Fifth Avenue, New York, NY 10010

Distributed exclusively in the USA by
St Martin's Press Inc.,
175 Fifth Avenue,
New York, NY 10010

British Library Cataloguing in Publication Data
A catalogue entry for this book is available from the British Library

Library of Congress Cataloguing-in-Publication Data
Sommerville, Margaret R., 1956–
   Sex and subjection : attitudes to women in early-modern society /
Margaret R. Sommerville.
      p.   cm.
   Includes bibliographical references and index.
   ISBN 0–340–64573–3. – ISBN 0–340–64574–1 (pbk.)
   1. Sex role–Europe–History.   2. Women–Europe–Social
conditions.   3. Sex discrimination against women–Europe–History.
4. Man-woman relationships–Europe–History.   I. Title.
HQ1075.5.E85S66   1995
305.3'094–dc20   95–14733
                                                                CIP

ISBN 0 340 64574 1(Pb)
ISBN 0 340 64573 3 (Hb)

1 2 3 4 5   95 96 97 98 99

Filmset by Wearset, Boldon, Tyne and Wear
Printed and bound in Great Britain by
J W Arrowsmith Ltd, Bristol

Contents

Preface and acknowledgements

This book would never have appeared without the help of a great many friends and scholars. Space permits me to thank only a few of them here, though particular debts on many points are recorded in the endnotes. I am indebted to the staffs of the Folger Shakespeare Library, Washington DC, and the Memorial Library of the University of Wisconsin at Madison – and in particular to Ed Duesterhoeft – for their unfailing courtesy and efficiency. John Morrill, Quentin Skinner, and Richard Tuck long ago struggled manfully to teach me basic historical techniques, and deserve credit for any merits this study possesses. Without David McDonald's and Patricia Springborg's criticisms of earlier versions of this work, the book would be far less coherent and comprehensible than it is; and without their encouragement it would never have reached the press. All the errors and misconceptions that remain are, of course, my own.

The spelling and punctuation of all English quotations have been modernized and contractions silently expanded. French and Latin quotations are unchanged except that, where appropriate, the letter i has been substituted for j, and some contractions expanded.

Who peyntede the leon, tel me who?
By God, if wommen hadde writen stories,
As clerkes han withinne hire oratories,
They wolde han writen of men moore wikkednesse
Than al the mark of Adam may redresse.

1

Introduction

This book describes the beliefs expressed about women by moral and political theorists of the sixteenth and seventeenth centuries. It concentrates upon the period from 1500 to 1700 for a number of reasons. First, these centuries witnessed the Renaissance and the Reformation – movements of profound importance in intellectual history. Second, this was a time of important social and economic change. Third, it was during these centuries that more women began to write and publish, and historians of women have detected the beginning of feminist thought in this period. Finally, the early-modern period saw the development of political theories that even today influence the way most people regard the nature of the state and the political obligations of citizens. In each of these cases it is simply impossible to understand the impact of social and intellectual revolution on attitudes to women without a sensitive and historical analysis of the standard male view of what women were (or should be).

Major claims have been made about the impact of the Renaissance on the position of women. The historiographical tradition that assumed (almost without examination) that humanism was a 'good thing' for males and therefore must have been good for females has been forcefully attacked by historians of women. They argue that rekindled interest in misogynistic classical literature increased hostility to women, that the growing importance of centralizing governments imposed a new division between public and private life which limited women's opportunities and horizons, and that a new stress on female chastity further restricted their freedom.[1] The assumption that the Protestant Reformation must have benefited women has also been subject to attack. It is now often argued that Luther, Calvin, and the other reformed divines were no better than their Roman Catholic predecessors (and quite possibly worse) when it came to women's status and their role in the church.[2] Doubts that Protestantism advanced women's cause have led historians to reconsider the assumption that Puritanism took it a stage further.[3] Sociological hypotheses have enticed us to regard Puritans as the panders of nascent capitalism, the prophets of the newly developing 'affective' family, or the apostles of the liberal bourgeoisie.[4] There may be some truth in any or all of these assertions but it cannot be established without careful study of where Puritans diverged from inherited orthodoxy.

This book deals with the problem of the impact of the Renaissance and Reformation by treating together the theoretical writings on woman of Catholics and Protestants, humanists and scholastics, and Englishmen and Continental Europeans. It will become apparent in the following pages that these theorists did *not* differ significantly from one another in their views of the nature and role of women. They shared a common intellectual heritage, read largely the same books, and respected the same authorities. Theological disputes between Catholics and Protestants, disagreements on philology, and variations in law and government between England and other European states, only rarely led to different opinions about the nature and position of women. This study concentrates on English protestant works of the sixteenth and seventeenth centuries, but there were very few issues on which these diverged significantly from medieval, Roman Catholic and Continental European theory. Similarly, puritan preachers generally just repeated ancient platitudes. Much of puritan teaching on women and the family was utterly conventional. Only by remaining wilfully ignorant of the Latin language, the classics, Scripture, the Church Fathers, medieval scholasticism, and writing on the European Continent can anyone make sweeping claims for the novelty of the early-modern English Protestants' view of women.

The sixteenth and seventeenth centuries were a time of profound social and economic change. How far the early-modern period saw the establishment of capitalism and the first steps towards the Industrial Revolution is a matter of debate, but that Europe's population increased enormously, that agricultural methods changed greatly, and that urbanization reached new levels cannot be doubted. These changes affected women's lives as well as males'; did they also affect theory about women? Some historians have argued that the growth of capitalism caused women to lose their earlier roles in economic production and to take on 'new and subordinated roles in economic consumption'. In turn, they maintain, this led to a new ideology in which 'seventeenth-century men consistently denigrated what they left to women of their traditional work'. Others have seen the sixteenth century as a period of 'particularly intense' 'gender conflict' where new economic stresses on patriarchal control led to 'an altered ideology of gender and work'.[5] Early-modern theorists *did* insist that woman's place was in the home, that her proper role was the reproduction and nurture of children, and that husbands and fathers should be the primary providers for their families and should control their families' goods. Yet the pages which follow show that these were far from being distinctively 'bourgeois' or 'capitalist' ideologies; virtually all classical and medieval theorists had argued in exactly the same way and early-modern theorists were consciously following them. Sixteenth- and seventeenth- century theorists repeated platitudes about woman's role inherited from antiquity and the Middle Ages, but they also placed significant qualifications and limitations on them. This study shows that the exceptions they made to traditional rules are frequently more revealing than their repetition of them.

The problem of how attitudes to women changed between 1500 and 1700 is an important one, to which I shall refer in passing at many points in the pages that follow, and return in my conclusion. The undermining of the clerical monopoly of learning that the Renaissance signalled, and the Reformation confirmed, did alter views about the female sex, sexuality, and sexual relations. The intense political debates sparked by the establishment of centralized states, and contests over who should rule them, had repercussions for theories of society and the family, and women's place within these. But change came very gradually. The roles and relationships of wives, mothers, and daughters were fundamental to society. Patterns of thought and conduct about them, established over hundreds of years, were not altered as readily and rapidly as forms of government, regimes, or dynasties. The chapters that follow – particularly the opening ones – bear testimony to the resilience and stability of certain dogmas about women and their nature throughout the early-modern period. Indeed, some of the beliefs described in the following pages (for example, that women are more emotional and less rational than males) have arguably still not been abandoned in many quarters.

In the course of the early-modern period, the spread of literacy and the cheapness of printed books meant that the laity wrote and read enormously more than in the Middle Ages. Almost for the first time, books were written for and by women. Amongst this output, some historians have detected the first stirrings of feminism.[6] But we cannot identify, let alone understand, feminist theory except by first describing the traditional theory of women. It is meaningless to assert that beliefs such as that male and female are equally made in the image of God, that they are spiritually equal, or that women should be treated fairly are 'feminist', when virtually every European Christian (no matter how determined a supporter of male superiority in secular spheres) had always maintained this. This book is not a work of women's history. Women do not find their voices in these pages. It is a study of males' attitudes to women.[7] In early-modern Europe, a few women did overcome the obstacles that deterred most of their sex from writing and publishing about their position in society, but their works will not be considered here. I shall not attempt to reveal the roots of feminism.[8] Instead, this book outlines the views to which feminists were responding; without understanding these, we cannot grasp what feminists wrote. It describes the conventional male views of women, their nature and their role. More specifically, this work is a study of *some* males' attitudes to women: males belonging to the educated élite. I outline theoretical attitudes to women in 'official culture'. Theologians, moralists, lawyers, and doctors were a tiny minority, but their views are the best evidenced. As the privileged experts of the day, they had the time and inclination to expound their views and the greatest access to pen and press.

Because I am interested in setting out the ordinary and commonplace views of woman and her role, I do not attempt to describe the views of radical, unrepresentative minorities except where their writings, or the responses to them, throw light on orthodox views. Quakers and Ranters, tub-preachers and inspired prophets, held beliefs significantly different from those opinions that I shall often (loosely) describe as '*the* early-modern' view. Their tracts show what divergence from the standard range of views was possible within the early-modern intellectual framework of thought, but they were marginal and generally regarded as such.[9] For similar reasons, I concentrate upon theories expounded in works of religious, ethical, and political theory, aimed at informing and educating, rather than entertaining. I say little about rhetorical exercises. When Renaissance rhetoricians debated woman's character, their main purpose was to amuse; they accepted the conventions of literary genres and employed conventional arguments. The rhetorical debate about womankind, which Professor Woodbridge has illuminatingly called 'the formal controversy', revealed many of the same attitudes to women as serious works of theory, but it did not seriously affect the status of women in official culture, nor does it seem that it was intended to do so. Similarly, the central focus of this study is not fiction. This study does not analyse literary conventions or dramatic genres. John Milton's political writings are cited frequently, his poetry only rarely. Although the seam is rich, Shakespeare's plays are not mined for references that elucidate early-modern attitudes to women. The role of woman as perceived by dramatists and poets has been expounded by many better qualified than myself. In this study, I aim rather to describe commonplace assumptions that provided the background to literary expression.[10]

The sixteenth and seventeenth centuries are of particular importance in the history of political thought because social contract theories of government (both Hobbes' absolutist and Locke's liberal versions) remain influential on modern political philosophy.[11] Recently, a great deal has been written about the place of women within early-modern contract theory. These analyses often give the impression that sixteenth- and seventeenth-century theorists were motivated by an unspoken (perhaps indeed unconscious) desire to oppress women and so were incoherent or even hypocritical in their theoretical treatment of them.[12] In fact, women were not at the centre of early-modern political philosophers' debates and concerns; theorists usually just applied to the case of women the principles they held about political power and obligation in general. It is easy to see early-modern theory as inconsistent and women as an anomalous exception if we look at the 'great texts' of political thought without understanding the ideological milieu in which they were written.

Hobbes, Locke, and other theorists shared most of their contemporaries' beliefs about biology, sociology, and anthropology, even as they developed novel political theories. They did not write in splendid isolation and we cannot understand these works torn from their intellectual setting. This book analyses early-modern theorists' attitudes to women in the context of their whole system of beliefs. It shows that most of what early-modern political philosophers said about women and their role in the state stemmed perfectly logically from the convergence of their views on woman's nature with their beliefs about mankind, government, and political power.[13]

Dealing with 'great' political and moral theorists in their context involves citing and analysing extremely 'minor' writers. Not all early-modern theorists' works should be treated as of equal weight and influence. Anyone wishing to understand how orthodoxy was established, traditions transmitted, and propaganda internalized must deal with the relative importance of moral and political theorists. Clearly, Thomas Hobbes' writings exercised more influence than an anonymous broadsheet. Cardinal Robert Bellarmine's pronouncements published under the Church's *imprimatur* in multiple expensive folio volumes were more authoritative than the wedding sermon of a country parson printed for a few friends of the bride and groom. Nonetheless, I often consider the opinions of Hobbes and Bellarmine alongside those of unknown pamphleteers and obscure curates. The great and the good did not live on another planet. All lived in the same society and shared – to a large extent – the same intellectual milieu. Not all these thinkers were equally important, but it *is* important that virtually all of them concurred on many aspects of the role of women. And it is important that when they differed, the lines of divide cannot be drawn between 'great' thinkers and 'minor' ones.

The primrose path to 'cultural imperialism' is avoided with even more difficulty when historians study their own past than when they investigate wholly foreign civilizations. It is all too easy to judge on our own terms arguments written in terminology we think we understand. When modern analysts look at the sixteenth and seventeenth centuries through the prism of current Western attitudes, they may find it tempting to suppose that only the desire to exploit or 'sexploit' women could have impelled early-modern thinkers to argue for woman's inferiority in the face of the obvious facts. In this book, I try to suspend judgement on whether the 'facts' were 'obvious', and recognize that all facts are 'theory-laden' in the case of my English-speaking, highly educated, intellectual forbears, just as in the writings of any alien culture. It is not inherently impossible that early-modern theorists listened to their teachers and preachers, read the books then available, looked at the world around them, considered the evidence, and sincerely decided that, by nature, women were physically weaker, less intelligent, and more emotional than males. In this book, I attempt to acknowledge this. I do so (not because I am certain that early-modern theorists sincerely believed in female inferiority) but because I believe that intellectual history must take as its starting point what people said they thought, not what we now believe they should have thought.

This book attempts to be historical, also, in employing the terminology of early-modern theory. This is part of a conscious attempt to deal only in concepts that were available to these writers. John Locke might have written better political theory if he had used the concept of gender on many of the occasions he employed that of 'sex'.[14] Isaac Newton might have written better physics if he had grasped Einstein's concept of relativity. Neither did so. The political philosopher can criticize Locke for his failure to perceive that 'gender' is a superior category of analysis, just as the physicist can criticize Newton's *Principia* for its errors. The historian who aims to describe what Locke or Newton thought must work with the categories and concepts available to them. For this reason, my text flouts the shibboleths of liberated language. This is aimed not at shocking, but at describing the views of the authors as accurately as possible. Where their views were ambiguous or equivocal because of (say) the dual meaning of 'man', an accurate portrayal must preserve these two senses. Throughout what I write, therefore, 'mankind', 'man', and 'men', include all human beings – male and female. On those occasions where it is important that there should be no ambiguity, I term as 'male(s)' those men who are not women. Similarly, women are a sex and not a gender. Where a statement about the female sex is clearly meant to apply to a social role, rather than to all those people who have female genitalia, I use the term appropriate to that role – 'wife', 'mother', 'queen', and so on.

The book is a study of the sixteenth and seventeenth centuries, but it is littered with references to the Bible, the classics, the Fathers, and medieval scholastics. These authors are interpreted as meaning what early-modern theorists believed them to have meant, although this sometimes involves mangling the original authors' views in a way that would turn them in their graves and sicken modern biblical, classical, and medieval scholars. Likewise, I accept early-modern theorists' beliefs on the authorship of texts. For the purposes of this work, for example, Moses wrote the first five books of the Bible, the Evangelists penned the Gospels, and Saint Paul composed all the Pauline Epistles. The same caveat applies to my treatment of early-modern scientific, historical, and anthropological opinions. The risible views of early-modern doctors on human physiology are treated seriously. Early-modern ignorance of the diverse possible forms of animal and human communities is not questioned. Caricatures of Jewish and Islamic faith and conduct are reported verbatim without the qualifications that a historian of the Mediterranean world would demand.

Notes

1. These assertions can be found implicitly or explicitly in almost all modern writing on the impact of the Renaissance on women, but the best statement of them remains Joan Kelly's seminal article, 'Did women have a Renaissance?' (Kelly 1984, 19–50).
2. For example, Greaves 1985 (especially Blaisdell, 35 and Greaves 3, therein).
3. One of the foremost scholars of Puritanism has argued defensively that we should not 'consign the question of puritanism's impact on the experience of women or on relations between the sexes to the historiographical scrap-heap' and accepts that it is questionable whether Puritans did possess 'a novel or distinctive set of attitudes to domestic relations' (Lake 1987, 143, 159). For the traditional view that Puritanism improved women's status, see Wright 1935, especially 202–26 and Schuecking 1969, especially 45–7.
4. The belief that Protestantism in general, and Puritanism in particular, established a new view of domestic and social relations that prepared the ground for capitalism was first argued forcefully by Weber (1930) and his views still influence the assumptions of many more recent historical works. It is, for example, implicit in much of Stone (1977).
5. Cahn 1987, 4, 121; Honeyman and Goodman 1991, 608, 613.
6. See, for example, Butler 1978, Nadelhaft 1982, Smith 1982, and Jordan 1990. One of the earliest and best expositions of the belief that the political upheavals of the seventeenth century influenced thought on women and the family can be found in Thomas 1958.
7. Patrick Collinson (1988, 83) has rightly insisted that '"women's history" is often nothing of the sort but rather the history of male attitudes and behaviour towards women'; I hope to avert such criticism by disclaiming at the outset any pretension to write women's history. However, Tilly (1989, 447) has argued that the 'best women's history ... endeavors to relate [women's] lives to other historical themes, such as the power of ideas or the forces of structural change', and my study does aspire to do this.
8. Jordan 1990 and Smith 1982 provide two of the best introductions to early-modern feminist thought.
9. Mack 1992 provides a scholarly and nuanced discussion of Quaker women.
10. Woodbridge 1984, 6. This work provides a brilliantly stimulating and thorough introduction to the place of women in literature in Renaissance England. See also Dusinberre 1975.
11. For instance, Rawls 1971 and Nozick 1974.
12. Clark 1977, Okin 1979, Nicholson 1986, and Pateman 1988 are outstanding examples.
13. For a complete account of the methodology to which I aspire, see Tully 1988, 29–132.
14. For an illuminating exposition of the concept of gender, see Scott 1986.

2

The basis of subjection

The inferior sex

During the sixteenth and seventeenth centuries, Europe changed profoundly. The power of central governments increased as nation-states emerged. Voyages of exploration to Africa, Asia, and the New World opened vast new territories to European exploitation. The Reformation divided Catholic from Protestant and led to religious wars not only between but within states. Unprecedented population growth and changes in agricultural methods fundamentally altered social and economic structures. By 1700, Europe was very different politically, socially, and economically from what it had been in 1500. The intellectual world also changed. During the Renaissance the revival of classical study led men to adopt new ways of thinking about the old problems of philosophy and science. In their political theory, early-modern Europeans employed Renaissance learning in an attempt to understand political change and its impact on their lives. They did not forget or abandon classical and medieval theories but modified and adapted them to respond to new challenges. As the power of governments grew, political theorists debated the proper role of government, the legitimate extent of its power, and the duties and rights of the individual towards it.

Some theorists saw government as receiving its power directly from God, 'for there is no power but of God: the powers that be are ordained of God' (Romans 13:1). Sovereigns ruled by divine right and were answerable to God alone for the way they used the absolute power He had given them. Mere men could not dispute the prince's power nor call him or her to task for the way (s)he used it. Active resistance to the prince was always immoral; disobedience was only legitimate when the prince commanded actions against God's laws, and even then the subjects must meekly accept any punishment imposed by the prince for their disobedience. Many absolutist theorists accepted that the people nominated their ruler but still insisted that his or her power was no less God-given and irresistible. One group of absolutists may be called patriarchalists because they argued that political power could best be understood as fatherly power. Children did not choose their father since they were born in subjection, but they were naturally obliged to obey him. Because God commanded children to obey their parents, paternal power was also divinely ordained.* Patriarchalists argued that political power was exactly like paternal power: people were born in subjection to their prince, whom they were bound by divine and natural law always to obey.[1]

Other political theorists attacked these beliefs. They, too, accepted that power was given by God but they argued that He had given power to the whole political community. The community appointed a ruler because of the difficulty of making every political decision democratically. 'A king is a thing men have made for their own sakes, for quietness sake. Just as in a family one man is appointed to buy the meat.' Governments received their powers (indirectly from God but) directly from the people. The political community transferred its power only on certain conditions; it made a contract with its prince. If the prince failed to abide by the terms of the contract, the people were released from their obligations. They need not obey when (s)he misused the power they had delegated and if (s)he attempted to compel them, they could resist force with force. This contractarian theory of government tended to allow the people more power against their prince than did patriarchalism, but there were also conservative contractarians who strictly limited the political community's right to take back its power.[2]

Patriarchalists and contractarians disagreed on some of the most fundamental questions of political philosophy, but they also shared some beliefs. They all believed that legitimate power stemmed ultimately from God and must be used in accordance with His laws. As most early-modern European theorists were Christians, they accepted that God's law was revealed in the Bible. They also believed God made binding natural laws. God was the creator of the world; He was the author of nature and its laws. Political systems would only operate well if they accorded with men's natural inclinations and abilities. It was essential to understand the nature of mankind in order to establish just and practical government. Contractarians and patriarchalists agreed that men were naturally unequal; some men were stronger, more intelligent, and more prudent than others. These inequalities must be considered when deciding who should govern, or government would not function efficiently. Wise men should rule, foolish ones should obey. However, agreeing that some men are wiser than others is a great deal easier than agreeing *which* men are wiser. Early-modern Europeans could not even agree that the 'barbarous' natives of Africa and the New World were their inferiors, let alone concur on which of their own number were superior. Yet consensus was achieved in one area.

It was axiomatic among all serious[3] early-modern philosophers, political theorists, and theologians that woman was inferior to the male in

---

* Honour thy father and thy mother' (Exodus 20:12).

many significant ways. She was held to be physically weaker, less intellectually capable, and less competent at controlling her emotions.
Renaissance thinkers were far from original in holding these beliefs.
Most early-modern thinkers did not aspire to develop novel moral and
political thought, but to apply traditional doctrine to contemporary problems. Two traditions were of overriding authority and importance: the
Bible (as interpreted by the Church Fathers and generations of theologians) and the moral philosophy of classical antiquity. By the sixteenth
century, classical morality cannot be separated from scriptural since each
had been interpreted in the light of the other for many centuries.
However, it is worth noting the differing sources, since tensions between
them still occasionally produced ambiguities and problems. Although
emphasis differed, the belief in female inferiority was one common to all
the classical, scriptural, patristic, and medieval authorities. So fundamental was the belief that women were inferior that it was rarely considered
necessary to provide arguments or evidence in its support. Female inferiority was simply obvious to everyone. It was more often asserted as an
axiom than defended as a proposition.

The nature of woman was discussed in serious early-modern literature
to determine the relevance – or irrelevance – of her natural inferiority to
particular philosophical, political, social, and ecclesiastical problems. It is
with these debates, rather than with *why* women were thought inferior,
that this study will be concerned. It is, however, useful to sketch out in
what respects this inferiority was believed to manifest itself. Early-modern theorists did not simply believe that all women were worse than all
males at everything. Rather they maintained that sexual difference led
the female to display different and unequal abilities, propensities, and
character traits. In the Renaissance, and beyond, it was accepted that
(even apart from obviously distinct reproductive roles) the sexes naturally differed in body, mind, and character.[4] Three interrelated beliefs were
particularly firmly rooted and widely held: first, that woman was physically weaker; second, that her intellect was not as developed; and third,
that her passions or emotions were stronger and less susceptible to the
control of reason. All three beliefs rested on the authority of ancient and
medieval thinkers. Early-modern theorists also believed that daily experience witnessed their truth.

Aristotle was the most influential of all classical philosophers. He
maintained that nature itself had designed the two sexes differently and
he described these differences between the sexes (both animal and
human). Aristotelian physiology characterized woman as a partly
formed male, 'as it were a mutilated male', whose whole system functioned at a lower metabolic level of efficiency: in his terms she was 'colder', 'moister', and 'more humid'. Medieval theorists translated, from
Greek into Latin, Aristotle's ideas on women and adopted them wholesale.[5] In their turn, Renaissance physicians accepted Aristotelian teaching

about the 'coldness' and weakness of women's bodies. One sixteenth-century French textbook on physiology, for example, argued that the female was 'colder' than the male and deduced, from this, women's greater needs and lower physical efficiency. An Italian sixteenth-century physician, Girolamo Mercuriale, called on medieval medicine to support his assertion that any particular male, however cold, was always hotter than even the hottest woman; it was their coldness that meant women were more often the culprits when a couple could not have children. A French midwife, Louyse Bourgeois, was arguing for the same conclusion when she asserted that 'the female sex is extremely humid and yet bilious'. Male attributes, such as amorousness, were often attributed to the fact that they were 'of hotter complexion'. English physicians adopted the same position – women were colder, moister, and weaker.[6]

During the Renaissance, some medical writers questioned the classical platitude that women were colder and the deduction that this resulted in a metabolism that functioned less efficiently.[7] But most writers continued to maintain that the male was naturally endowed with 'greater strength of body'. Nature 'disarmed' woman 'of corporal strength'. Early-modern thinkers generally agreed that woman, by nature, had a finer and more delicate body, while the male was stronger and more muscular, and that this was a natural, rather than an environmentally produced, characteristic. In harsh climatic conditions or in primitive societies where women were forced to perform hard manual labour, the differences in physical strength between the sexes were less apparent. Nonetheless, it was considered natural for woman to be a physically inferior specimen.[8]

Many classical writers had also taught that women were less intelligent than males; indeed, some had doubted whether women could truly be described as rational beings.[9] Although occasionally challenged, the belief that women were intellectually inferior to males was adopted by most Renaissance thinkers. A judge deciding a case in the King's Bench in 1566 asserted that women were rational creatures but (citing Aristotle) argued that the male was superior. Husbands should rule wives because 'men are for the greatest part more reasonable than women'. He clearly did not regard this as a contentious statement, but as the universally accepted maxim on which he could rest a judgement. In 1701, another theorist argued that women's intellectual abilities were 'naturally inferior to males'. He admitted that some women were possessed of an 'acute wit' but argued that in every field of endeavour – political, military, scientific, and artistic – women had achieved less. Mentally, just as physically, women were second-division material.[10]

This view was commonplace all over Europe throughout the sixteenth and seventeenth centuries. The Italian Catholic friar, Tommaso Campanella, adopted many heterodox views but agreed that husbands should rule wives because natural law made them generally wiser and stronger. A humanist author in the Netherlands noted that 'memory and inventive power' were more perfect in the male. A Milanese physician (a faithful neoscholastic) cited Aristotle that woman's 'intellectual actions' were inferior to the male's. The chaplain to Philip IV of Spain taught that, by nature, the male was more perspicacious, acute, and prudent; woman's mind was simply 'feebler' than the male's. The Jesuit commentator, Cornelius a Lapide, explained that the ministry was naturally as well as scripturally a male preserve, because males excelled women in judgement and in rational and critical discourse. The nonconformist Puritan, John Bunyan, was at odds with most of Lapide's theology, but stood shoulder to shoulder with him on this. The ministry should be 'performed by the most able'; no one could 'be so silly as to turn a company of weak women loose' for worship. Neither academic affiliation, nationality, nor religious denomination disturbed the tranquil agreement on woman's intellectual inadequacy.[11]

In early-modern theory woman's intellectual imperfection was seen as almost inevitably associated with poor judgement and poor self-control. Just as her reason was weaker so her emotions were stronger, and this combination produced a dangerous cocktail. Early-modern psychology postulated an internal division in each person between 'reason' and 'emotion'; people were thought to be living out an eternal battle between two halves of their nature. Reason was forever struggling to assert its supremacy over the baser passions. Since women were less reasonable, they lost the fight far more often than males did. One English puritan preacher bemoaned that just as women 'are weaker in understanding so they are stronger in passion'. Italian Catholic opinion was identical: women were less controlled by reason and more often overcome by their desire and appetite. Women were inferior because nature had loaded the scales against them, giving them stronger passions and weaker intellects than males.[12] Aristotle had argued that the female of any species was 'softer in character', 'more compassionate', 'more easily moved to tears', 'more jealous, more querulous, more apt to scold and to strike'. His view found acceptance amongst both medieval and early-modern theorists. In the seventeenth century, an Italian Catholic cited Aristotle and Seneca to prove that woman was more easily overcome by her affections. An English Protestant described women ('the inferior and weaker sex') as 'more subject to jealousies and groundless suspicions'. Others echoed the ancient commonplaces that women were more easily moved to anger, and that males were more firm in all affairs and 'more constant and of better persuasion in love'.[13]

Excessive passion entailed poor judgement; women were seen as less balanced and less moderate. Their inconstancy sprang from 'lack of prudence and judgement'. The male has more 'courage and authority' and 'more sharpness and quickness of wit, with greater insight and foresight than the woman'. One of woman's avowed defenders explained that 'the female sex is to be pardoned if their passions of love, anger, fear, and the rest are more predominant than man's' because Scripture 'does confirm common experience that their judgements are not so fixed, nor their reason so elated'. Afflicted with too much emotion and too little reason, women started with a handicap in life's race; therefore, special allowance had to be made for their poor performance. Husbands must show patience in the face of 'the natural imbecility of the female sex'. Most women were 'easily cast into anger or jealousy or discontent', and because of their 'weak understandings' were 'unable to reform themselves'. A woman was 'betwixt a man and a child'. The belief that women were 'childlike' was another commonplace of classical, medieval, and early-modern thought. Women's defenders attacked the 'abject, outcast kestrels' who indiscreetly revealed women's 'natural imperfections' but only because they spoke as if women could 'be rid of them'. Nature made women less perfect than males and there was nothing much they could do about it.[14]

Classical, medieval, and early-modern theorists regarded woman as inferior in physique, intelligence, judgement, and self-control. Since all these theorists viewed nature in teleological terms, refusing to see any aspect of the natural world as the product of random chance, the question immediately arose of why nature had produced such an inferior specimen. Aristotle provided part of the explanation by arguing that the female was simply a partially formed male; women were males arrested in an early stage of development. Nature stopped women developing before the point where they could emit semen and beget offspring so that they would be receptive to the male and incubate his.[15] Aristotle also argued that nature made females with another purpose than reproduction. In the case of humans, it was to provide the males with housekeepers. Ever attentive to male needs, nature had custom-built women to do all those trivial tasks in the home that made life comfortable. This allowed males to get on with the serious tasks like agriculture, commerce, politics, and philosophy without having to worry about looking after screaming children or cooking the dinner. In Aristotle's opinion, nature had designed males to accumulate goods by hunting and farming, and females to preserve them by careful housekeeping. Women were too weak for tough work outside and males too restless to sit and sew. Nature 'made one sex to lead a sedentary life and not strong enough to endure exposure, the other less adapted for quiet pursuits but well constituted for outdoor activities'. Aristotle ridiculed Plato's idea in the *Republic* that women should hold public office for if women were out working, 'the men will see to the fields, but who will see to the house?' Women were made to do the housework, males to do fieldwork; 'from the start their functions are divided'. Other classical authors, like Xenophon and Plutarch, also argued for a natural division of labour between the sexes.[16]

During the thirteenth century, Thomas Aquinas played a vital role in adapting Aristotle's thought to render it acceptable to Christian audiences;

and he readily accepted the natural, sexual division of labour. Unlike
other animals, humans mated not only in order to produce children but
also to share a common domestic life in which some tasks were
appropriate to the wife and others to the husband. Aristotle and Aquinas
bequeathed to early-modern theorists the belief that 'the woman should
take care of domestic affairs'. Nature had designed women to be house-
keepers and they should devote their energies to fulfilling this task properly.[17] Aristotle's characterization of woman as a natural housewife
proved equally acceptable in the Renaissance when expounded and
interpreted by a new generation of classical scholars. Orthodox theory
still held sway and Cardinal Cajetan agreed with Aristotle's *Economics*
that many tasks in domestic life were not only better done by the wife
but were simply unfitting for the husband. Males were no more suited
by nature for doing the dishes than women were for heavy physical
labour or intellectual endeavours. The classical truism was still generally
accepted: 'the wife hath sufficiently performed her duty, if she do safely
preserve and keep those things which her husband hath brought in'.
Even those who believed that the wife could and should work to supplement the family's income believed her labour should be performed within the home.[18]

The correlative of the belief that women were made to run the home
was that they were unsuited for work outside it. Their mental and physical handicaps rendered them unfit for public business, which might
moreover expose them to (possibly irresistible) attacks on their modesty
and chastity. Classical Greek authors had supported the notion that
respectable women should never leave the home. Another source for
Renaissance attitudes was Roman or civil law, and this, too, aimed to
restrict women to domestic activities. Civil law was administered by few
courts in England but many of its concepts were influential; Roman civilization was deeply admired and its laws carried authority. 'The force of
the Roman arms and extent of their empire made them formidable
throughout the whole world; but how much more were they so by the
wisdom of their laws?'[19] The Roman law tradition firmly upheld the
notion that woman's weakness, silliness, and emotionalism were all part
of nature's way of saying that she should stay at home and leave the difficult and demanding tasks of making money and law to males.
Medieval exponents of civil law accepted the exclusion of women from
all public offices.[20] Early-modern theorists did the same. The French civilian, Hugues Doneau, admitted that the law recognized women as able to
make legally binding actions and decisions; they were no less *sui iuris*
than men. But he argued that civil law was right in limiting some public
offices to males. Woman was excluded from these both by her ignorance
and by the indecency of a woman taking part in such public activities.
Other expositions of the civil law also accepted the propriety of these
restrictions.[21]

Most early-modern theorists accepted the natural division of labour between the sexes. All woman's defenders (and many of her defamers) agreed that she was very useful about the house. In the ages before consumer electronics, woman was the original, supremely adaptable, household gadget. *The batchelor's directory: being a treatise of the excellence of marriage* commended a wife to any potential husband as 'an admirable means to ease him in his domestic cares'. Few could have been found to dispute this opinion. When the monasteries were dissolved and celibacy abandoned, protestant clerics came to appreciate the usefulness of woman, the housekeeper. Marriage was a 'help' to ministers of religion, since it gave them 'one at home that careth for domestical affairs'. A married minister would have more time for study and contemplation, because a batchelor was obliged 'to play Martha's part' and continually run the household, 'which if he had a trusty yoke-fellow he need not to trouble himself withal, but give himself quietly over to his book'. Learned scholars, discussing the knotty problem of whether they should marry, given how foolish women were, readily conceded that women were thoroughly useful for taking care of domestic matters. The eminent humanist scholar, Isaac Casaubon, witnessed this from actual experience. He bemoaned when his wife was away:

> Deliver me, my heavenly Father, from these miseries, which the absence of my wife and the management of my household create for me. Not being used to keep our accounts, I am perfectly aghast when I see the expenditure of this family.

It was difficult both to be at the cutting edge of classical scholarship and to deal with tradesmen and household bills.[22]

Throughout the early-modern period, the sexual division of labour continued to be seen as stemming naturally from the particular strengths of males and weaknesses of females. English divines – puritan and Anglican alike – argued that the male's greater physical strength, higher intelligence, and better judgement fitted him for public business. Woman's physical weakness, natural tidiness, modesty, timidity, and mental dullness rendered her fit only to run the home.[23] It was deemed 'contrary to the law of nature' that women should handle commerce 'and commonwealth affairs, and men to keep within doors'.[24] The refrain that home was the proper sphere for woman echoes throughout the literature of Europe from Plato to the present day.[25] Of course, conceptions of what this involved varied over time. Many of those who later cited Aristotle would probably have been surprised (and quite possibly appalled) if they had been interpreted to mean (as Aristotle did) that women should be totally secluded from all males who were not close relatives, and strictly confined within the house. Theorists who commenced their advice with the recommendation that women should stay at home frequently proceeded to list numerous exceptions to the rule. Any young woman who went out 'every day with her mother and haunteth feasts and banquets' risked the loss of her 'maidenly modesty', but 'not to suffer a maid to go abroad but once or twice in a year is the way to make her become foolish, fearful, and out of countenance in company'. No one doubted that women's lives should be centred on domestic duties, but a woman 'who can do nothing but sow and spin' was deprecated as 'the very figure of a country milkmaid'. The recommendation that women should stay at home was frequently coupled with the complaint that, in practice, they did not do so and that many accepted and even encouraged this. Fathers should be 'more careful over daughters and servants, and not to employ them so commonly as they do in journeys and travail and solitary business'. Custom was far more important than theory in determining what 'staying at home' actually entailed.[26]

In north-western Europe, women were not strictly confined to the home like their counterparts in the Mediterranean basin. Studies of medieval England show that women's work was centred on the home, but that it was also commonplace for them to go to market and undertake seasonal agricultural labour in the fields. Similarly, one seventeenth-century historian observed 'that English women among the lower orders in [the Stuart] period ... enjoyed considerably more social freedom than do their counterparts in more modern Mediterranean peasant societies'.[27] Preachers often bemoaned that women hobnobbed with males at alehouses and dances. Theory postulated the ideal of a clear-cut division between women (active within the home) and males (active outside it), but practice was far more complicated. Nonetheless, throughout Europe, women's proper place was thought to be in their homes, and greater restrictions were placed on their freedom of movement and association than on that of males.[28]

Throughout the sixteenth and seventeenth centuries, sexual differences were believed to extend well beyond those physiological differences related to reproduction. The precise categorization of these differences varied from theorist to theorist and to some extent over time. Nevertheless, virtually every early-modern theorist believed that males and females were different in many important respects. Those differences made woman more fit for managing household affairs and for having children, and the male better equipped for virtually everything else. Physically, intellectually, and psychologically, woman was formed for specifically female functions; the sexual division of labour was not merely customary, it was natural.

Natural subjection

The major consequence that early-modern thinkers drew from woman's inferiority was that she should be in subjection to the male. In arguing for female subjection they consciously followed numerous earlier theorists who had reached the same conclusion. Once again, the obvious starting-point is Aristotle – for his biology and morality, and therefore his beliefs about the sexes, were enormously influential on later European thinkers. Aristotle had not asserted that sexual roles were separate but equal, but instead that 'the male is by nature superior, and the female inferior; and the one rules, and the other is ruled; this principle, of necessity, extends to all mankind'. This belief was in keeping with the rest of his thought. He held that natural differences should result in political inequality between males as well as between the sexes. In answer to the question of whether anyone is 'intended by nature to be a slave', Aristotle answered in the affirmative. He argued that 'from the hour of their birth, some are marked out for subjection, others for rule'. The best men were ruled by reasonable souls, 'but the lower sort are by nature slaves, and it is better for them as for all inferiors that they should be under the rule of a master'. Aristotle admitted that it was not always obvious which men were which, because 'the beauty of the soul is not seen'. Nonetheless, he was certain that some were born for rule and some for subjection.[29]

Aristotle's views on natural political inequality were not accepted in their entirety by medieval or early-modern political thinkers. A master's powers over a slave were very extensive; a slave could be used by an owner for his or her own purposes without regard to the slave's will or welfare. Even Aristotle's most faithful followers generally denied that *slavery* was natural. Both Christian and Stoic ideas were inimical to the notion that some men were created simply for the benefit of others. However, early-modern political theorists *did* accept that political differences should be based on natural ones. It was, they argued, both natural and moral that the more intelligent should rule the less, the stronger the weaker, and the braver the more cowardly. Thomas Aquinas followed Aristotle, in part, and maintained that there would have been political rule even if man had not fallen into sin. He gave two reasons for this: first, that there can be no society without some authority. Second, those with greater talents for rule should use their gifts to benefit others. Aquinas was in no doubt that there had always been natural distinctions of sex, age, and 'spiritual differences' regarding both justice and knowledge amongst men. Other medieval philosophers adopted the same position, denying Aristotle's thesis of natural slavery, but accepting natural political subordination. Natural inequality in mental powers, Duns Scotus reasoned, should not lead to 'extreme servitude' but should lead to political subordination.[30]

In the early-modern period, as in the medieval, even those theorists who believed that naturally all men were born politically free, accepted that there was a natural subordination of 'the foolish to the prudent'. They distinguished between natural and legal/political inferiority. Legal subjection arose when one person contracted to obey another, and it was quite possible for a wise man to be employed by a foolish one. In contrast, natural subjects were those unable properly to order their own affairs without another's guidance. It was obvious that someone lacking talent, shrewdness, and prudence was designed by nature to serve, 'since he knows neither how to rule himself nor others'. This was not slavery in Aristotle's sense, as it was for the subject's good not the ruler's. No one was made for the good of another man, nor could he be obliged to obey except by a legitimate judicial or legal act. All men were equal: 'The king, the subject, the lord, the slave, the infant recently born, the senile and decrepit old man, the stupid, the wise, the female, the male – all are equally human beings'. None was better or worse than another in respect of their common humanity, but some were naturally more able than others. And it was in this sense, early-modern theorists asserted, that Aristotle had stated that some were slaves by nature, that is of servile spirit and designed to obey rather than to rule. Although born free and equal *qua* human beings, the weak and stupid were naturally fated to subjection.[31]

The Bible taught that 'the fool shall be servant to the wise of heart' (Proverbs 11:29) and early-modern theorists cited this passage in support of Aristotelian élitism. A strong, wise man should not *impose* his rule on a weak, foolish one, but it would benefit both if it arose by consent. It was 'the intention of nature' that foolish people should be servants; the rule of the wise was a natural and commendable way of organizing human relations. The foolish 'cannot be compelled to serve', but society would function better if they did. Natural inferiority did not itself place someone in a position of political subjection but did indicate that (s)he should accept it. When dealing with wifely obedience, theorists referred to just this idea that some were born to liberty and others to service, and called on Aristotle's authority in their support. Aristotle's argument that some 'are made for servitude, others for direction' continued to be prevalent in the late seventeenth century. Theorists admitted that 'some have undertaken to censure Aristotle for saying that there are slaves by nature', but held that this condemnation was based on confusion. Natural inferiority did not justify 'the condition of the conquered under the lash of the victor'. However, anyone afflicted with 'impotency of the mind ... naturally requires the tuition of another'. Like their intellectual predecessors, theorists of the later seventeenth century were convinced that it was entirely unobjectionable to believe that those 'of the clearest merit ought naturally to preside over the more unworthy'. Meritocracy was a reasonable and natural form of government.[32]

Early-modern theorists bluntly dismissed as 'a fancy which in absurdity hath no equal' the suggestion that there could ever have been equality between 'fathers and children, husbands and wives, older and younger'. Inferiors would suffer if they did not submit to superiors; they must be subject for their 'own good'. Any community needed some hierarchy in order to function, and it was thought natural and reasonable that the older and the wiser should take precedence. Patriarchalists argued that inequality was as natural as government was necessary,

> for there has been always an inequality amongst men, and as we differ in strength and proportion or other natural abilities, so do we in reference to authority and subjection. Some are born to govern and some are born to be governed, and that by the prime laws of nature antecedent to any positive pact or contract between governors and the governed.

Superior beings would simply use power better than inferior ones.[33] The belief that it was entirely natural for the wiser to rule was adopted by contractarian theorists just as much as by patriarchalists. In the mid-seventeenth century, the Scottish Presbyterian, Samuel Rutherford, wrote forcefully against patriarchal theory. He argued that all political power arose from the consent of subjects. But Rutherford, too, accepted that subjection was perfectly natural when it resulted from greater abilities or from the wisdom that came with age. His belief that all were born free and equal with regard to political subjection did not stem from a belief that all were naturally equal in strength, intelligence, and judgement.[34]

In constructing his absolute state, Thomas Hobbes insisted on the approximate natural equality of all men in all respects, but his views were unusual. Even Hobbes only upheld natural equality in the very limited sense that almost any person (even the weakest) was capable of killing another (even the strongest).[35] Hobbes' ideas of natural equality were dismissed by many as 'evidently false'. The learned lunatic, Johannes Lyser, was often at odds with his contemporaries, but he was merely describing accepted views with exceptional vigour when he insisted that men varied greatly in intelligence – human intellectual abilities differed so greatly that there was often 'more difference between one man and another, than between man and animal'.[36] There was almost no trace in Renaissance thought of the modern notion that everyone is born approximately or potentially equal in ability, and that only environmental, social, or cultural factors make some significantly less able and accomplished than others. Early-modern thinkers believed that physical and mental differences were natural. Some men were more intelligent, prudent, and strong than others and it was natural for them to take positions of authority and rule. The weak and stupid could not be forced into subjection, but it would be better for everyone (including themselves), if they recognized this natural order of merit.[37]

Virtually everyone accepted that men were not equal in ability. The problem in creating an aristocracy of merit was that of obtaining agreement on who were the natural aristocrats and who ought to obey. As Thomas Hobbes commented acidly,

such is the nature of men, that howsoever they may acknowledge many others to be more witty, or more eloquent, or more learned; yet they will hardly
believe there be many so wise as themselves.

Aristotle's difficulty remained – 'the beauty of the soul is not seen'. It
was one thing to agree that the fool should obey the wise man, another
to identify each, even in the case of individuals let alone groups. The discovery of the New World and the prospects of seizing land and slaves
provoked some to argue that the aboriginal peoples were such a group.
'Defenders of the conquest such as Sepulveda' it has been noted, 'argued
on Aristotelian grounds that the inhabitants of the Americas were inferi-
or to Europeans, and were, therefore, clearly designed by nature to be
slaves'. Even in this case – where it seemed in the interest of all
Europeans to agree on the inferiority of native Americans – Vitoria, Las
Casas, and others rejected and attacked Sepulveda's arguments.[38]

However, there was one section of the population that all the theorists
could agree was naturally, obviously, and indisputably inferior –
women. Centuries of learned discourse maintained it. Everyone agreed
that those weaker in body, mind, and character, should be governed by
the stronger; everyone agreed that women were weaker than males in
body, mind, and character. The conclusion that women should be governed by males followed inescapably.

The most influential of the Church Fathers, Augustine of Hippo,
described woman as 'the weaker part of that human alliance', designed
for subjection to her husband. Thomas Aquinas believed that even in the
state of innocence before the Fall of Adam and Eve, difference of sex
would have been a determinant of subjection. Where natural inferiority
was writ so large, difference in power must surely follow. Early-modern
divines and theorists adopted the same logic. Male government followed
naturally from woman's imperfection and weakness. Physical differences pointed to moral subjection more clearly in the case of the two
sexes than in any other area. 'Nature hath formed the lineaments of [the
male's] body to superiority.' No male should 'suffer this order of nature
to be inverted'. Nature showed by 'the weakness of all females' that they
were 'inferior to the males'. Theorists could not agree that Africans or
native Americans were naturally inferior to Europeans but they could
and did reach a virtually universal consensus that women were naturally
inferior to males. Nature had marked out women with a badge of inferiority so obvious that, for a change, all the theorists could agree on its
presence.[39]

Instituting an Aristotelian meritocracy amongst adult males was difficult because 'nature, which distinguishes the husband from the wife and
parents from children, does not in like manner distinguish other men but
has made them equal'. But nature made female inferiority perfectly obvious and so showed where rule should lie; in the case of male and female
it was thought completely clear who had greater gifts. In the middle of

the seventeenth century, the Scots patriarchalist bishop, John Maxwell, readily linked Aristotelian notions of natural inequality with female subordination:

> Is not the female sex by the ordinance of God and nature inferior and subordinate to the male? Doth not nature teach that the wife by the law of nature is subject to the husband? If you will believe Aristotle in his *Politics*, he telleth you that a man of weak understanding is subject to him who is more intelligent and prudent and ... that he is *natura servus* [by nature a slave].

However, it was not only conservative patriarchalists who thought in these terms. In many areas, seventeenth-century Whig theorists were the intellectual precursors of modern liberals, but they certainly did not believe in the natural equality of all men. William Fleetwood, Bishop of Ely, was a plain example of the tenaciousness of the belief in natural inequality, even amongst Whigs. One of Fleetwood's attacks on Tory principles was so outspoken that it was condemned to be burnt, but he held contractarian ideas alongside the conviction that 'where people are in other respects equal, strength of body and capacity of mind will undoubtedly make them superior'. He admitted that training might have some effect on women's ability to handle difficult and demanding tasks, 'yet all the use and education in the world would never fit them for the performance of the great businesses of trade and merchandize, and making wars abroad, and executing justice at home'. It was 'demonstrably certain' that women had less strength and ability. Nature had made males superior to females in all areas but housework. Nature did nothing in vain; it was pointing a lesson when it made women inferior; nature had 'designed this superiority'.[40]

It was commonly believed that what was natural could be discovered by looking at the customs and conventions of most nations. Men were quite capable of acting unnaturally and unreasonably, but as such unnatural modes of conduct would be unprofitable and self-destructive they would not find general acceptance. The *ius gentium* [law of nations] was often defined as that which natural reason establishes 'among all peoples', and which is equally kept and observed by all nations. The 'light of nature' was equivalent to 'the declared sense of mankind' on moral matters. Widespread acceptance of a law or institution at all periods of history was good evidence that it was natural and reasonable. Early-modern theorists believed that 'there must certainly be good reason in a practice that all the different nations of the world would agree and centre in'.[41] Of course, evil and corrupt customs could be found among some pagan peoples both in the past and the present, but it was assumed that these would not last for long nor be generally accepted, simply because they were contrary to human nature. Natural law should not be deduced from the actions of a few primitive tribes, for 'the practice of some vicious and barbarous nations is no proof that a thing is good, and the law of nature must be learned from the soberest and most civilized nations'. Early-modern anthropological observation almost always led to the confirmation of pre-existing beliefs on the law of nature. If theorists approved of a practice found in other non-Christian societies, they cited this as proof of its naturalness; if they disapproved, the custom was merely a vicious exception. The content of the natural law's edicts was not discovered by empirical research; observation merely confirmed rational deduction.[42]

Nonetheless, the general acceptance of a custom in all nations and periods of history was *prima facie* evidence that it stemmed from nature. Few customs could boast such universal acceptance as the subjection of women. It would do discontented women 'little service' to look to 'the usages and customs of all countries in the world', since at 'all times' women have been placed 'but a little above their slaves or menial servants'. Even the most civilized nations have 'allowed them to be only made for the solace of mankind, the care of some domestic matters, and the continuance of the world'. Early-modern theorists thought that universal practice made clear that woman's role was to serve males, stay at home, and have babies. Pagans did not have the Bible to instruct them; they were guided only by natural reason, but 'the very heathen philosophers which knew not the word of God, yet could teach that the rule for the wife to live by was her husband'. Since all peoples had established institutions that subjected women to males, this must be the natural way of doing things. Whenever 'they are linked together in one yoke, it is given by nature that he should govern, she obey. This did the heathen by the light of nature observe'. Men did not need the Bible to tell them that, as woman's superior, the male should control her. The Bible gave Christians an edge that enabled them to organize their affairs better than others, but when it came to husband and wife, 'even heathens understood' this natural subjection.[43]

The seventeenth-century English philosopher, John Locke, thought that God's providence was shown in the widespread subjection of wives, for 'generally the laws of mankind and customs of nature have ordered it so; and there is, I grant, a foundation in nature for it'. The French historian, Pierre Petit, devoted hundreds of pages to arguing that the classical legend of the Amazons was based in truth, and that it was not inherently impossible for a society to exist where women ruled over males. Nonetheless, even he accepted that male rule was more natural. Just as biological 'sports of nature' (such as animals with two heads) were occasionally produced, so at infrequent intervals arose an oddity like a society that allowed females to rule. These rare exceptions did nothing to undermine the general law of nature; instead, their failure to survive or spread bolstered the reasonableness of more natural arrangements. Benedict de Spinoza, the great Dutch and Jewish philosopher, considered the possibility that women were excluded from government only by custom (not nature), but rejected it, arguing that 'actual experience' proved that women's exclusion from power was 'due to their weakness'. Nature simply had not given women an equal ration of that 'strength of mind and intellectual ability, in which human power and therefore human right mainly consists'. Women's subjection could not simply result from convention, for then

> surely among so many different nations some would be found where both sexes ruled on equal terms, and others where the men were ruled by the women and brought up in such a manner that they had less ability. But since this has nowhere happened, I am fully entitled to assert that women have not the same right as men by nature, but are necessarily inferior to them.

Women were inferior, not because of social conditioning, but because nature made them so.[44]

To most early-modern thinkers, the naturalness of female inferiority and subjection was clearly evidenced by the absence of any significant exceptions to the universality of male rule. For much of the early-modern period, most theorists accepted that the legend of the Amazons had some basis in historical fact, but it caused them no concern. Some argued that Amazonian women had gained the upper hand because Amazonian males were feeble and effeminate. Others maintained that the Amazonian women only succeeded in self-rule by excluding males from their community; if males had remained, women would soon have been subordinate housekeepers. History, classical scholarship, and anthropology were all martialled during the sixteenth and seventeenth centuries to support the belief that women were physically, mentally, and emotionally second-rate, suited only to a role within the home, and designed for subordination. Early-modern thinkers believed that right reason led even heathens to recognize female inferiority and subjection. But these theorists were not heathens. And as Christians they had an infallible source of knowledge to aid their reason – the Bible. Did they believe that God's will revealed in Scripture abolished, modified, reinforced, or extended woman's subjection? It is to this question that I shall now turn.[45]

Original subjection

In the sixteenth and seventeenth centuries, virtually all moral and political thinkers regarded the Bible as the most compelling authority on ethics. Early-modern theorists believed they found female inferiority and subjection writ large in Scripture just as in the classical tradition and anthropological observation. The book of Genesis was the most important source, for the Creation story provided the hard evidence of God's original intentions about the human condition and the relationship that should obtain between male and female. Virtually everyone thought the

book of Genesis was a literally true record of the world's history. God's
creation of Adam and Eve was the beginning of humanity and of human
nature. Natural law followed from that human nature; it showed the best
way for men to live in accordance with their nature. Arguments from the
Creation story were not regarded as religious. The Gospel, recognized as
binding by Christians, came much later and supplemented, *not* undermined, the natural law established by God when he created the world
and its inhabitants. God made nature, and the Bible was the word of
God. The Bible told truths that man could not discover by reason alone
but it never contradicted deductions from natural reason for God is not
inconsistent. God's laws in the Bible and His natural laws were believed
to be in perfect accord. Early-modern theorists saw arguments from
nature and from the Bible as complementary props of the legitimacy of
woman's subjection.

Both the Bible and nature were seen as giving evidence of God's
design and precepts, and each drew strength from the other. One example of this was the early-modern attitude to the pain and difficulty of
childbirth in the human female. God had pronounced painful childbirth
to be part of Eve's punishment, and daily observation showed that parturition was indeed more difficult for women than for other female animals; these were regarded as mutually supportive facts. Physicians
discussing the problem of painful childbirth explained that all theologians and Christians recognized that it resulted from Eve's sin, because
God wanted to torment the sex that caused mankind's death. They
admitted there were more immediate causes, such as woman's weakness, her sedentary life, and 'the size of the foetus' head'. Whatever the
natural mechanisms, however, the relative sizes of the female pelvis and
the foetus' head were not a morally neutral physical accident, but part of
God's plan and woman's punishment. The 'pains and weakness and
indisposition of body' of women during pregnancy and parturition 'arise
from a cause in nature, yet it is ordered and appointed by God himself in
a course of justice'.[46] Early-modern commentaries on Genesis cited
Aristotle's observation 'that women bring forth young with more pain
than any other creatures' to show that God was still punishing women as
He had promised, and thereby giving independent, scientific evidence
(visible even to pagans) of His providential justice. Exegetes noted with
pleasure that infidel scientists were able to see this 'evident demonstration' of divine retribution. With godly economy, agonizing childbirth fulfilled two functions: it reminded women of their shared responsibility for
Eve's sin, and provided empirical proof of God's providence. When in
childbirth 'nature seems to outstrip herself, and to torment the poor
woman beyond all natural causes'; it was obvious 'that their danger and
sorrow is the just punishment of sin, for the first offence of the woman'.
To divines, labour pains provided a recurrent, satisfactory proof of God's
justice in the world.[47]

The key biblical text for female subjection was the story of the Creation and Fall of Adam and Eve as told in Genesis. Because of its fundamental importance, it is worth repeating that the use of this biblical text did not, in early-modern theorists' eyes, involve them in viewing the assertions, made on the basis thereof, as true only for those who accepted Scripture as the word of God, such as Jews and Christians. Adam and Eve were thought to be (literally) the parents of all mankind. The human nature that God created on the sixth day was shared by everyone – Jew and Gentile, Christian and infidel. All shared, too, in the repercussions of the Fall. Without Scripture to guide them, pagans would not perceive clearly all the consequences of the arrangements God made in Eden, but they could not escape them; human nature, and the laws governing that nature, were irreversibly fashioned then. God 'spoke to the persons of Adam and Eve as the protoplasts of mankind, to relate to all the males and females that ever should be born'. God's edicts in Genesis were 'not to be understood of Eve alone, but of all women'. Because the Bible was seen as a factual narrative of history, no incongruity was perceived in citing Genesis alongside Aristotle; both contained truths about human nature – truths largely deducible by natural reason.[48]

Genesis described the Creation of Eve in the following terms: 'And the Lord God said, It is not good that the man should be alone; I will make him an help meet for him'. God then brought the animals to Adam who named them.

> But for man there was not found an help meet for him. And the Lord God caused a deep sleep to fall upon Adam, and he slept; and He took one of his ribs, and closed up the flesh instead thereof: and the rib which the Lord God had taken from man, made He a woman, and brought her unto the man. And Adam said, This is now bone of my bones, and flesh of my flesh: she shall be called Woman because she was taken out of Man. (Genesis 2:18; 2:20–23)

The story of Eve's creation – analysed in minute detail by early-modern biblical commentators – was used to provide many props for belief in the inferiority of woman and her subjection to man. Three were of overriding importance. First, Eve is made *after* Adam; woman is an afterthought – Adam gets there first. Second, Eve is made as a 'help meet' *for* Adam when God notices that it is not good for Adam to be all alone. Third, Eve is made *from* Adam, not formed *de novo* from the dust of the ground (as Adam was, back in verse 7). Each of these facts was used to support the idea that woman was naturally subject to the male (and note that this is *before* Eve has slipped up by eating the apple and tempting her husband to do the same).

The first and most influential Christian commentator to ground the subjection of woman on her later creation was Saint Paul, who stated 'I suffer not a woman to teach, nor to usurp authority over the man, but to be in silence. For Adam was first formed, then Eve'(I Timothy 2:12–13). Paul was generally referred to as *'the'* Apostle (just as Aristotle was *'the'* philosopher); nobody casually disagreed with a doctrine taught by him. Early-modern divines often turned to Paul in support of their arguments for the subjection of women. Protestant ministers were less enthusiastic than the (avowedly celibate) Catholic clergy about those passages of Paul where he recommended virginity over married life. However, theorists of all denominations found in Paul's Epistles reassuringly wholehearted support for the belief that woman was inferior and should be subject. Paul's interpretation of the significance of woman being created second was accepted by virtually all later Christian writers. They noted approvingly his argument from 'priority of creation' and were perfectly persuaded by his logic; 'God who made other creatures male and female together, did not so in mankind: which Paul observeth ... making it one reason of the woman's subjection'. God wanted to show that the male was made 'first, and as the more principal' so He did not create Adam and Eve simultaneously. Eve was the runner-up in creation and – just as she had taken second place then – she should take it for ever afterwards. 'It is true from the beginning the woman was subjected, as in order of time she was created after man.'[49]

Occasionally, an apologist for woman would note that there was a certain inconsistency in maintaining that man was made after all the other animals in order to show that they were made for humans' benefit, while failing to apply the same logic to the order of Adam and Eve's creation.[50] In general, however, Paul's view triumphed. Early-modern moralists used the argument from first creation and looked to Paul's words where he did 'prove this subjection to be ... due ... from man's being first created'. Adam was 'first created as a perfect creature, and not the woman with him at the same instant'; Eve was made only after God had put first things first, after Adam's 'constitution and frame ... then was she thought of'. The majority of theorists opted for Paul's version of events: 'God first created man and by so creating him, gave him the power and dominion over the woman'. God gave Adam pride of place as a metaphorical lesson to every other married couple, to teach 'every wife, uprightly to fulfil the law or command of her husband to the benefit of him or another'. Eve came second to Adam; wives should come second to husbands.[51]

God made Eve as a 'meet help'. Saint Paul also took the lead in pointing out the importance of woman being made *for* the male. When exhorting male Corinthians (whose women were getting a bit above themselves) to exert proper authority, he stated 'neither was the man created for the woman; but the woman for the man'(I Corinthians 11:9–10). God's purpose in making woman was her usefulness to man. Eve was 'an assistant or helper'.[52] To Renaissance thinkers, it was obvious that a helper did not have the same importance as the helped. They saw nothing in Genesis to suggest that God had had Eve's desires and needs in mind when He made Adam, but the text was explicit that He was thinking of Adam's when He made Eve. John Milton took this attitude to its logical extreme in his divorce tracts, but even in more moderate and charitable commentators, the same perspective was apparent. Women were 'vessels ... for use: they are helps to piety, helps to society, helps to house-government, and helps to propagation'. Woman was the earliest handy, all-purpose utensil; 'man's best moveable'. Woman was 'God's gift, the first and best that God gave to man'. Eve was God's birthday-present to Adam. To early-modern commentators, it seemed clear that woman's divinely ordained role was an ancillary one.[53]

Paul's view was repeated by both puritan and Anglican divines. Extreme separatists pointed to Paul's words 'and gathered another reason of the woman's subjection'. The Bishop of Exeter adopted Paul's argument. Woman was made to help the male; home-helps should not try and run affairs:

> that the husband hath power over the wife is so clear, both in nature and reason, that I shall willingly save the labour of a proof. It is enough that, by her Creator, she was made for a helper; and a helper doth necessarily argue a principal.

Woman was not the principal, she was the auxiliary. Early-modern theorists thought that God had taken particular pains to make woman's ancillary status clear:

> God might have said barely, I will make him a woman and no more; but God chose ... to describe the woman's duty, 'I will make a help meet for him'.

Woman's obligation to minister to males' wants was spelt out as her original function. The point of God's actions seemed to them so obvious that even a woman should grasp it (which was of course deemed to have been God's intention).[54] Antony Hickey was an Irish Catholic, but on this point he did not differ from English Protestants. In early-modern eyes, Genesis stated clearly that God made woman in order to improve things for Adam. This was the purpose of her existence. 'She owes her being to the man's necessities and convenience.' Woman was there to be 'helpful and profitable to man', so naturally she must take second place to him. Adam was made for God, but Eve was made for Adam. 'He for God only, she for God in him.' 'Who can be ignorant that woman was created for man, and not man for woman?' asked John Milton rhetorically; certainly, no seventeenth-century theorist was. To early-modern commentators, God's reasoning was obvious: Eve was made to help man, not to do things on her own.[55]

The significance of Eve being made *from* Adam, not created independently, was also not neglected by theorists of female subjection. Normally males are born from women. On this important first occasion,

God made a special effort to ensure that the norm would not give any false impressions about sexual priority: the woman was 'born' from the male. In Martin Luther's words,

> This is a reversal of the pattern of the entire creation. Whatever is born alive ... is brought forth into the world by the female. Here a woman herself is created from the man ...

Although every living male came from a female, divine creation was thought to show that in a much more profound sense the female was derived from the male. Commentators stressed the significance of this reversal in proving male superiority: the male was the true source of the female, and as a mere offshoot she counted for less, 'because that which is derived cannot be equal to that from which it was derived'. In Renaissance thought, it seemed obvious that the by-product of a rib should be in subjection to her perfect forbear; her secondary derivation showed 'her usefulness by obedience; and this is not an human (as some women would have it) but a mere natural, or rather a divine, imposition upon the sex'. Since woman 'had her beginning' from the male, it must be 'the duty of a wife to be obedient to her husband'. At the end of the seventeenth century, divines used precisely the same logic: 'woman was made out of the man, and therefore ought to be subject to him'. Adam was the author of Eve – naturally he held authority over her.[56]

These three points – Eve being made *after*, *for*, and *from* Adam – constituted the main proofs of the subjection of woman even before the Fall. It was almost universally agreed that woman's subjection followed 'from the rationale of creation itself' – for woman was created after the male, from the male and on account of the male. Early-modern exegetes were in no doubt that this was the message that God was trying to convey in Genesis, because Saint Paul supported this interpretation. Paul was God's Apostle, personally converted by Him on the road to Damascus: 'There is no doubt that Paul of all men understood the meaning of Christ most correctly'. God would not have allowed him to err on such an important matter. Because Eve was made from the man, 'Paul admonished' women to be subject and obedient to their husbands; because the woman was made after the man, and on account of him, and from him, 'and not *vice versa*'. Being omnipotent, God could have created mankind differently and it was significant that He did not do so. God, the Creator, was pointing a lesson or He 'might have done otherwise, that is, yielded to the woman coequal beginning, sameness of generation, or relation of usefulness'.[57]

Adam's priority in the order, origin, and purpose of creation was seen by early-modern biblical interpreters of woman's role as God's major message in Genesis, but other lessons were also drawn. In particular, they noted that initially God spoke only to Adam, issuing to him the command not to eat of the forbidden tree (Genesis 2:16–17). God had no direct speech with Eve until he called her to task for apple-eating (Genesis 3:13). The fact that God did not speak to Eve directly was rarely held to excuse her. God had told her husband, and it was obviously Eve's duty to accept whatever Adam said.[58] No early-modern biblical exegete attributed God's silence to mere absent-mindedness. God's taciturnity where Eve was concerned was aimed at conveying the message that in religious matters (as in almost all others) woman should take her lead from the male. Again, this was a traditional view; Hugh of Saint Victor had argued this way in the twelfth century and early-modern theologians agreed: 'The wife's obedience (implicitly) depending upon the husband's: he is to obey God and she him'. Radical Protestants like John Bunyan were just as insistent on this as medieval Catholic monks: 'worship was ordained before the woman was made, wherefore the word of God at the first did not immediately come to her'. God had made man male and female with the object of establishing the first hierarchical power relationship; 'man was made to this end and purpose, that God's glory should appear in his rule and authority: on the other side the woman was made, that by profession of her obedience she might the more honour her husband'. When males wanted to glorify God they should order their wives about; when females wanted to honour Him, they should obey these orders.[59] God had not spoken to Eve until the time came to condemn her misbehaviour because she did not need to be told very much. As a good wife she could learn from her husband, who would handle all the difficult matters while she took care of the housework.

> It is probable Eve had less knowledge than Adam and yet had as much as was required to the perfection of a woman . . . .. It was not necessary for her to know things by their causes . . . but she was skilled in those things that concerned her family and related to her husband and children.

Until she sinned, Eve was perfect; the perfect woman stays at home, looks after the children, and obeys her husband. Before the Fall, Eve was an ideal, immortal, sinless second fiddle.[60]

Fallen woman

Early-modern theory maintained that Eve was created after, from, and for Adam. She was an ancillary subordinate even before she was tempted into sin by having 'tittle tattle too long and too much . . . with Satan in the serpent'. Before the Fall, Eve was a subject, fertile housekeeper; because she was the first to sin, her position deteriorated; her subjection was confirmed and extended. Women should never be allowed to forget that Eve sinned first. It was their fault that mankind was no longer sporting happily in an idyllic garden. 'Let it cause modesty in women that she was the first sinner, and who having cast man down in Paradise, did cast him out of Paradise.'[61] Fortunately for the Renaissance biblical scholars who argued that women were naturally and divinely ordained for obedience, Genesis left them in little doubt that the Fall was definitely Eve's fault and that her punishment (along with labour pains) was being subjected to her husband's rule. God penalized Eve for her disobedience in these words:

> Unto the woman He said, I will greatly multiply thy sorrow and thy conception; in sorrow thou shalt bring forth children; and thy desire shall be to thy husband, and *he shall rule over thee*. And unto Adam he said, Because thou hast hearkened unto the voice of thy wife, and hast eaten of the tree, of which I commanded thee, saying, Thou shalt not eat of it: cursed is the ground for thy sake; in sorrow shalt thou eat of it all the days of thy life. (Genesis 3:16–17 [my italics])

To the early-modern theorist, it seemed apparent that God viewed Eve as the fall guy for the whole affair. Saint Paul dispelled any doubt on this point: 'and Adam was not deceived, but the woman being deceived was in the transgression' (I Timothy 2:14). It was significant that Eve sinned first, just as it was important that she was created second. 'The woman's being then (as often since) before man in the transgression and fall'. It was commonplace to assert that 'Eve's sin was greater than Adam's because she was first in the transgression'. God's message seemed perfectly clear to early-modern eyes. Eve was the instigator of the crime and could not be trusted afterwards. As a penalty for her insolence God placed her (and all future wives) under an obligation to obey: 'because woman was first in the transgression, God layeth this sentence upon her of a painful subjection to her husband'.[62]

Saint Augustine's writings persuaded some theologians to deny that Eve's subjection antedated sin. Martin Luther argued for this. Other German Protestants adopted the same position. Cornelis Jansen, Catholic bishop of Ypres in the early seventeenth century, was another enthusiastic Augustinian and he, too, maintained that the husband's dominion over the wife resulted from Eve's sin. The early English reformer, Hugh Latimer, argued that Adam and Eve were equal at first but that after 'she had given credit to the serpent', God commanded her to live in subjection to her husband. Wives, he reasoned, 'being before at liberty, must now be obedient to their husbands'. The Lutheran influence on the English Reformation soon waned but this view continued to find a few adherents. Had Eve not sinned,

> it may be supposed she had been equal in authority with him, checkmate, and no diversity of superiority had been betwixt them, but now the case is altered, there is no equality in this nature but the man hath wholly the prerogative.

Others, too, taught that: 'The first subjection of the woman began at sin'.

Augustinians had a low view of male nature after the Fall, but saw woman as even more abject.[63] By far the more common opinion among Catholics and Protestants was that woman's subjection preceded the Fall, but that it was made more onerous thereafter, both in its extent and in her new unwillingness to accept it. One exegete put some additional words into God's mouth for women who could not follow His terse style in Genesis. God was telling Eve (and every wife) that her husband

> shall rule thee, not with that sweet and gentle hand which he formerly used as a guide and counsellor only, but by an higher and harder hand as a lord and governor . . .

Jean Calvin argued that Eve had 'previously been subject to her husband' but that, after she had 'perversely exceeded her proper bounds', 'servitude' replaced her former 'liberal and gentle subjection'. The biblical scholar, Isaac LaPeyrere, devised some strangely heterodox views on Genesis in his touching attempts to reconcile biblical chronology with historical evidence, but he was wholly orthodox on this point. Eve's subjection before the Fall was pleasant, after the Fall it was penal. For the vast majority of commentators, the Genesis 3:16 text 'he shall rule over thee' was capable of only one interpretation: it embodied God's will that for ever after the Fall wives should obey their husbands.[64]

God's statement that Eve's desire should be to her husband was not interpreted as meaning that she would obey him willingly. On the contrary, fallen woman's subjection was expected to be an endless struggle because of her proud and rebellious nature. Her very distaste for subjection was part of the punishment; innocent woman would have loved doing what her husband told her. Any wife who balked at her husband's authority over her was simply suffering proper punishment for her part in Eve's sin; this must be the case because *only* a sinful woman would want to resist it. To early-modern divines, Eve's taste in fruit had shown woman's unreliability and proved that her antelapsarian subjection was just; it simply had not gone far enough. Unless curbed by their husbands' authority, wives could not be trusted to avoid dangerously sinful ways: 'good reason it is that she who first drew man into sin, should be now subject to him, lest by the like womanish weakness she fall again'.[65]

One of the ticklish points that theologians and biblical commentators enjoyed debating was whether Adam or Eve sinned more.[66] This debate was not about apportioning responsibility for mankind's damnation. For soteriological purposes, it was agreed that mankind did not Fall until Adam also ate. Saint Paul had asserted that 'by one man sin entered into the world' (Romans 5:12). The Greek was 'henos anthropou', the Latin 'unum hominem' – both mean one human being, the term is not sex-specific. But one excluded two, and Adam was always assumed to be the one in question.[67] Renaissance commentators rarely considered the possibility that a mere woman could alone have been responsible for such a significant change in mankind's state, particularly as Paul had also stated that 'in Adam all die' (I Corinthians 15:22). As Calvin commented, 'truly, Paul . . . states that sin came not by the woman, but by Adam himself'. Soteriological discussions of the nature of original sin often talked solely about 'Adam's sin' with barely a reference to Eve. The sixteenth-century Italian Catholic, Lorenzo da Brindisi, was orthodox in arguing that if Eve alone had eaten the apple, the whole human race would not have fallen. In contrast, had Adam alone eaten, all humanity would have been subject to the penalties of original sin, 'and so truly it is on account of Adam's sin that we are lost'. At the end of the seventeenth century, a protestant Yorkshireman taught the same view: 'Eve's eating would not have spoiled mankind had not Adam eaten also'. Established opinion held that Adam's sin was the really significant one in bringing about the Fall of all; he was the proper 'beginning of mankind, . . . not Eve'. Original sin was 'consummated' in Adam and descended through the male not the female; 'it is written that we sin in Adam not in Eve, because the man being the principal agent in generation, sin is rather derived from him into his offspring'. When crucial theological points were at issue, Adam's sin was the decisive one. Even when sinning – the only thing she was remembered for – Eve had less impact than Adam.[68]

Discussions about Adam's and Eve's comparative sinfulness are interesting because of the arguments employed and the attitudes these embodied. As we have seen, God criticized Adam for listening to his wife; 'the cause of the first sentence God passed upon man was "because thou obeyest the voice of thy wife"' (Genesis 3:17). This opinion was regarded as authoritative and frequently found its way into theoretical discussions of the Fall. Saint Augustine had characterized Adam's fault as standing by Eve even when 'this involved a partnership in sin', and medieval thinkers also saw Adam as the first of many husbands to fall from grace by loving his mate too much.[69] Adam's uxoriousness found as many critics in the early-modern period; by succumbing to his wife's persuasion he overturned the proper order in human relations – Eve 'was given rather to help than advise'. In some accounts of the Fall it became difficult to see why God punished such a nice fellow as Adam at all. For Adam 'was not deceived by the serpent or tempted by the fruit, but only he was drawn by the love of his wife out of his kind nature'. God requited Adam's kind, uxorial nature with lifelong toil. Adam had to choose between doing the right thing and excessive affection for his wife, and he chose wrongly. Since he was in authority over her, the choice definitely lay with him. 'Adam's sin was greater than Eve's in this respect, because he was her head and governor.' Adam had been given the responsibility of governing his wife and endowed with more intelligence so that he could do the job properly. Because Adam merely followed her suggestion, when he had been 'appointed to be the woman's head' and given these 'gifts of knowledge and wisdom, the man was more faulty than the woman'. When Eve suggested eating, Adam 'should have rebuked her .... Then had her sin been personal, rested upon herself and gone no further, had he not hearkened to her voice'. Adam's sinfulness in taking notice of what his wife said to him needed stressing because, as Hugh Latimer complained, 'there be now many Adams that will not displease their wives, and do as them listeth'.[70]

The other side of the coin was Eve's pride, disloyalty, and disobedience in making the decision to eat the apple without first consulting her husband. Disobedience was often portrayed as Eve's crucial offence. Robert Parsons argued that God had given Adam charge of religious matters but that Eve 'became his doctor and mistress in this ecclesiastical function and thereby turned upside down, to both their ruins and to the ruin of us all, the whole order of subordination'. As a Jesuit attacking Elizabeth I's governorship of the Church of England, Parsons had a particular axe to grind against female supremacy in ecclesiastical matters, but his argument that Eve should have followed Adam's lead (not vice versa) was generally accepted. The Roman Catholic polemicist, Nicholas Sander, used the same argument against Elizabeth I's supremacy; female authority in ecclesiastical matters was something 'the Devil instituted in paradise when he made Eve Adam's mistress in God's matters'. Standing at the opposite end of the confessional spectrum, John Bunyan disagreed with almost all of Sander's opinions on church government, but he concurred on this. Even the devil, Bunyan insisted, knew 'that the man was made the head in worship'. He drew the same conclusion as Sander's: women must never be allowed any part in the ministry or rule of God's church.[71]

It was commonplace to argue that Eve's sinful conduct encompassed the faults of 'disloyalty', pride, and 'breach of that social love, which by order of creation was due unto her husband'. Many thought it 'an usurpation' simply for Eve to speak to the serpent, 'when her husband who was more able than she was at hand to whom ... the law was given as chief'. Eve 'should have known his consent before she presumed to have passed hers'. Eve must have known that any matter of importance should be decided by her husband; she was unwarrantably presumptuous in daring to meddle with 'the great and weighty matters that concern eternity'. Inevitably disaster followed. Not only did Eve – without permission – pass the time of day with a snake in the grass and eat the apple, she tempted Adam to follow in her sinful ways. She was not only disobedient, but (as Saint Paul had complained) a seductress. Adam was not misled by the devil, 'but only enticed and deceived by the woman, who was the tempter's agent'. Every woman since Eve should bear in mind the dreadful consequences of this first instance of independent female action so she would remain 'humble [and] in a low opinion of herself and that lower order wherein God hath fixed her'. Eve's pride was the root of the whole problem; she sinned more because she was a woman, who should be humbler, 'for pride is a greater sin in woman than in the male'.[72]

There was, of course, a great deal to fault in both's conduct. Indeed, 'it would not have been an easy matter to have determined whose sin had been greater, had not God done it ... by inflicting a greater punishment on Eve than on Adam'. In case Eve's greater sinfulness was not obvious to anyone reading holy writ, God had removed all doubt by 'His order and gradation in the punishment'; Adam was punished 'last and least' because he 'only listened to his wife'. Most early-modern exegetes were moved by the same logic. A life of toiling in the sweat of one's brow to grow food in thorny ground was fairly unpleasant, but not as bad as one of excruciating childbearing ruled over by a fallen husband. 'The punishment of man was great, of woman greater ... for the degrees of punishment were equally proportioned to the degrees of sin'. Since most early-modern commentators on Genesis believed God to be just, they reached the same conclusion. Woman's punishment was greater than the male's, because she was the greater sinner.[73]

The debate over Adam's and Eve's relative sinfulness served to reinforce the conventional interpretation of woman's subjection: Eve was meant to obey Adam even before sin entered the world. He held authority over her in both secular and religious matters. The overturning of this order led to calamity in Paradise. Adam's big mistake was to hearken to what his wife said. Adam rebelled solely against God; Eve rebelled against her husband too. In early-modern eyes, the Creation story seemed to provide divine endorsement of the belief in the natural subjection of woman expressed in classical texts. Genesis, and Saint Paul's exegesis of it, seemed in complete accord with their observations of natural, physical, and mental differences between the sexes, and of the customs of all the societies they encountered. Given such a comforting consensus of all authoritative sources, early-modern theorists were satisfied that women were naturally inferior to males. They were equally convinced that this inferiority was intended by God to result in woman's subjection.[74]

Notes

1. For a brief introduction to the elements of early-modern political patriarchalism in England, see Sommerville 1986a, 27–34.
2. Selden in Thornton 1934, 55. For the best introduction to European political theory of the Renaissance and Reformation, see Skinner 1978, and to English seventeenth-century political thought, see Sommerville 1986a.
3. A few pamphlets arguing for woman's superiority were written either as *jeux d'esprit*, or to display the writer's rhetorical skill. (A comparable exercise was maintaining that dogs can reason.) One of the most famous and entertaining is Henry Cornelius Agrippa's 'Female Pre-eminence' (1509) reprinted in 1980.

Poullain de la Barre argued for feminine equality (apparently seriously) but admitted that no belief was 'more ancient nor more universal' than that in women's inferiority (1989, 15).

4. The sexes' different role in sexual reproduction will be considered in Chapter 5.

5. Aristotle 1984, I, 'History of animals', 948–9 (608), 'Generation of animals', 1144 (737a), 1186 (766b). See, for example, Hugh of Saint Victor 1951, 151; Albertus Magnus 1890–9, 30, 101 (IV.XXVI.3), 207 (IV.XXIX.3); Weyer 1991, 181 ff.; Allestree 1673, 29; Nicholls 1701, 98–9.

6. Dubois 1596, 5; Mercuriale 1591, 16; Bourgeois 1626, 2–3 ('le sexe feminin est extremement humide, et neantmoins colerique'); Chartier 1596, 23; Crooke 1615, 198–9.

7. For example, Lotichius 1630, 59–60; and Primrose 1655, 'Destructio', 51–2.

8. Dod and Hinde 1614, 33 [greater]; Heale 1609, 2 [disarmed]; Sydenham 1844, 371; Petit (1687, 106–7) cited as examples of societies where harsh living conditions led to similarity between the sexes: the Lapps, Siberians, and Norwegians.

9. See Chapter 3, Spiritual equality.

10. Plowden 1816, I, f. 305r; [The case was *Sharrington v. Strotton*, etc. in B.R., I, ff. 297–307]; Nicholls 1701, 87–8.

11. Campanella 1968, 200; Baudius 1638, 321 [memoria et ingenium]; Settala 1626, 57 ('actionibus intellectus'); Turturetti 1629, 386 [prudentior, infirmius]; Lapide 1854–9, IX, 267; Bunyan 1860, 666. Only the Quakers were 'so silly' as to establish a meeting system that allowed 'a formal collective identity for women as women' (Mack 1992, 344). See also Ludlow in Greaves 1985, 98.

12. Preston 1635, 206; Settala 1626, 288. See Maclean's outstanding study (1980, 40–5) for early-modern notions on the relationship between women's physiology and psychology.

13. Aristotle 1984, I, 'History of animals', 948–9 (608b); Prandoni 1649, 88; Poole 1696, I, sig. Dd2v [jealousies]; Turturetti 1629, 318 ('ad iram proclive'); Chartier 1596, 16. On medieval views, see, for example, Gerson 1606, II, 621.

14. Wright 1621, 41 [lack]; Pricke 1609, sig. K2r–v [insight]; *Marriage asserted* 1674, 16–17 [pardoned]; Baxter 1830, 15 [betwixt]; Pritchard 1579, 37 [imperfections]. Compare Batty 1581, f. 5r; *A discourse of the married and single life* 1621, sig. A8v; Crompton 1632, 31; Ken 1682, 12; Rogers 1642, 34; Weyer 1991, 540.

15. Aristotle 1984, I, 'Generation of animals', 1184–6, (765b–6b). Galen stated that woman was made 'less perfect' than the male, because this gave

> no small advantage for the race; for there needs must be a female. Indeed you ought not to think that our creator would purposely make half the whole race imperfect, and, as it were, mutilated, unless there was to be some great advantage in such a mutilation. (Lefkowitz and Fant 1977, 155)

See Crooke 1615, 217–18, for a faithful repetition of this view.

16. Aristotle 1984, II, 'Economics', 2131 (1344a); ibid., 2007 (1264b); 'Ethics', 1836 (1162a); Xenophon 1923, 419–23; Plutarch 1603, 315–25.

17. Aquinas 1872–82, I (1872), col. 1378 (I.92.2.3); Raulin 1512, f. 26v ('mulier debet habere curam domesticarum rerum').

18. Cajetan 1581, 413–14; Dod and Hinde 1614, 34. Compare Camerarius 1581, 17–19; Settala 1626, 44; Tilney 1568, sig. C5v.

19. Ferriere 1724, sig. iii. Compare Coke in Sommerville 1986b, 253.

20. On medieval civil law, see, for example, Bartolus 1570, I, f. 15r.

21. Doneau 1828–33, I, 630 (III.xxiv.12); Zouche 1682, 124–5; Harpsfield 1878, 214–15.

22. *The batchelor's directory* 1694, 25; Willet 1634, 291 [domestical]; Baudius 1638, 293–4; Veron 1562, sig. B5r [Martha]; Casaubon quoted in Pattison 1892, 94–5. Compare Houlbrooke 1984, 109–10.

23. See, for example: Smith 1982, 58; Smith 1866, 22; *The court of good counsell* 1607, sig. E2v; Crompton 1632, 11–12, 4; Dillingham 1609, f. 10v; Cawdry 1656, 99; Cosin 1843–51, I, 56; Dickson 1659, 162; Nicholls 1701, 101–3; Pricke 1609, sig. L3r–v; Tilney 1568, sig. E1r, C2v–r; Goodman 1629, 258.

24. Finch 1627, 75.

25. The consensus of all theoreticians, throughout the centuries, that woman's function was to run the household must necessarily make one sceptical about the belief that this idea was coined in early-modern Europe to uphold the infrastructure of nascent capitalism. S. Findley and E. Hobby (in Baker *et al.* 1981, 13), for example, argue that '[t]he seventeenth century is marked by the development of a sharp division between two separate spheres – the private where the woman belonged, and the public, which belonged to the man'. It is far from clear that there was any theoretical *development* of this belief; the classical doctrine was simply restated as a truism.

The implicit assumption in almost all modern works of women's history is that for women to stay at home is a 'bad thing', whilst for them to work outside the home, earning wages by paid employment or by producing profits through capitalistic enterprise is a 'good thing'. It is presumed that a woman dependent for her means of support on pleasing an employer or consumer is more autonomous and prestigious than one dependent on her husband since she is more 'empowered'. Probably, no more explanation for this historiographical phenomenon is required than that most works of women's history are written by professional women who have consciously rejected the role of dependent housewife. However, it is important to note that sixteenth- and seventeenth-century theorists did not see matters in this way – rather the opposite; a woman supporting herself by work outside the home was seen as unfortunate, not successful.

26. *The court of good counsell* 1607, sig. H3r; Topsell 1597, 150–1. For example, Pricke 1609, sig. L3v; Abbot 1653, 77–8 both made so many exceptions to the rule as to render it almost meaningless. For complaints of women 'gadding' out 'to all spectacles and shows, to speak in all discourse . . . and to hazard upon all companies', see Durham 1675, 360.

27. Sharpe 1980, 18. For medieval practice, see Hanawalt 1986, 113–16, 150–3, and Goldberg 1992, 108–25. On early-modern English practice, see also Collinson 1988, 77; Wall 1990, and especially Willen 1988, whose scholarly account shows how much women's activities intruded into the public sphere.

28. For example, Topsell 1597, 151; Rogers 1642, 32. See Clark (1983, 131–2) on the conventions governing women's visits to alehouses, and compare Smith in Goldberg 1992, 38, on medieval practice.

29. Aristotle 1984, II, 'Politics', 1990–1 (1254b–5a).

30. Aquinas 1872–82, I (1872), col. 1414–15 (I.96.3–4 'secundum animam diversitas'); Duns Scotus 1639, IX, 756 (XXXVI.I.2 'servitute extrema').

31. Cipolla 1547, 3 ('insipientes prudentibus'); Bañez 1594, 125–6; Settala 1626, 303 ('neque se, ne alios gubernare novit') and 310–13; Saravia 1593, 36–7 ('Rex,

subiectus, dominus, servus, infans recens natus, decrepitus senex, stultus, sapiens, foemina, mas, pariter aeque sunt homines').
32. Wemyss 1633, II, 55 [intention]; Baudius 1638, 319; *Treason, popery* 1680, 8–10 [impotency].
33. Gomersall 1634, 17–18 [absurdity], 4; Sedgwick 1624, 72 [own]; Foreness 1683, 13–14 [inequality].
34. Rutherford 1843, 51.
35. Hobbes 1991, 86–7 (Chapter 13).
36. Lowde 1694, 154; Lyser 1682, 104–5 ('adeo ut saepe magis differat homo ab homine alio, quam homo a bruto').
37. K.F. Koerner (1985, 70) has rightly argued that Locke's assertion that all men were sufficiently rational to form a state, did not mean he regarded all as equally intelligent. 'Locke simply never started with a postulate that all men are equal in terms of understanding.' Locke was not unusual; virtually all early-modern political theorists started from the same assumption. John Maxwell (1644, 137) for example insisted:

> To be rational in the first capacity and natural power is essential to all men and equal in all; but in the use, the exercise of the rational faculty there is vast disparity because of a great latitude in different actual ability.

38. Hobbes 1991, 87 (13 [61]); Aristotle 1984, II, 'Politics', 1990–1 (1254b–5a); Muldoon 1979, 143.
39. Augustine 1977, 272 (XIV.11); Augustine 1955, 5 (c.1); Aquinas 1872–82, I (1872), col. 1379 (I.92.1); Davenant 1627, 427; Whately 1617, 18; Cawdry 1656, 103.
40. Domas 1705, 49 [distinction]; Maxwell 1644, 112; Fleetwood 1716, 133–5.
41. Lagus 1566, 10 ('apud omnes populos'); Whaley 1698, 6; Fleetwood 1716, 22 [reason]. For similar arguments, see Wiseman 1657, 31; Taylor 1828, XII, 312, (II.II.iii); Suarez 1872, 125 (II.XIII.6).
42. Comber 1679, 59.
43. Fleetwood 1716, 135 [service]; Carter 1627, 48–9 [heathen]; Gouge 1622, 270 [yoke]; Doneau 1828–33, III, 979 (XIII.xxi.15 'Etiam ... ethnici intellexerunt').
44. Locke 1988, 174 (I.47); Petit 1687, 128; Spinoza 1958, 442–5.
45. Prandoni 1649, 141–3; Petit 1687, 113; Spinoza 1958, 442. Contemporary drama also expressed the belief that an Amazonian gynocracy was not really feasible, and that women would naturally surrender the freedom of self-rule for the joys of virile love. These themes have been discussed in Gagen 1954, 177.
46. Mercuriale 1591, 60 ('capitis foetus magnitudo'); White 1656, III, 201 [course].
47. Poole 1696, sig. B4r [pain]; Willet 1605, 51 [demonstration]; Goodman 1629, 315, 324 [sorrow]. Herbert's *Child-bearing woman* gave an exemplary exposition of the view that women deserve the pain of childbirth for their guilt in their 'first mother's transgression' (1648b, 14). For similar arguments, see Needler 1655, 105–6; Patrick 1695, 77; Becon 1844, 28; Dering 1597, sig. F6r–v; Topsell 1597, 181–2; Howe 1822, VII, 542; Ness 1696, 52.
48. Tyrrell 1681, 102 [a protoplast is a model or archetype]; Brenz 1584, sig. D5r.
49. Jackson 1617, 48 [priority]; Ainsworth 1627, 12 [reason]; Willet 1605, 15 [principal]; Torshell 1650, 8–9 [order].
50. See, for example, Littleton 1680, 57, and Agrippa in Bornstein 1980, 9. Catharinus (1551–2, 249) dismissed this point.

51. Coke 1662, 34 [created], 124 [uprightly]. Compare Richardson 1655, sig. B3r; Vauts 1650, 46; Abbot 1653, 41; Whately 1647, 3.

52. Blount 1911, 320.

53. Abbot 1653, 45 [vessels]; Heydon 1658, 72 [moveable]; King 1614, 6 [best].

54. Ainsworth 1627, 12 [another]; Cawdry 1656, 104; Hall 1837, 463-4 [labour]; Swinnock 1868, 516 [duty - he was citing Ferus].

55. Duns Scotus 1639, IX, 571 (XXV.II [Hickey's commentary]); Hopkins 1701, 166 [necessities]; Milton Paradise Lost, book iv, line 299; Milton 1959, 324. See also Lorenzo da Brindisi 1935, 237; Dickson 1659, 57; Wemyss 1633, II, 16; Torshell 1650, 8-9; White 1656, II, 87; Hughes 1672, 21.

56. Luther 1958, 123; Pettus 1674, 65-9 [derived]; Willet 1605, 42; Hopkins 1701, 166 [subject]. See also Ainsworth 1627, 13 [as Ainsworth notes, St Paul had once again been the first to spot this reason for woman's subjection]; Babington 1615, 14-15; Wemyss 1633, I, 264; Dickson 1659, 58; Tyrrell, 1691/2, 24.

57. Davenant 1627, 428 ('ab ipsa creationis ratione'); Chemnitz 1978, 748 [Paul]; Hunnius 1604, 71, ('admonet hinc Paulus', 'et non econtra'); Rogers 1642, 255. See also Maxwell 1644, 84; Lawson 1659, 191; Foreness 1683, 3.

58. See, for example, Catharinus 1551-2, 197; and for the opposite view Charnock 1866, 402.

59. Hugh of Saint Victor 1951, 122 (I.7.v); Bunyan 1860, 664-5; Pettus 1674, 82; Attersoll 1618, 1169-70.

60. Needler 1655, 61-2.

61. Babington 1615, 17 [tattle]; Jermin 1639, 244 [shame].

62. Vauts 1650, 46 [since]; Needler 1655, 72; Hughes 1672, 39 [layeth]. See Willet 1605, 51; Davenant 1627, 428; Abbot 1653, 41; Swan 1635, 500; Cawdry 1656, 102; Bovet 1684, 51, for the same view.

63. Luther 1958, 115; Lotichius 1630, 137; Jansen 1641, 45; Latimer 1844-5 I (1844), (252); 1845, II, 161; Carter 1627, 154-55 [checkmate]; Cleaver 1598, 231 [subjection]. See also Feltham 1631, 101; Heydon 1658, 75-6.

64. Poole 1696, I, sig. B4r [gentle]; Menochio 1678, 5; Calvin 1948, 172; LaPeyrere 1656, 15. On the long tradition of this view, see Maclean 1980, 18 ff. For the same account, see, for instance, Catharinus 1551-2, 248; Fisher 1597, 1292-3; Lorenzo da Brindisi 1935, 292; Lapide 1854-9, I, 71; Baynes 1866, 338; Bentley 1582, 24; Saravia 1593, 6; Tyrrell 1691/2, 12.

65. Gouge 1622, 268-9. See, for example, Salkeld 1617, 129-30; Ainsworth 1627, 17; Wemyss 1633, II, 17; Needler 1655, 106; Richardson 1655, sig. B3r; White 1656, III, 201; Lawson 1659, 69; Leighton 1844, 147; Bunyan 1860, 438; Hughes 1672, 39; Pettus 1674, 69-70; Hopkins 1701, 167.

66. This debate dated back to 'the ancient fathers of the Church' (Willet 1634, 868). As Joseph Hall summarized it:

> Saint Chrysostom directly teacheth, that the sin of Eve was more heinous than Adam's; which opinion most of the Schoolmen follow, upon Sentences 2 Distinction 22. Contrarily, St Ambrose by many arguments proves, that the man sinned worse than the woman. St Au[gu]stin[e] so speaks as if he thought the sins of both equal. (Hall 1837, 100)

['Sentences' refers to Peter Lombard's Sententiae in IV libris distinctae, one of the most frequently annotated and massively influential works of theology ever written].

67. See *Disputatio perjucunda* 1641, 15, for an entertainingly logical extrapolation of this view to prove that only Adam, not Eve, was a human being.

68. Calvin 1948, 172; Lorenzo da Brindisi 1935, 273 ('sicut vere propter solum Adae peccatum amisimus'); Ness 1696, 43 [spoiled]; Ames 1968, 116 [beginning]; Wilson 1615, 29–30 [principal].

69. Ness 1695, II, 164; Augustine 1977, 272 (XIV.11); Hugh of Saint Victor 1951, 125 (I.7.x).

70. Pettus 1674, 123 [help]; Mayer 1631, II, 495 [kind]; Needler 1655, 71 [governor]; Willet 1605, 47 [faulty]; Trapp 1662, 24 [rebuked]; Latimer 1844–5, 254. Similar criticism is found in, for example, Topsell 1597, 40; B.,Ste. 1608, 57; Caryll 1647, 273; White 1656, III, 209; Waker 1664, 7; Swinnock 1868, 507; Bunyan 1860, 438.

71. Parsons 1606, 80; Strype 1711, Appendix, LXXVII, 119 [mistress]; Bunyan 1860, 664–5.

72. Jackson 1617, 55 [breach]; Bunyan 1860, 428–9 [usurpation]; Poole 1696, II, sig. Bbbb4r; Maldonado 1677, I, 57. ('maior enim est superbia in muliere, quam in viro'). Compare Patrick 1695, 77; *The Fall of Adam and Eve* 1702, 11; Grotius 1952, 204.

73. Needler 1655, 72 [easy]; Vauts 1650, 46 [least]; Lawson 1659, 67 [greater].

74. Turner (1987, 118) has ridiculed early-modern thought for the 'hopelessly heterogeneous collection of factors' on which they based female subjection. In fact, early-modern theorists perceived this diversity of sources as a strength rather than a weakness: the convergence of religious, historical, anthropological, and biological evidence reassured (not troubled) them.

3

Exceptional women

Restrictions on inferiority

Sixteenth- and seventeenth-century theorists' bald assertion that woman was naturally inferior was subject to two important exceptions. First, it was a generalization not a universally binding law of nature. Virtually no early-modern theorist ever attempted to argue that all women were inferior to all males in every respect. Second, they believed that in certain matters male and female were equal; in particular, woman was deemed naturally – not spiritually – inferior. Each of these qualifications was significant.

To early-modern thinkers, the axiom that women were weaker did not entail that *every* woman was necessarily less strong, less intelligent, and less balanced than *any* male. Today, a zoologist might generalize that 'the female is shorter than the male' – meaning that most females are shorter than most males – without intending to assert the absurd proposition that no female is taller than any male. Someone can maintain it is 'natural' for females to be shorter than males yet admit that genetics and environment produce exceptions to the rule. In the same spirit, no serious early-modern theorist denied that there were some wise and rational women. Subjecting wives to their husbands was approved because males were believed to be generally superior, but many freely admitted that, 'some wives may be of a stronger mind and judgement than the husbands'; the normal 'rules respect the general condition of the sexes and speak of the female as ordinarily weaker'.[1]

The admission that some women excelled was most readily forthcoming when early-modern theorists discussed queens. Bishop John Aylmer argued that 'you can never show in all England since the conquest so learned a king as we have now a queen'. It may not surprise the reader that Elizabeth I was frequently the object of fulsome praise, even after her death. In France, Salic law excluded women from the throne, but powerful regents like Catherine and Marie de Medici provoked similar accolades.[2] Nor was it only when discussing royal women that such admissions were made. Authors receiving largesse from wealthy patronesses rarely omitted to mention how exceptionally virtuous were some women. One commentary on the Book of Ruth was typical in repeating many of the commonplaces about woman's inferiority in the text, while heaping praise on the author's benefactress (Frances, Countess of Warwick) in its epistle dedicatory. John Donne's poem *To the Countess of Huntingdon* (Elizabeth Stanley) lamented that virtuous women were unusual, but decided that his patroness was just such an exception.[3]

Early-modern moralists conceded that even amongst commoners the wife was sometimes cleverer than the husband, although this concession was often tied to the insistence that the characteristic indication of a wise and discreet wife was that she obeyed her husband.[4] Disapproval was evident when theorists admitted that it 'sometimes falls out, though unfitly' that wives 'have more knowledge or graces' than their husbands. The puritan divine, Nicholas Byfield, attempted to answer the question of why the Apostle Peter so stressed wives' obligation to obey their husbands. He hardly complimented males with his answer:

> Because many times their provocations from absurd husbands are so great as, if God did not speak very much to them, they could never endure it with subjection.

Byfield was eager to call husbands to task for their failure to be the shining examples to their wives that God and nature intended. 'It is a great dishonour to many men in this age', he complained, 'that women excel them in knowledge, both for the measure of it and care to use the means to get it'. Early-modern theorists recognized that just as husbands were sometimes more ignorant so they were often less prudent. The husband would 'oftentimes dissipate and waste his estate and patrimony in sensuality and riot'. Catholic casuists sympathized with the wife who had to stand by while her husband

> spends her wealth and substance at dice, at play, in drinking, banqueting, prodigal giving, improfitable bargains, improvident contentions ....
> Moreover he leaveth the charge of providing for the whole household to her and yet alloweth not wherewithal sufficiently to do it.

Such husbands were not 'rare', 'the whole world is full of them'. The desire of preachers and teachers to bemoan the vices of both sexes often produced condemnations of ungodly husbands.[5]

Many theorists admitted that 'it oft falleth out that a wife, a virtuous and gracious woman, is married to an husband destitute of understanding'. The contractarian political theorist, James Tyrrell, was eager to point this out to gain polemical advantage against patriarchalists. John Milton was anything but feminist in his beliefs and defended the justice of husbands' power over wives with zealous enthusiasm, but conceded that 'particular exceptions may have place'. If a wife 'exceed her husband in prudence and dexterity, and he contentedly yield', she should direct him. Male superiority was a general, not a universal, rule and theorists conceded that not *all* women were stupid and temperamental. On the Continent, Roman Catholic priests proffered advice to temperate women with dissolute husbands. Works of domestic counsel often repeated the classical truism that husbands should not seek their wives' advice but frequently subjected it to the qualification that, if the husband knew his wife to be 'prudent, firm, and sensible', he should take her advice. A virtuous woman's advice was worth having 'for godly men may hear their godly wives', and 'many have found much profit in following their wives' counsel'. Although theorists dismissed many women as foolish, they admitted that prudent wives were occasionally the salvation of stupid husbands.[6] Women's inferiority had not prevented certain women equalling 'the perfectest of men': 'whether we take them national or singular, we shall find them to parallel men, as well in the liberal arts as in high facinorous acts'. The Elizabethan philosopher, John Case, expounded (and subtly distorted) the ideas of Aristotle to suggest that although Aristotle had regarded it as natural for the male sex to be superior to the female, he had not meant to include every case in the general category. Others insisted that women varied amongst themselves just as males did; some were weak characters, yet as 'diverse women have sundry complexions, so they be subject to sundry passions'. Women were generally but not invariably inferior to males. This was a minor concession but 'the rule is, *What is, may be.* If some of the sex have been so, the proposition is firm that the sex is capable and may be so'.[7]

Moral and political theorists were convinced that these outstanding women proved the rule rather than undermined it. Women whose abilities exceeded most males' were oddities and therefore insignificant. William Fleetwood admitted that there were 'a great many instances' of such superior women,

> yet the number of such neither is, nor ever was, nor ever will be great enough to show that nature intended to give that sex superiority over the men...

Exceptional women were just exceptions to the normal rule.[8]

The recognition that some women outdid some males often found expression when piety was discussed, for it was commonly accepted that, whatever talents women lacked, many were profoundly religious. It was a medieval platitude that the female sex was the more pious. Women's emotional nature was thought to incline them more to 'mercy', 'pity', 'compassion', 'piety, and devotion' than males. Widows were extolled for their generosity to charitable causes. Divines preaching sermons at the funerals of godly women often used the occasion to praise them and censure others by emphasizing that 'many women do shame a great many men that are ignorant, idle, or profane'. Renaissance theorists admitted to few areas where the whole female sex excelled the male but a number held that devotion, prayer, and church attendance was one:

many women 'dignify their sex, leading lives worthy of the vocation wherewith they are called, outdoing men in actions of piety, humility, and charity'. Preachers explained this unusual female excellence by pointing out that

> men have more advantages of aspiring to honour in all public stations of the church, the court, the camp, the Bar, and the city than women have, and the only way for a woman to gain honour is an exemplary holiness.

Not all women were inferior to all males; males were naturally dominant in all areas of power and material achievement, but extraordinary females might compete even there. And all women might comfort themselves with the belief that they could outdo males in piety.[9]

Spiritual equality

Piety naturally leads to the second limitation on the doctrine of woman's inferiority – that woman was not inferior in all respects. In particular, religious dogma held that woman was equal to the male as a spiritual being. Theologians insisted that women had souls; philosophers maintained that woman, too, was rational. Divines had no interest in excluding women from religion and showed no inclination to do so. 'In this wicked age it is the numbers of that sex (which are constantly present at the public prayers and communions of the church) ... which make those ordinances in many places so well attended.' Throughout early-modern Europe religion and the religious were supported by hosts of pious women. In the godless immorality of Restoration England 'the reputation of religion is more kept up by women than men; many of the one countenancing it by their practice, whereas more of the other do not only neglect but decry it'. Ministers of religion did not want to lose at least half of their constituency by excluding women from their ministrations.[10]

Despite the clear interest of the Church in insisting that women should also be seeking salvation, occasionally a theologian broke ranks. In 1594, an anonymous *Disputatio perjucunda* [Jovial discussion] was published arguing that women were not human beings and were incapable of salvation. In fact, the pamphlet was not aimed against women at all. It was an extremely witty attack on Scriptural literalism. By pointing to the difficulty of proving through direct quotation from the Bible the belief that God is interested in saving women, it aimed at a *reductio ad absurdum* of the doctrine that every religious dogma must be demonstrable by explicit Scriptural citation. The local German church did not see the funny side of this tract at all. They set about prosecuting the printer and pursuing the (anonymous) author. Protestant worthies wrote long, dull responses which addressed the putative point that women were not intended for salvation, rather than the underlying attack on strict adherence to the letter of Scripture. The fury of their response was a telling indicator of the strict limits that Christian theorists placed on the doctrine of woman's inferiority. Early-modern theologians wished to exclude women from any position of power in the ecclesiastical hierarchy, but found dreadfully alarming the suggestion that Christianity offered nothing to women.[11]

Occasionally, theologians felt obliged to set forth convincing proofs that woman, too, had a soul, and that she was intended for salvation or damnation. Explicit biblical texts were thin on the ground for 'mention of women in Scripture is rare'. The Bible provided some examples of pious and godly women, but Old Testament law was addressed to the male, and it was easier to find passages bemoaning women's tendency to vice and excluding them from priestly functions than any explicitly asserting woman's spiritual equality.[12] Nonetheless, biblical exegetes teased out some Scriptural support for the doctrine that God also gave women souls which Christ was concerned to redeem. Two Scriptural texts particularly were cited to prove this: Genesis 1:27 and Galatians 3:28. The first stated: 'God created man in His own image, in the image of God created He him; male and female created He them'. Despite the confusing change of pronoun suggesting that God's image was ascribed directly only to 'him', the standard exegesis of this passage was that woman resembled God through possession of a rational soul.[13] 'There is neither Jew nor Greek, there is neither bond nor free, there is neither male nor female; for ye are all one in Jesus Christ' (Galatians 3:28) was glossed as showing that women, too, could be saved by faith in Christ. Basil the Great had argued for this interpretation in the fourth century, and Catholic commentators followed it in the seventeenth: Saint Paul's words in Galatians were held to show how Moses' confusing grammar in Genesis 1:27 should be understood.[14]

Standardly, early-modern commentators on Genesis insisted that 'without question the woman was made after the image of God as well as the man'. Divines attributed to women complete spiritual equality in the same breath as they insisted on their natural subordination. Although God made women 'inferior ... in outward condition, yet he hath made them equal in the inheritance of life'. Anglicans and Puritans agreed wholeheartedly that 'woman as well as man was created after the image of God' and as both were 'equal in creation, so there is no difference between them in state of grace'. Women were inferior to males in virtually all natural respects: they were blessed with less strength, less intelligence, and less self-control, but 'Moses tells us, that both male and female ... had an immortal soul'. God gave women stingy portions of natural talents, but in the one supremely important area – their invisible, immortal souls – He had given them as ample rations as males. Jews were criticized for 'not knowing the Scriptures' because they held 'a base conceit of women, as that they have not so divine a soul as men, that they are of a lower creation, made only for the propagation and pleasure of man, etc.' The Jewish practice of excluding women from worshipping in the synagogue was characterized as another error. Any suggestion that women were not spiritual beings was condemned out of hand. The Bible's account of Creation 'condemneth the folly of those that use to say that women have no souls; which though it be a most sottish speech, yet no opinion is so sottish which hath not some to embrace it'.[15] The fourth-century bishop, Ambrose of Milan, and the fifth-century theologian, Theodoret, were two of those said to have doubted the spiritual equality of women.[16] Their beliefs were almost universally rejected. Genesis 1:27 was typically expounded by arguing that 'him' referred to mankind as a species and 'them' to the two sexes. By using the singular pronoun, God had not meant to exclude women from sharing his image. Those adopting (pseudo-)Ambrose's and Theodoret's interpretation were accused of 'manifest error'.[17]

John Donne vigorously rejected as heretical the notion that woman had no soul. She was made in God's image and at the resurrection 'her sex shall not diminish her glory'. Donne discussed (pseudo-)Ambrose's commentary and its doubts about whether woman was made in God's image. He argued that this opinion alone showed Ambrose was probably not the real writer.

> No author of gravity, of piety, of conversation in the Scriptures could admit that doubt whether woman were created in the image of God, that is, in possession of a reasonable and an immortal soul.

Donne's sermons propounded the orthodox theological belief that male and female souls were essentially alike. In insisting that women, too, had souls, Protestants were following in the ancient traditions of the Church. Augustine and Aquinas had taught that woman 'was also made to the image of God', that 'the soul has no sex', and that 'in those things that relate to the soul, the woman does not differ from the male'. Their views established orthodoxy and Catholics continued to maintain that women were spiritually equal from the Middle Ages to the Counter-Reformation and beyond. Catholics welcomed the adherence of women 'under the title of the devout sex, which is given you by the Church'. Woman's spiritual equality was accepted by Roman Catholic theologians and protestant Reformers alike.[18]

Some Scriptural texts seemed to imply that woman was not on a par with the male spiritually, but this implication was never sanctioned by the orthodox. One such text was where Saint Paul taught woman 'shall be saved in child-bearing, if they continue in faith and charity and holiness with sobriety' (I Timothy 2:15). An obvious interpretation of his words was that woman's salvation depends on the reproduction of well-behaved offspring. The outrageous *Disputatio perjucunda* seized on this point.

What sane man has ever taught that living faith is found in a woman? Indeed the Apostle proclaims the contrary and takes away faith from every woman, writing: the woman is saved not through faith but through child-bearing.

In their enthusiasm for keeping women in their reproductive place, a few more conformist theologians teetered on the brink of endorsing this view.

Yea, they shall be (saith Saint Paul) saved by bearing of children and bring-ing them up in faith, love of God, and in holiness. For this is their office, this their function and calling wherein they ought to please God and to attain everlasting bliss.

Woman's role was not to meddle with religious doctrine, but to bear and rear children; she could do nothing else that would please God as much. For Protestants supposedly convinced that 'justification was by faith alone', this was dangerous ground.[19]

The orthodox interpretation of I Timothy 2:15 was that 'they' referred to the mother(s) not to children; that is, a childbearing woman would be saved if she herself persevered in the faith. The shift in pronouns (from 'she/her' to 'they') was ignored or glossed away. 'For who could believe that Paul wanted to make the salvation of the mother depend on the chil-dren's perseverance?' Theologians insisted that Paul only meant women would be 'more blessed' on account of bearing children, and 'attributeth the cause of salvation to faith and to love etc., not to the procreation of children'. Others employed the argument that women were here being exhorted to virtuous conduct which was impossible without faith. Women who saw the text as 'heightening their perplexities' by suggest-ing that 'if a woman continue barren in a state of marriage there is no hopes of her salvation' were ridiculed. Any such reading of the text was a

false interpretation – so remote from the wisdom, goodness, and truth of God that it is strange the fancies even of women (which it is confessed are very forcible) should be able by any chemistry of the imagination to extract such a disproportional meaning from that text.

On the contrary, Paul was merely trying to cheer and console women after 'having fixed that sex in their due place of subordination in the church'. The text was almost always glossed to stress that, in relation to salvation, the woman's 'faith, love, and sanctification' were more impor-tant than dutiful acceptance of her reproductive function.[20]

Paul's Epistle to the Corinthians caused further difficulties, for it implied that woman was not made in God's image. He taught that 'a man ... is the image and glory of God: but the woman is the glory of the man' (I Corinthians 11:7). This text was frequently cited to justify the subjection of the female to the male in both secular and religious affairs.

John Milton came close to expounding this text as (pseudo-)Ambrose had done, arguing 'that the woman is not primarily and immediately in the image of God but in reference to the man'. Yet even Milton allowed that a woman could be 'herself the redeemed of Christ'. More orthodox theologians than Milton, when it suited their purposes, would point to Paul's words as proof that women were not males' equals in ecclesiastical matters. The Jesuit, Robert Parsons, used them to argue against Elizabeth I's supremacy of the Church of England. The protestant zealot, John Bunyan, used the text in much the same way. The Holy Spirit forbad women 'to teach, yea, to speak in the church of God'. They 'are not builded to manage such worship, "they are not the image and glory of God, as the men are"'.[21]

However, in purely spiritual matters, when ecclesiastical place and preferment were not at issue, Saint Paul's words in I Corinthians 11:7 were normally explained away. The Catholic commentator, Cornelius a Lapide, argued that Paul only called woman the image of man 'improperly and analogically', because Paul must have known woman was equal and equivalent to the man in her rational soul since both sexes were made in the image of God. Like males, women had souls; natural inferiority disqualified woman from secular offices, but spiritual equality entitled her to a place in heaven. Paul never intended to deny this; he only wished to argue that God's 'authority' is seen better in the male. In the more important area of 'righteousness and holiness', woman resembled God just as much as the male did; 'the Apostle denieth not the woman, as she is a creature, to be made in the image of God'. Paul was just explaining that the husband is 'more honourable and must have the pre-eminence; in which the woman is rightly called the glory of the man, because she was made for him and put in subjection to him'.[22]

Early-modern divines placed a strict limit on the literal interpretation of Scripture in the case of the Timothy and Corinthians texts. The established Christian tradition (that women did have souls) was used to interpret Paul's meaning; it was not deduced from the apparent sense of his words in isolation. Scripture was the word of God and the foremost authority for these theologians, but they represented its sense in a manner consonant with existing beliefs, and dismissed, ignored, or simply did not see alternative readings wholly out of keeping with these views. Christians of all denominations insisted that women, too, must grasp the 'means of salvation'; their nature, too, was 'immortal'. They also should hear and read the word of God and instruct their children in it. 'Neither doth Saint Paul reprove women for talking of the Scripture ... so it be with modesty seemly to the sex.' Woman must have religious knowledge so 'she may learn to know her own sins' and 'reading of spiritual books also is commended by the Fathers'. When insisting on the need for religious instruction, woman's weakness, foolishness, and emotionalism were ignored and her spiritual equality was accented.[23]

Nonetheless, the belief that males more resembled God was latent in the writings of many early-modern divines; Paul's teaching to the Corinthians was not completely ignored. A 'good woman' was described as 'in her soul the same that a man is, and she is a woman only in her body; that she may have the excellency of the one, and the usefulness of the other': this assertion of woman's spiritual equality was phrased to suggest not that 'souls have no sexes' but that all souls were male. The same ambivalence could be detected in many commentators; they taught explicitly that in the spiritual realm male and female were equal, but clearly felt that some were more equal than others. Divines accepted women as spiritual beings but could rarely refrain from implying that God had a closer relationship with males. Women were lucky that Paul had not meant to deny that they could share in God's grace, for 'I assure you, if God had had respect to the sex, your estate of women hath been miserable'.[24] God's image was 'common to both the man and woman' but in the male 'after a more special manner'. Women shared God's image only 'in the second degree'. Biblical commentators rejected the dangerous suggestion that women were not spiritual beings, but happily quoted Paul in order to uphold male authority: 'the husband is to be honoured of the wife, because he beareth before the woman the image of the glory of God ... I Corinthians 11:7'. Women should 'rejoice' because they 'resemble man' in having faith. Even in spiritual matters, women tagged along behind. Most theologians accepted the doctrine of God's salvation of women but their tone suggested that, like the peace of God, this was a mystery passing all understanding.[25]

Even more than Scripture, much of the classical tradition suggested that women were incapable of virtue to the same extent as males. The word 'virtue' comes from *vir*, a male – the Latin word, *virtus*, meant 'manliness'. There was something inherently odd about saying that the female could be as 'manly' as the male. Before Christianity shifted its definition, virtue referred to the strength, bravery, toughness, and self-control that allowed males to assert themselves over outrageous fortune. No classical author readily thought of these as female characteristics. The Latin word for woman was *mulier*; it was derived from *mollior* – meaning softer, gentler, and more pliable. In classical moral philosophy it was the male, *vir*, who should have *virtus*, force, vigour and courage. Only when Christianity offered a less macho description of moral perfection (faith, hope, and charity) did it become slightly easier to ascribe 'virtue' to women.

Classical authors had been sceptical about whether the female was possessed of rationality in the same way as the male. Plato doubted 'whether women be reasonable or unreasonable creatures', and Aristotle maintained that the 'deliberative faculty' of the woman was inferior to that of the male. However, the classical tradition was not univocal. Xenophon had said that God granted 'to both [sexes] impartially memory and attention' and 'the power to exercise self-control'. Plato did not always dismiss women's abilities; in his *Republic* he allowed that women could rule and philosophize and maintained 'that if there be any distinction between their sufficiency and ours, it is not essential but accidental and such a one as is grounded merely upon use'. Plutarch thought 'the virtue of man and woman ... all one and the very same', and Seneca that women had 'the same vigour and free faculty of mind as men have, to apprehend that which is honest'. Their views made it easier for early-modern theorists to argue against Aristotle.[26] His assertion that women could not be truly virtuous or rational was one to which most Christian apologists gave short shrift. They standardly attacked 'that heathenish conceit of the Philosopher that virtue is not properly said to be in women'. Early-modern theorists knew that Aristotle thought woman a mere 'aberration of nature' but they rejected this, insisting that 'Scripture, the word of truth, dignifieth them as consisting of the same essential parts and capable of the same celestial perfections with men'. Theorists resorted to *ad hominem* insults when considering Aristotle's view of women's rationality and virtue, dismissing it as stemming from 'cankered malice'.[27]

When accepting Aristotle entailed the possibility that women might be deprived of the hope of heaven and freed from the threat of hell, his authority could not compete with Christian tradition. Were the female sex to adopt the 'absurd' and 'mad conceit' that 'women had no souls', they might

> commit all licentiousness with boldness, for if they have no souls it could be no fault in them more than in brute creatures to give over themselves to all sensuality and libidinousness.

Christian divines, eager to exclude any pretext for such conduct, insisted that the immortal soul and 'rational mind' were equally in the male and female sexes. Sex was crucial in determining familial, social, and political roles, but 'sexual distinction should make no difference at all to virtue'. In women, too, 'virtues dwell, or should, at the leastwise'. Woman's natural flaws would not exclude or excuse her on the Day of Judgement; God would sit in judgement of her sexless soul. God had instilled in women 'as well as in men a feeling of that which is honest', and He treated all the elect – male and female – alike. 'For grace and God's fear make as great a difference among men as amongst women.' All divines concurred that 'wives are as straightly bound to be religious as husbands'. However naturally defective woman was, she was obliged to conform to religious precepts and moral standards just like the male.[28]

The implications of spiritual equality in early-modern thought should not be overemphasized. In particular, it is important to remember that spiritual equality was *not* held to involve ecclesiastical equality. Woman's soul might be equal to the male's in the eyes of God, but this was not construed to mean that she should play the same part in teaching, preaching, or administering the sacraments. Christ welcomed all men into His Church but only males to positions of authority within it. Scripture and the Fathers were cited to show that women's natural inferiority and wives' subjection to their husbands excluded them from the difficult and authoritative role of church ministry. Woman's unsuitability for the ministry was one issue on which Catholics and Protestants were in sweet agreement. In their polemical literature, one denomination would sometimes malign another with the insulting accusation that they allowed women to exercise ministerial functions, but only the most outrageous sects in fact did so. Catholics taught 'that all male Christians universally, even boys or madmen may become priests' – their 'impediment' was 'merely contingent, unlike that of women'. Calvinists chorused that 'the sex of womankind is not capable of spiritual government and therefore may not ordinarily exercise any ecclesiastical function of authority in the church'. Anglicans concurred that God had made woman for housework and child-rearing, 'her hands (saith Solomon) must handle the spindle ... or the cradle, but neither the altar nor the temple'. All orthodox early-modern Christians accepted that God had decided to save females. They also insisted that He had decided to redeem them through an exclusively male ministry.[29]

Like medieval theologians, sixteenth- and seventeenth-century theorists thought the Gospel preached a message of spiritual equality but denied that Christ's teaching undermined natural inequality. Early-modern divines were certain that Christian religion should in no way subvert or overthrow natural hierarchies. Rather, it would establish and strengthen 'those interests that arise from natural relations or from voluntary contracts (either domestical or civil) betwixt man and man'. They enlightened any women so confused as to doubt this:

> Many women might think by reason of religion all were equal, as Christian servants likewise might imagine that religion should bring in an equality. But we must know that Christ hath freed men and women from the bondage of sin and death, and not from outward subjection.

The equality of Jew and Greek, slave and free, male and female was strictly confined to the spiritual sphere. The distinctions of sex and servitude were 'taken away from the spiritual body of Christ which is the Church, but not from the outward bodies and societies of men here upon earth'. Saint Paul's words to the Galatians* should not be taken to extend beyond the supernatural domain, 'for then there should be likewise no distinction of countries or sexes'. Christ offered 'everlasting happiness' to all, 'but in regard of the outward man they still abide masters and

---

*'There is neither Jew nor Greek ... there is neither male nor female; for ye are all one man in Jesus Christ' (Galatians 3:28).

servants, prince and people, bond and free, noble and ignoble'. Women should not expect Christ to lighten the load of natural subjection; instead they should anticipate equality in the life everlasting. 'Look not at things at present but at the estate to come when they shall – both husband and wife, without any subordination – be both heirs together of the grace of life.' Male and female were spiritual equals – one in Jesus Christ, but two for all other purposes.[30]

Spiritual equality was virtually the only equality that early-modern theorists were willing to accord to woman, and in their eyes it had no necessary implications for social and political relations. The assertion that prince and subject, lord and commoner, master and servant were all equal in Christ was regarded as quite compatible with the belief that there could (and should) be significant differences in power between them. Just the same logic was applied in the case of the sexes. The possession of an immortal soul opened up for women the possibility of eternal salvation not earthly liberation. Theorists admitted the existence of exceptions, but believed that as a rule women were physically, mentally, and psychologically inferior products. The exceptions proved the rule and did not undermine it. God and nature did nothing in vain, but had made woman in this way in order to show her proper place and purpose, which was as a help to the male. Woman's inferiority implied woman's subjection.

Subjection to women

Just as there were limits on the belief that woman was inferior, so also the doctrine of woman's subjection was qualified. Woman was naturally subject, but not all females were subject to all males. No one thought Elizabeth I was subject to her stable-boy. Wives were subject to husbands and daughters to fathers, but mothers were not subject to sons – on the contrary, they possessed authority over them. Likewise, queens held legitimate power over subjects and mistresses over servants. Sex was merely one factor in deciding a person's position in the hierarchy of legitimate power. Because it was such an important factor, theorists felt uncomfortable when faced by women in positions of power over males, particularly in the highest office – hereditary monarch. But even in this case, other determinants of legitimate authority, in particular the laws of succession, were generally believed to overcome the disability of sex.

In the middle of the sixteenth century, English theorists were positively eager to describe the limits of female subjection. They were motivated first, by opposition to French Salic law and second, by the accession of Mary and then Elizabeth Tudor to the English throne. In both cases, Englishmen wanted to show that females could succeed to the throne and exercise royal power and prerogatives when they did. The Salic law of succession dictated that a woman (and a male whose claim came only through a woman) could never accede to the throne; the French thought it so fundamental a law that even an absolute king could not will the crown to a female. Early-modern English theorists alleged that the French had only adopted Salic law in order to exclude from the French throne the rightful claimant, the English king Edward III. More than two centuries after 'the exclusion of King Edward the third from the crown of France upon this pretence', English irritation was still evident.[31]

William Shakespeare included a vigorous attack on Salic law in *Henry V*. He poured scorn on the poor historical and legal basis of Salic law and ended by concluding that all French kings

> ... appear
> To hold in right and title of the female;
> So do the kings of France unto this day.
> Howbeit, they would hold up this Salique law
> To bar your Highness claiming from the female,
> And rather choose to hide them in a net
> Than amply to imbar their crooked titles
> Usurped from you and your progenitors.

Shakespeare put forward the standard English view that Salic law was a sleazy French trick perpetrated to prevent English monarchs from exercising their proper rights. On the matter of Salic law, Englishmen felt honour-bound to disagree with the French. Richard Knolles translated Jean Bodin's *The six bookes of a commonweale* into English, but he mangled the translation where it involved ridiculing female rule. Knolles clearly approved most of Bodin's thought but he added marginal notes strongly dissenting from defences of Salic law, for example: 'Poor French shifts, for the avouching and proof of the Salique law'. English political theorists attacked at length Bodin's exposition of Salic law and frequently accused the French of using Salic law merely as a pretext to exclude the English monarchs from their entitlement.[32]

French theorists, of course, defended Salic law with as much enthusiasm as English ones attacked it. Although many French defences insisted that Salic law was in accordance with natural law, most maintained that it was a particular custom of France, not generally binding on all nations. The discussion of this point by the French civil lawyer André Tiraqueau was typical; he argued that it was unnatural for a female to succeed before a male of the same degree and this was why primogeniture passed over daughters in favour of their younger brothers. Yet he also admitted that customs varied about female succession to the crown, and that at many times and in many realms women had succeeded. In other kingdoms, such as France, reasonable customs excluded the female. Even Frenchmen rarely argued that natural law forbad woman's rule at all times and in all places.[33] Nicholas Harpsfield was a Catholic defender of Mary Tudor who attacked the pamphlet, *A Glasse of the truth*, which defended Henry VIII's unorthodox attempts to obtain a male heir on the grounds of the disastrous consequences of female succession. Harpsfield complained that the pamphlet's author spoke 'neither like a good Englishman nor like a good Christian man. Somewhat tolerable this talk had been in the mouth of some Frenchman' but for an Englishman 'this discourse was not very seemly'. Well aware of his audience's anti-French prejudice, Harpsfield wanted to smear the rejection of female rule as an 'un-English activity'.[34]

The English tradition of hostility to Salic law worked in favour of female rights to succeed to the throne, but the case of Henry VII served royal women's cause less well. Henry VII was crowned king of England after his army had defeated and killed in battle Richard III. Apart from any claim that might stem from conquest, Henry could only claim to be legitimate ruler as the representative of either the House of York or the House of Lancaster. In fact, the best Yorkist claim was held by Elizabeth, daughter of Edward IV, to whom Henry was betrothed. Any Lancastrian claim Henry might make must theoretically have been subordinate to that of his mother, Margaret Beaufort, from whom it stemmed. With the acquiescence of Parliament, neither woman was mentioned in Henry's assertion of his right to the crown. Writs ran only in Henry's name, coins bore only his head, the heirs of his body (even if not of Elizabeth's) were established in the succession. Although few doubted that Elizabeth's claim was theoretically better, Henry (not their son Henry VIII, who inherited his mother's titles) continued to rule for 6 years after her death. In 1485, tracts were not published denying women's right to rule in order to justify Henry's actions; female claims to the throne went apparently unnoticed.[35]

In the 1550s, the case was altered. The accession of (the Catholic) Mary Tudor, and then (the protestant) Elizabeth Tudor to the throne of England gave Englishmen of both religious denominations reason to defend female rule. After the death of Mary's and Elizabeth's younger brother, Edward VI, no male had a plausible claim.[36] The Wars of the Roses had shown how disputes about the title to the crown disrupted and endangered all Englishmen's lives. Reluctance to see the crown on a woman's head was nowhere near as intense as the fear of seeing it fought for across England's green and pleasant fields. The policies of the zealously Catholic Mary Tudor helped provoke the fervently protestant John Knox's intemperate attack on her (and any woman's) right to rule in his *First blast of the trumpet against the monstrous regiment of women*. This was published (with singularly poor timing) in 1558, very shortly before the protestant Elizabeth's accession in November of that year. Obviously, under these circumstances almost no Protestants felt it in their interest to support Knox's stance. Another protestant exile, Christopher Goodman, expressed views similar to Knox's in some respects, but he abjectly withdrew them after Elizabeth acceded. As one Anglican apologist commented, there were so few protestant attacks on women's right to

rule that 'if there had been but one less, ... there had been but one at all'.
Moreover, until the execution of Mary, Queen of Scots in 1587, the main
Catholic claimant and heir presumptive to Elizabeth was also a woman.
Living in such glass houses, neither faction threw stones at female rule.[37]

Theorists mounted various arguments in defence of queens' rights to
accede and rule. One was that God in His providence disposed of kingdoms and if He saw fit to place a woman on the throne it was not for
mere mortals to resist His ruling. The accession of a woman would probably lead to trouble, but it was God's judgement and meekly to be
accepted. It was by 'God's will and order' that a woman was rightful
heir, for 'He could have provided other'. Men simply had no right to
resist God's providential disposition of the crown. God gave kingdoms
to good and bad without 'any distinction of sex', age, or condition, and
men must accept His provisions. Defenders of Elizabeth used Romans
13* – the favourite text of theorists preaching political obedience – to
show 'that the queen of England is queen by God's ordinance'.[38] Even
Mary's determined opponents believed they were obliged to submit
when the Lord took 'away empire from a man' and gave it to a woman,
'whom nature hath formed to be in subjection unto man'. God's action
was 'an evident token of [His] anger towards us Englishmen', but it was
unwise to defy an angry God. Those who had survived many years of
female rule adopted a calmer tone, but their argument was essentially
the same. The same obedience owed to kings 'is due also to queens when
they have, by God's special providence, sovereignty of government'.
Defenders of female rule did not feel obliged to abandon the orthodox
view that women (queens included) were inferior, they simply asserted:
'The inferiority of their sex must not in this case prejudice their authority
over their subjects'. Succession to the throne was in God's hands, 'when
God doth give a child ... or a female, subjects must be content and submit themselves to God's pleasure'.[39]

God's providential disposition of kingdoms was invoked specifically
in the case of the crown, but queens' rights were simply one instance of
all women's rights to inherit property and any associated power. Female
inferiority rendered women bad choices for elective offices, but their title
to 'hereditary dignities' was indisputable. As usual, early-modern theorists found a biblical authority to support their position: Numbers
27:6–8.† In default of male heirs, females should inherit. English exegesis

*'Let every soul be subject unto the higher powers: for there is no power but of God: the
powers that be are ordained of God. Whosoever therefore resisteth the power, resisteth the
ordinance of God' (Romans 13:1–2).
†Zelophehad died on the journey to Canaan leaving no sons; his daughters petitioned
Moses for an allotment of land, and Moses consulted God.

  And the Lord spake unto Moses saying, The daughters of Zelophehad speak right: thou shalt surely give
  them a possession of an inheritance among their father's brethren; and thou shalt cause the inheritance of
  their father to pass unto them. And thou shalt speak unto the children of Israel, saying, If a man die, and
  have no son, then ye shall cause his inheritance to pass unto his daughter. (Numbers 27:6–8)

of Numbers 27 explicitly pointed out the connection between divine providence and female succession to any Frenchman or Knoxian who might have missed it.

> We may note how grossly and grievously they err that condemn the government of women, when crown and kingdoms by lawful descent in the all-guiding providence of God fall unto them.

The sex of the heir to the throne was an expression of God's will not a genetic accident. The 'divine law' given in Numbers was 'diametrically opposite to the Salic law of the French'. God's will worked through rules of inheritance; He 'disposeth kingdoms,' and only 'busy heads ... unlawfully seek to withstand the law of nature in succession'.[40] Writing in defence of James VI and I, Thomas Craig also had good reason to defend the right of succession through the female, for James largely derived his title to the throne of England from his mother Mary, Queen of Scots. Craig insisted there was 'certainly no manner of reason to despoil women of, or to exclude them from their patrimonial and paternal inheritance'. He added that the precept in Numbers was one of those 'which do continually bind and oblige us'. God had not been speaking only to Jews about the laws of succession they should apply in the Promised Land; daughters' rights of succession were eternal. Male primogeniture diminished daughter's rights in inheritance but could not legitimately abolish them.[41]

In the book of Numbers, God had given 'a general law for inheritances that they should pass to the female, if the father died without male issue'. The rule was natural as well as divine and applied not only to queens but all female governors: 'in royal rule and other inferior hereditary government, it is a common rule by the law of nations that, in the absence of sons, daughters succeed'. Jean Calvin protested that he had tried to discourage John Knox from publishing the *First blast*, by arguing that 'both by custom and public consent and long practice it has been established that realms and principalities may descend to females by hereditary right'. Calvin's affirmations of support for women's rule were probably not without ulterior motives, for in 1559 it was clearly in the interests of the Church at Geneva to establish good relations with Elizabeth. Nonetheless, it was an argument frequently employed and he clearly thought it plausible.[42]

Theorists commonly defended daughters' right to inherit, even apart from upholding woman's rule for political reasons. In civil law, the dowry given to a daughter at marriage was seen as her share of the family estate. Catholic casuists taught that fathers could be legally compelled to endow their daughters (even, some argued, illegitimate ones). Legal theorists characterized 'law which entitles the daughters together with the males to the inheritance of a father' as 'perfectly natural'. Civil lawyers accepted that by 'natural equity' and 'common right' females should succeed equally with males to their fathers' possessions. They recognized that feudal law customarily excluded the female, and maintained that only necessity and the public good could justify such 'apparently wicked' custom. Natural equity meant that, whatever written law laid down, daughters should not be 'altogether excluded from inheritance'.[43]

Because they regarded it as natural that daughters, too, should inherit, early-modern theorists condemned Roman laws that excluded women from inheriting goods, and exhorted parents to 'provide for their daughters as well as for their sons and not to leave them to the wide world'. Those fathers who disinherited their daughters 'break both God's law and the law of nature too'. There was a general consensus that daughters had rights to inheritance like their brothers, although not such extensive ones.[44] Male primogeniture obtained in most parts of England, but daughters inherited when there were no legitimate sons. One legal historian describes the Court of Chancery as 'very sympathetic towards the young single girl who came before the Court seeking her inheritance'. Another recent study

> clearly indicates that in early modern England daughters inherited from their parents on a remarkable equitable basis with their brothers. ... When it came to distributing their patrimony, fathers normally gave their daughters shares comparable in value with those of their brothers, although girls usually inherited personal property and boys more often real property.

English medieval practice amongst the lower orders had also generally been that daughters inherited in default of sons, and were entitled to dowry even where there were brothers. It was generally accepted that heiresses should not be denied their rights, and defenders of female rule based their arguments on this consensus.[45]

The laws of inheritance overrode the normal rule that women should be subject, and placed some women in positions of power over males; a power exercised alone and in her own right, if the woman were a single adult. In many discussions of female subjection, theorists slipped without comment or conscious ambiguity between the terms woman and wife. It was a tag of English common law that all women 'are understood [as] either married or to be married'.[46] The status of both males and females improved at marriage; the married took precedence over the single. Social theorists argued that bachelors behaved immaturely, but 'taking a wife induces dignity'. Most women married. The position in France has been described as one where '[a]part from the religious, almost all adult urban women in the first half of the sixteenth century were married or had once been so'. Although England had a higher proportion of spinsters than some countries, it did not significantly differ. When one divine noted of Eve that 'no sooner was she a woman, but presently a wife; so that woman and wife are of the same standing; and the first vocation of man was *maritari*, to be a husband', he was describing the facts of early-modern life. Yet in both common and civil law, a single woman had the same legal rights as a male in private law.[47]

Despite the ease with which it was assumed that all mature women were wives, Renaissance theorists readily distinguished, when it served their polemical purposes, between speaking about the wife in comparison with the husband and speaking 'absolutely' about woman and the male. Discussions of adult women were usually – but not necessarily – about wives. The subjection of the female sex no more entailed that every woman should obey any male than the subjection of servants implied each servant should obey every master. Saint Peter had exhorted wives to 'be in subjection to your *own* husbands', not to any and every male (I Peter 3:1 [my italics]). Peter had added 'own', commentators maintained, 'to bound and circumscribe their obedience that it was to be only to their own husbands, not to others'. If a woman had to obey *any* husband, the world would soon be a brothel, 'and so when [Peter] persuades them to subjection, he cautions them against unchastity'. A woman's obedience was due only to her own husband or her own father.[48] English supporters of woman's rule were particularly motivated to establish that women and wives differed, and they leapt to spell out the distinction between a wife's subjection and a woman's freedom: 'Neither oweth every woman obedience to every man, but to her own husband'. The belief that the female sex was framed for subjection was accepted as generally as the belief that children needed adults to guide them; but – just as children were thought only obliged to obey their own parents – women need only obey their own husbands. '[N]either the divine law, nor the law of nature, nay, nor yet the Civil Law put any other woman under the subjection of men but only such as have husbands.' Notwithstanding the preferment of 'the male kind', wives need not be obedient to any but 'their own husbands'. The Swiss Reformer, Henry Bullinger, insisted that apostolic exhortations 'concerning the obedience of wives' were not to be utilized to exclude women from political power: they were 'not to be wrested unto reigning'. Another of Elizabeth's champions, Bishop John Bridges, used the same argument: Saint Paul forbad the authoritative rule 'of women not simply, but of such women as are wives. Neither of all authority, but of authority over the husband'. Women 'must be subject to their husbands, not to their family; therefore this reason doth not infringe the government of women'.[49]

Although wives were subject to their husbands, not all women were subject to any male. Some males were subject to some women; indeed, in the case of a queen, all the males in the realm were subject to her. When a queen regnant married, these two principles seemed to come into conflict. Saints Peter and Paul had not inserted exemption clauses for the case of married queens. 'God requires subjection of all wives .... A queen hath no more privilege than the poorest cottager's wife.' The married

queen must obey her husband; the subject husband must obey his sovereign. Could each obey the other at one and the same time? If not, did the
claim to obedience of the queen 'trump' that of the husband, or vice
versa? Renaissance thinkers could not comfortably leave the problem
unresolved. In 1531, *A Glasse of the truth* complained that Henry VIII
needed a male heir (despite Mary Tudor's 'many dotes and gifts')
because 'if the female heir shall chance to rule, she cannot continue long
without a husband, which by God's law must then be her governor and
head, and so finally shall direct this realm'. Henry Howard, later Earl of
Northampton, was a defender of Elizabeth I. Yet even he feared that anyone Elizabeth married would probably try to seize control of government
from her, because 'the course of nature and human ordinances do yield a
pre-eminence to the husband as head of his spouse'. This was true,
'albeit the order of government of the realm requireth a superiority in
her Majesty as chief of the state'. A spinster queen caused problems for
the succession; a married queen gave excessive influence to a male not
born to be king.[50]

Early-modern defenders of female rule did their best to resolve the
conflict. One solution was the providential one. God had placed a
woman on the throne and joined married couples together. His creatures
should leave God to resolve any tricky points that might arise: 'we must
commit all this to God', Bishop John Aylmer urged, 'let us not meddle
with bridling of queens in marriage this way or that way; we are none of
those to whom it shall be said, Who shall give this woman?' Biblical history showed that God had handled the problem before. Deborah,
'though a married woman and subject to a husband, reigned over the
Lord's people notwithstanding'.[51]

Another solution was for theorists to distinguish between political and
conjugal power. The belief that there were different kinds of power had a
long and respectable history. Aristotle himself had distinguished the
power held by a political ruler from that of a father, and both from that
of a slave-owner. If a son were a magistrate, he had to show his father filial obedience; the father had to show him civil obedience. Each obeyed
the other in their respective spheres of power.[52] Defenders of married
queens regnant continued to maintain that a *wife's* subjection was divinely and naturally ordained, but distinguished wives from women and the
political rule of queens from the domestic rule of husbands. Political and
conjugal power were distinct. A queen could be 'her husband's inferior
in matters of wedlock, and his head in guiding of the commonwealth'.
One of the champions of Mary, Queen of Scots reasoned that a married
queen 'may without any impairing or maiming of her duty to God or to
her wedlock, repress her husband's misdemeanours'. If the public good
required, a queen could discipline her consort and yet still not be
'exempted from such duties as the matrimonial conjunction craveth of
the wife toward her husband'. The Scots Presbyterian, Samuel

Rutherford, adopted the same position about Mary: 'Our late queen, being supreme magistrate, might by law have put to death her own husband for adultery or murder'. A queen regnant's marital subjection was thought quite compatible with executing her husband.[53]

The Spanish Catholic theorist, Francisco Suarez, also insisted that a queen regnant was not subject to her husband in political matters. Her consort did *not* control the administration of the realm, because this office belonged to the queen's 'own person' and therefore was not, like other goods, transferred as part of her dowry. On the contrary, the queen's husband was her subject in all matters relating to political jurisdiction. He denied that this form of subjection was contrary to reason or Paul's teaching. Like Scots and English political theorists, Suarez accepted that a queen could be subject to her husband in relation to marriage and the family, whilst being 'superior in relation to political government'. One German civilian argued that the relationship between a queen regnant and her husband simply depended on the custom and law of each particular country and on the agreement or treaty reached between them. He asserted that when Ferdinand of Aragon married Isabella of Castile, the two became joint rulers of both countries; in contrast, Philip II of Spain did not become ruler of England on marrying Mary Tudor, for he was not the partner of his wife either 'in rule or administration'. Later in the century, the German theorist of natural law, Samuel Pufendorf, echoed these arguments and examples when discussing 'women who have crowns of their own, [and who] take to themselves husbands without relinquishing their full authority over the state'.[54]

No mature woman was subject to any male except her own husband. Even a wife's subjection was only to her husband's conjugal power and this differed essentially from political power. A wife could be conjugally subject while being politically superior. A queen regnant's consort had to obey her in political matters even as he ruled her in domestic ones. Mary Tudor and Mary Stuart were subject *qua* wives, but superior *qua* queens; 'a woman being a prince hath as great authority as a man'. The principle of hereditary power eclipsed that of female subjection.[55]

It is interesting to compare the debate on female rule in the mid-sixteenth century with the apparent indifference to the issue in the later seventeenth century. When, in 1688, James II was ejected from the English throne and his newborn son labelled supposititious, the next heir on hereditary principles was James's daughter Mary. Her husband, William of Orange, held a better title to the throne than any other male, and in addition was on English soil at the head of an army. Nonetheless, even William's 'most extreme supporters' accepted that he could expect no more than joint sovereignty with Mary. While she lived, Mary II's name was used in all governmental acts. Jacobites bitterly (and to some extent truthfully) complained that 'all this was but pageantry and she in the mean time a Queen of Clouts, while the executive power was lodged solely in [William]'. Mary's absolute refusal to countenance any settlement that involved her holding power independently of her husband neutered her supporters, but it was clear that there were many ready to assert Mary's rights had she allowed them to do so. William's continued reign after Mary's death involved 'postponing and jostling out of its natural, lineal, and due place the right of the Princess of Denmark' (Anne – Mary's younger sister). However, the right of Princess Anne and her offspring to succeed ahead of any children William might have by another wife was accepted universally; female title in the succession had come a long way since Henry VII nonchalantly ignored the claims of his mother and fiancée.[56]

When Anne acceded to the throne of England in 1702, there was deafening silence on the matter both of women's capacity for rule and of the subjection of married queens regnant to their husbands. In Anne's reign, the political nation was less worried by the influence of her husband than by that of her ladies-in-waiting. Female rule did not provoke controversy in pulpits or press. One of the few theorists who did mention Anne's sex, complacently pointed to Elizabeth's reign and piously commented 'such another queen we trust God has now given us'. Women were still perceived as inferior but queens now were not. By the end of Anne's reign, a woman's rule perturbed Englishmen so little that it was merely an occasion for elegantly polite humour. In 1714, Bishop Thomas Sherlock preached that:

> To such a height of glory has this female reign arrived, such honour and such triumphs has it brought our nation, that should any future king prove unfortunate, Britain (perhaps grown superstitious upon the successes of her Queen) will wish, *He had been a woman.*

Whigs and Tories alike accepted a queen as readily as a king.[57]

Mothers and mistresses

The great majority of early-modern theorists accepted that a queen regnant held legitimate political power over all her subjects (even her husband). The female sex was inferior, but this was quite compatible with a woman holding political power over males. There was also a far more common case where sixteenth- and seventeenth-century theorists supported male subjection to the female – that of sons to mothers. No one in early-modern Europe doubted that children were morally bound to obey their father. The whole system of political patriarchalism was based upon this conviction, and even those who rejected it accepted paternal authority within the family. The father's authority was believed to be natural. Just as all the societies of which early-modern theorists were aware subjected wives to their husbands, so all subjected children to their fathers. Nonetheless, the theorists looked to the Bible for added support. There they found the fifth commandment: 'Honour thy father and thy mother' (Exodus 20:12). Because of the need to interpret 'honour' as involving 'obey' (in order to uphold fathers' power) and because of her explicit inclusion, commentators concluded that children owed obedience to their mother also. God did 'not say "obey your fathers" but your "parents" ... under which word both fathers and mothers are equally comprehended and that not without just cause'.[58]

Many discussions of children's duties simply stated that obedience was owed to parents without distinguishing between father and mother. They stated that 'a child is bound to obey his parents in all things in the Lord' and that 'the command of our parents' must be obeyed as 'being a natural effect of that authority they have over us'. Often theorists began by saying that parents should be obeyed and then slipped into talking about the father alone, since he was regarded as the key figure. However, no theorist ever suggested that mothers could be disobeyed with impunity.[59] The paradigm case of a disorderly family was one where 'the boy ruled the mother' and the mother ruled the father. Mothers should govern children just as husbands should govern wives. The mother was subject to the father, 'yet she is superior to her child as she is a mother and may command and must in no wise be neglected or disobeyed'. Samuel Rutherford opposed political patriarchalism but believed that parental authority was based on God's word: 'The mother hath a parental power as the father hath ... so the fifth commandment saith "Honour thy father and thy mother"'. The subjection of her sex did not undermine a mother's authority as a parent.[60]

God had so explicitly included the mother in the fifth commandment as a corrective, because 'the mother is the weaker sex and commonly most doteth upon her children, which maketh her look for less honour and them less to esteem her'. Early-modern texts often suggested that there was generally more affection between mother and child, and that this led mothers to exercise less rigid control than fathers. God stressed the need to obey mothers 'that none might take exceptions against his mother or think himself exempted from her jurisdiction'. Theorists insisted explicitly that children must not think that the mother's subjection to the father diminished her authority over them; she must be given 'equal honour with their father'.[61] When John Locke stressed the obedience owed to mothers, he was motivated by his desire to attack the political patriarchalism that formed the foundations of Sir Robert Filmer's absolutism. Parents had 'one common right belonging so equally to them both, that neither can claim it wholly, neither can be excluded'. Locke had a polemical axe to grind, and knew that Filmerians could not reject this commonplace of moral theory. No Christian could ignore explicit biblical injunctions; the mother must be obeyed. In Leviticus 19:3, God

'putteth her before' the father, 'that it might be the rather observed'.*
Biblical commentators compared the two key texts. The father was 'put
first' in Exodus 20:12, the mother in Leviticus 'to show that we owe this
duty promiscuously and indifferently to both of them'.[62]

The fifth commandment was interpreted as endorsing political as well
as familial power. Popular catechisms taught the faithful to reason that
God 'comprehended all that are above me under the tender and venera-
ble names of father and mother'. God had intended to include magistrates and rulers of all kinds in his command to obey parents: 'in the
father and mother ... God at first settled all that power which since is
derived from them to our governors according to their several places and
degrees'. God had mentioned mothers by name not only to ensure that
the authority of 'natural mothers' was not 'neglected', but also to prevent
'any other kind of authority or eminence wherewith women are preferred above others' being 'despised'. Legitimate power could not rightly
be disdained even if held by a weak and feeble woman. English political
writers stressed the divine rights of mothers in their attacks on arguments against female rule. In Jean Bodin's theory, one wrote, 'that which
is most monstrous and impious is that it is inimical to the laws of nature
for any child to obey and honour his mother, because she hath not prudence, magnanimity, and force of command'. Women's possession of
inferior talents afforded no justification for disobedience to mothers or
female governors. A mother's divinely endorsed parental authority not
only trumped the normal subordination of the female to the male, it also
implied that women could hold political power.[63]

Many early-modern theorists argued that obedience to mothers was
not only divinely commanded but also stemmed from the law of nature.
It may seem obvious to the modern reader that whatever duties children
naturally have to their father they must *pari passu* have to the mother.
This was far less obvious to the Renaissance thinker. There were two reasons for this. First, early-modern medical knowledge implied that the
mother contributed less to the child's constitution than the father.
Second, Roman law had given extensive power to the father – *patria
potestas* – but none to the mother.

Aristotle's theories had enormous influence on early-modern medical
science. In his physiological theory, the woman contributed little or nothing to her child's composition; she merely incubated the male's seed. A
woman contributed as little to the child she bore as a field did to the
crops that grew in it. Each simply provided nourishment and a suitable
medium where the seed could grow; in all its essential characteristics the
seed was already perfect and entire when sowed. A mother could only
assist or stunt a child's development in the same way that rich or poor
soil facilitated or retarded a plant's growth. Early-modern medical theory

---

*'Ye shall fear every man his mother, and his father' (Leviticus 19:3).

was deeply permeated with the Aristotelian view that each child 'doth naturally draw his original and beginning from the sperm and seed of man, projected and cast forth into the womb of woman as into a field'. The child was formed in the male's image, but someone 'must supply the part and office that the earth does to vegetables, ... to contain, preserve, and supply it with fitting nourishment, which is done by the female'.[64] The Aristotelian view was not unchallenged. The most influential ancient dissenter was the Hellenistic physician Galen. He had argued that a child was formed by the union of the male's perfect semen with the female's 'imperfect semen'. The father's contribution was greater but the mother did contribute something, and this was why children sometimes resembled her. With great confidence but virtually no anatomical knowledge (by modern standards), various other classical medical theorists had taken one side or the other in this debate. Zeno and Hippocrates had adopted the same stance as Aristotle; Pythagoras, Epicurus, and Democritus, the same as Galen.[65]

Renaissance medical theorists continued to debate with equal enthusiasm and equal ignorance 'whether or not the female contributes any spermatical particles towards the formation of the foetus in coition'. The problem was difficult to resolve because ovulation was not understood in the early-modern period. Theorists could easily see the role of male sperm in reproduction but had difficulties finding the equivalent in the female. What 'seed' (if any) did woman add to the process? Their best guess was the lubricating mucus produced by Bartholin's gland during sexual intercourse. This assumption produced considerable uncertainty about whether woman's seed *was* necessary to reproduction, for observation showed that a woman during (brief) coitus might not 'seminate' (in this sense) and yet still become pregnant. Many reached the conclusion that 'to say woman has true seed is false and erroneous', or hesitantly said she had 'seed, or somewhat like seed'.[66] Whether they adopted Aristotle's or Galen's view, physicians generally agreed that the mother contributed less to the composition of the child than the father. Even if Galen were right and the woman did contribute 'seed', 'notwithstanding it is out of all doubt that the quality of them is not alike: for the seed of man doth exceed woman's seed in heat and thickness'. Woman was created for reproduction but even in this area it was argued that her physical inferiority led her to contribute less.[67]

It is interesting that moral and political theorists generally disregarded physiological theory when discussing children's duties to their mothers. Although biologically the male was thought 'the principal cause of the propagation of children', moral arguments commonly assumed that the mother's role in reproduction was significant. One theologian insisted 'that father and mother together generate the soul in children'. Some even reasoned that mothers' role in nourishing and carrying the foetus and breast-feeding the infant meant they contributed more to reproduction. In Aristotle's physiology of reproduction, every mother was just a living incubator, a 'surrogate', but for moral purposes this was ignored. Divines maintained that 'the power of a parent is by the bond of nature in that a child hath his being from his parents'. Medical theory indicated that had the father copulated with a different woman, essentially the same child would have been produced. Protestant and Catholic moral theory ignored contemporary science and said: 'Both parents are, under God, a like means of their children's being'. They based parental authority on joint contributions to the reproductive process; 'in relation to their children they are both as one, and have a like authority over them'. The fifth commandment's inclusion of the mother was defended on the grounds that she 'hath a great share in the begetting' of her children. Theorists argued from the mother's part in bearing the child to the assertion that 'the subjection and obedience of every child is alike due to father and mother'.[68]

A second reason for doubting the naturalness of maternal power was the example of Roman law. Just as the classics were often seen as embodying the philosophical knowledge that pagans could achieve by unaided natural reason, so Roman law was regarded as an epitome of a natural legal code. Ancient Roman law had afforded the mother virtually no legal authority over her child. Early-modern civil lawyers were well aware of this and defended confining *patria potestas* to the father. The Italian civilian, Scipio Gentili, was sure that restricting the *patria potestas* to the father alone was natural 'as Aristotle clearly testifies'. Other civil lawyers adopted the same stance.[69] Yet even civil lawyers did not deny that children should obey their mothers. The *patria potestas* consisted of a complex set of legal rights. Civil lawyers could not support the extension of all these to mothers also, but this did not mean that they believed that God had got the fifth commandment wrong. Scipio Gentili conceded that just as much respect and affection were due to the mother as to the father. Another Italian civilian, Giovanni Sordi, was typical in arguing that a mother, like a father, should be supported and maintained by her children; for parents,

> as I have often said, should not be treated unequally; nor is the mother to be laid under more obligations than the father, and wholly the same logic should be applied to the mother as to the father and therefore they should have the same rights.

Civil lawyers, like divines, thought that mothers had rights over their children.[70] One English theologian argued that Roman law was not wholly in accordance with the law of nature simply because it denied maternal authority: Roman law was unjust to afford the father power but 'the mother none; when as God's law always, as well as the law of Moses gives them equal interest'. Protestant legal theorists explicitly rejected the Roman civil law belief that mothers had no authority over their children.[71]

Roman law had made the father's (but not the mother's) consent necessary to a child's marriage. Most protestant theorists followed civil law in holding that parental consent was necessary to a child's marriage. They maintained that the mother's as well as the father's was required. 'It is the mother's duty as well as the father's to provide a godly husband for her daughter, or a virtuous wife for her son.' Divines often cited Saint Paul in support of their contention that both parents had legitimate authority over their children's marriages (I Corinthians 7:36–8). The Corinthians text was glossed as supportive of mother's rights despite the fact that Paul's words were obviously addressed to fathers, not mothers. Early-modern theorists held that a mother's authority over her children's marriages was subject to that of the husband if he were still alive. They often qualified it in the case of a mother who had remarried. Nonetheless, virtually all Protestants agreed that a child could not marry without a mother's consent.[72] English lawyers were similarly respectful of mothers' rights. In one case, a mother's consent to her daughter's marriage was deemed valuable 'consideration', because 'by the law of nature' she had 'a special strike' in this matter. The Court of Chancery disinherited an orphaned woman who had married with the consent of noble executors but not that of her grandmother, because she had ignored 'that due obedience [to] which law and nature oblige her; and she should have applied to her grandmother for consent'. In another case, Ecclesiastical High Commission 'allowed the act book to be used to record' one couple's request 'on their knees' for forgiveness for marrying without the mother's consent. In marriage, the sex of the forbear was not deemed relevant to his or her authority over the child. English divines in general, like Continental Protestants, insisted on the need for maternal, as well as paternal, consent to a child's marriage.[73]

The belief in maternal authority long predated the sixteenth century. The doctrine that mothers held authority over their children had been often rehearsed in the context of the ban on the marriage of parent and child. Theologians had always insisted that the most unnatural and immoral case of incest was that of a mother marrying her son.[74] The natural (as opposed to biblical)* argument mustered was that a mother who married her son would be subject to him *qua* wife, which was wholly incompatible with her authority over him *qua* mother. Thomas Aquinas and Duns Scotus were amongst the many who employed this argument; both insisted that a mother held authority over her son by natural law. Lecturing in the 1530s, the Scottish Catholic canonist, William Hay, continued to appeal to this principle: 'if a son marries his mother, the son would be above his mother in virtue of the marriage and more honourable, yet in virtue of his son-ship he would be inferior to her and less

---

*'The nakedness of thy father, or the nakedness of thy mother, shalt thou not uncover: she is thy mother; thou shalt not uncover her nakedness' (Leviticus 18:7).

honourable. This is impossible'. A son could not marry his mother without producing fundamental conflict in the natural hierarchy of authority. This was not a doctrine that the Protestant Reformers abandoned; Theodore Beza and Hugo Grotius both employed it. English theorists, too, continued to use this argument from the son's subjection to the mother to defend the prohibition of mother–son marriage. The English common lawyer, Sir John Vaughan, readily adopted the scholastics' argument and proscribed such marriages: 'For as a husband to her, the son is both to command and correct the mother as his wife, but as a son to be commanded and endure her correction as mother'. A son was subject to his mother's discipline just as much as a wife was to her husband's. English divines adopted exactly the same position; 'if a son marries his mother, she who is in authority greater by the right of geniture becomes *minor in matrimonio* [the lesser in marriage]'. It was simply 'a contradiction of rights that the same person should be the superior mother and the inferior wife'. The hierarchies of birth and sex should not be set at odds with one another; wives should be subject to husbands and sons to mothers.[75]

Mothers, like fathers, naturally possessed authority over their children; 'for concerning children, nature gives their mother power over them and God confirms it upon them jointly with their fathers who have the priority of that authority'. The mother's authority was subordinated to the father's because the wife was subject to her husband, not because a child (son or daughter) owed her no subjection. Those who discussed the problem were clear that the mother should not be obeyed against the father's command. Father and mother were equally entitled to reverence, 'sustentation and maintenance', but 'concerning obedience, because the man is the head of the woman and the master of the family, obedience ordinarily is rather to be given to the commandment of the father than of the mother'. By subjecting wives to husbands, God had established a hierarchy within the family; the mother was superior to the child, but the husband was superior to the wife – the child must obey the higher authority. 'God himself in the very beginning, subjected even the mother to the father and thereby plainly showed that he ought to be preferred.' God had not left children uncertain whom to obey when there was discord between parents; where the father and mother gave contradictory commands, 'I doubt not but the father must be obeyed'.[76]

Sixteenth- and seventeenth-century theorists believed there was a clear chain of command within the family; the husband and father stood at its pinnacle. They were equally clear that where the father did not countermand the mother's command or where the mother was a widow, the child must obey her. '[I]f the father's contrary authority be not interposed or if the father be dead, then is a mother simply and absolutely to be obeyed.' The mother possessed power over children in her own right; it was not simply derived from the father. A child was duty-bound to obey his father, but 'if the father be dead, we owe the same duty to our mother which is alive'. She held authority over her household even when not joined with the father, 'for sure it is nowhere against the law of God for a widow to govern her family'. In early-modern theory, on the death of the husband the headship of the household passed to the widow. While her husband lived, the mother's authority was overshadowed by his, but 'when the mother is the only parent, then her authority increases and she is then solely to be regarded'. The contractarian theorist, James Tyrrell, used the case of a widow's authority over her children to attack Robert Filmer's political patriarchalism. If Adam had predeceased Eve, he stated that 'her natural right of governing the children which she herself brought forth' would have come before 'any right of her eldest son'. A mother's authority was latent during her husband's life but it would 'revive' at his death. Filmer was wrong to state that the eldest son became governor at his father's death, because

> the mother of the family hath the best right to the government ... after the husband's death . . .. For if the commandment of *Honour thy father and mother* signify more than bare reverence and respect ... then this obedience which was due to the father belongs likewise to her when his power ceases.

For contractarians (unlike political patriarchalists), a parent's authority largely ceased when the child came of age, but until then the mother held full parental power in the absence of the father.[77]

The belief that the mother held authority over her children, independently of the father during his lifetime and exclusively after his death, was not simply a contractarian notion constructed to subvert political patriarchalism. It was taught in virtually all tracts of domestic moral theory and contractarians were appealing to this consensus. Mothers needed power in order to guide their children, so God and nature gave it to them; for it was 'the duty of parents in general and in special of mothers' to raise children. Mothers were frequently exhorted to chastise and correct their erring children, and in pious homilies disobedience to mothers was castigated just as much as disobedience to fathers. Since the woman's natural role was in the private sphere of the home, '[h]ow to govern well her family should be her chiefest study'. In moral teaching on a widow's duties, governing and educating children always featured large. Both parents jointly and severally held authority over their children.[78]

The female sex was inferior as a sex, but early-modern political theorists did not believe this excluded women from all positions of authority over males. Another example of males' subjection to women was found within the household. A mistress[79] rightly exercised authority over her household and servants. The household was an important unit of early-modern England's structures of social discipline.[80] Domestic servants were subject to 'governors of families, husbands and wives' and bound to obey the commands of either, for 'though the husband is called the guide of the wife, yet the wife is called the guide of the house'. Woman's proper domain was household management and she was believed to hold the power to rule it by both natural and divine law. The power of the wife and mistress was 'the ordinance of God, Who appointed that the wife next under the husband should bear the chiefest sway in the administration of the family'. Christian and classical tradition both accorded the woman rule within the household. Inherited ideas of the natural division of labour entailed the belief that it naturally belonged to 'the woman to take charge within'. In a properly ordered household, the wife's authority was subject to that of no one but her husband; 'the husband and his wife are the chiefest in a family, all under them single persons: they governors of all the rest in the house'. The wife was 'under her husband' but she was 'ruler of all other things'. The wife was like 'the master's mate' on a ship, 'a great commander in the house' though 'turned and ruled by her husband'. Early-modern Puritans prudishly insisted that the mistress should not corporally punish male servants any more than masters should do likewise to maid-servants, but this was on the grounds that it was 'not comely or beseeming' for each to do so, not because they lacked the authority. Husbands rather than wives should chastise 'menservants and ... sons that are of age', but the wife must notwithstanding 'inquire into the ways of the whole family'.[81]

The very definition of a household included the mistress's power; a family was 'property, where several servants obey the same master or mistress in the same house'. Wives and mistresses held authority as did husbands and masters, they were 'joint governors with them over all the inferiors'. Sixteenth- and seventeenth-century moralists believed wholeheartedly in the need for domestic discipline, and exhorted servants to obey mistresses. Every husband should preserve his wife's authority in this area, for 'she is the mistress of the house, just as you are its master'. The authority of the mistress within the household was so central that some theorists used it as the defining characteristic of a wife as against a concubine. A concubine was merely a 'half-wife ... betwixt a servant and a wife'. Concubines simply provided children, wives were joined in 'management of the family'. Mere concubines did not rule the family, they only 'served the first wife' who was mistress. It was part of the definition of a wife that she ruled everyone in the household except her husband.[82]

Mistresses were as common as masters, and generally exercised day-to-day control over domestic servants. Homilies on servants' duties often used the term 'master' but early-modern theorists explained that they intended this to include mistresses also. Even when only the term 'masters' was used 'most especially let it be noted that both sexes – mistresses as well as masters – are here meant'. Servants' duties existed without regard to sex and should be performed 'both by maid-servants and also by men-servants that are under mistresses'. When Scripture exhorted servants to obey their masters, early-modern exegetes glossed the text by adding,

we must understand it likewise without distinction of sex. For the head of the family – be it man or woman – must be obeyed ... No Salic law for the cutting off of this.

The obedience owed by a male servant to his mistress was as real as the obedience a wife owed to her husband. The hierarchy of sex was only one in a complex network of obligations. The wife's control of the household was underpinned by God's law; she 'was by God's ordinance ... admitted into the participation of household matters and government'. Scripture was thought to teach 'the servitude and entire subjection of the handmaid to her mistress'. Saint Peter had used Sarah (Abraham's wife) as an example of dutiful submission in a wife (I Peter 3:6). Early-modern theorists used Sarah equally often as an example of female rule in the household. Sarah was at first 'mistress of one family' and made by God 'independently a princess and mistress'. Biblical commentators insisted that even after the maid-servant, Hagar, had been 'suddenly made Sarah's partner in the privilege of Abraham's bed', she was acting sinfully in not showing due obedience to her mistress, for 'she was Sarah's maid, to whom she owed subjection and service'.[83] None of the theorists wished to undermine domestic discipline by suggesting that servants of either sex could disobey their mistresses. Each link in the hierarchy of power reinforced the others. The obedience owed to mistresses was just like that owed by wives – both were commanded by God. If wives 'would have their subjects obey them, they must make conscience to obey their husbands'. Social and domestic hierarchy was divinely endorsed; women were often inferiors but sometimes superiors. 'God deals fairly with the wife, in that he makes her subject but to one and lets her rule many.' God endorsed the mistress's power over her servants as He did all domestic hierarchy.[84]

Just as the power of mothers and widows proved that women could legitimately hold political authority, so also did that of a mistress over her household. Bishop John Bridges upheld female rule in the polity by pointing to the accepted fact that a 'wife hath inferior government in her household'. This case, like that of female regents, showed that it was 'false' to assert that a woman could not be a 'governor'. William Fulke was more puritan in his inclinations than Bridges but he employed the same argument on female rule.

The error of them that did write against the regiment of women is easily confuted by the fifth commandment, where civil authority and government is established as well to the mother, mistress, lady, and queen, as to the father, master, lord, and king ...

By commanding that mothers should be obeyed, God told men that some women held legitimate power from Him. Patriarchalists and contractarians alike believed that the fifth commandment endorsed all forms of legitimate power and that women were also included. Most early-modern political theorists did *not* believe that the inferiority and subjection of the female sex excluded women from positions of power over males. Women were not subject to all males, even though wives were subject to their own husbands. Moreover, the conjugal subjection of a wife was compatible with her holding authority in other areas. The authority of a mother over her son and of a mistress over her household existed independently from that of the father and master.[85]

Subjects and queens, sons and mothers, servants and mistresses – all of these were instances where the general rule of the subjection of woman was overridden by another principle: hereditary succession, parental authority, and contract/rank, respectively. The subjection of woman did not necessarily involve the subjection of all women, as Elizabeth I and many a wealthy widow well realized. However, most women were wives for much of their adult lives. Despite the limits described above, they *were* subject – subject conjugally to their husbands. What was the nature of this subjection? How did early-modern theorists characterize the relationship between husband and wife? The next chapter deals with this problem.

Notes

1. Leighton 1844, 152.
2. Aylmer 1559, sig. H2r–v; on France see, for example, Coste 1647, sig. o2r ff. For fawning praise of Elizabeth I, see also Whetstone 1586, 136; Tasso 1599, 'Defence or answer' sig. L2r; Vaughan 1600, sig. R6v; Tuvil 1616, 104–5; Lotichius 1630, 110, 123; Heywood 1640, 212; Rogers 1642, 306; Torshell 1650, 87; Trapp 1662, 384; Craig 1703, 84–5.
3. Bernard 1845, 1 [Topsell's commentary on Ruth, *The reward of religion* 1597, did just the same]; Donne 1971, 236–8. It was hardly surprising that Donne would not insult the 'noble Lady', 'in whose protection I am, since I have, nor desire other station, than a place in her good opinion'; (quoted in Bald 1970, 276). See also Lake 1987, 161.
4. For this argument see, for example, Baxter 1830, 143; Cleaver 1598, 221; Pricke 1609, sig. L1v–2r; Cottesford 1622, 11; Griffith 1633, 329; Elton 1637, 548; Abbot 1653, 38–9; *The house-keepers guide* 1706, 125.
5. Stock 1865, 174 [unfitly]; Byfield 1637, 579, 640; Anderton 1642, 9 [dissipate]; Lessius 1621, 94–6. Compare Feltham 1631, 264.
6. Gouge 1622, 287–8 [oft]; Tyrrell 1681, 109; Milton 1959, 589; Raulin 1512, f. 20r and 27r; Settala 1626, 150 ('uxorem prudentem, solidae mentis et constantem'); Topsell 1597, 40 [godly]; *The court of good counsell* 1607, sig. C4r–v [profit]; Nevizanus 1570, 316–17.
7. Whetstone 1586, 136 ['Facinorous' means extremely wicked]; Heywood 1640, sig. xxr; Case 1588, 32; Wright 1621, 42 [passions]; Torshell 1650, 16.

8. Fleetwood 1716, 134. The willingness of Renaissance theorists to admit that there were many instances of strong, balanced, intelligent women, and yet to dismiss them as mere exceptions, has been seen as paradoxical by some modern commentators (Woodbridge 1984, 214–15; Jordan 1990, 132–3). Perhaps the early-modern view can be best explained as the 'mirror-image' of one modern argument. Faced with the under-representation of women amongst (say) Nobel prize laureates, a possible response is that women are *naturally* equal, but that social, economic, and cultural factors sometimes (appear to) result in inferior attainments. The arguments of Renaissance male supremacists followed the same form but employed contradictory axioms. In each case, the proposition is: naturally women are inferior (equal) but environmental conditions sometimes makes them appear equal (inferior). The same point applies to woman's 'natural' character. In Renaissance thought, as Woodbridge has argued, 'if a vast majority of women had failed to conform to expectations about timidity, passivity and tenderness of heart, that would have proved only that a good many women were unnatural nowadays'. Today, if a vast majority of women *do* conform to expectations about timidity, passivity, and tenderness of heart, that merely proves that a good many women are oppressed nowadays. Neither then nor now did beliefs about woman's inherent natural abilities stem from a random sample of women whom a theorist had met. A Renaissance theorist felt no more obliged to abandon the belief that (generally) women were inferior because he met some outstanding ones, than the zoologist does to abandon his view that 'the female is (generally) shorter than the male' because Tutsi women are taller than Mbuti males.

9. Joannes de Sancto Geminiano 1585, f. 145v; Wright 1621, 40 [pity]; Lessius 1621, 327–8 [widows']; Geree 1639, 3 [shame]; Littleton 1680, 58–9; *Marriage asserted* 1674, sig. A6r–v [dignify]; Ken 1682, 12 [exemplary]. For the view that women exceed males in piety, see, for example, Raulin 1512, f. 20, 23; Nevizanus 1570, 319, 401; Salkeld 1617, 106; Lotichius 1630, 94–9; Byfield 1637, 621; Bernard 1845, 1–2; Rogers 1642, 284; Coste 1647, sig. i1r; Gauden 1656, 180; Dickson 1659, 162; Leighton 1844, 147; Guillaume 1665, *passim*; Bovet 1684, 51; Poullain de la Barre 1989, 53; Ness 1696, 157. See also Lake 1987, 143–4; Willen 1992, *passim*; Moran in Greaves 1985, 127–31; and for medieval antecedents, Cullum in Goldberg 1992, 182–211.

10. Nicholls 1701, 115; Allestree 1673, sig. b5v. See also *The house-keepers guide* 1706, 131; Thomas 1958, 45; Todd 1987, 101–17; Moran in Greaves 1985, 125–49.

11. The *Disputatio nova* or *Disputatio perjucunda* was reprinted frequently (generally with a response by Simon Gedike) over the next 150 years. For an account of its origins and impact, see Bay 1934.

12. Calvin 1948, 573 [mention]; Bird in Ruether 1974, 49 has stated: Old Testament laws were formulated to 'address the community *through its male members*' as 'only a male is judged a responsible person'.

13. As Milton (1959, 589) pointed out, 'had the image of God been equally common to them both, it had no doubt been said, In the image of God created He *them*'. Milton's gloss was heterodox.

14. Lorenzo da Brindisi 1935, 200. [Compare Augustine 1982, 99]. For other texts sometimes cited to prove female salvation, see Luther 1958, 201; Lotichius 1630, 47–8.

15. Needler 1655, 14–15 [rational]; Byfield 1637, 667; Torshell 1650, 5, 10; Trapp 1662, 36 [Jews]; Maynard 1668, 180–1 [sottish]. At this time, it was still almost

universally accepted that Moses himself wrote the first five books of the Bible. Hobbes and Spinoza were amongst those who questioned Mosaic authorship, but in the late seventeenth century, orthodox divines such as Kidder (1694, 'Preface') still staunchly upheld it.

16. A commentary on St Paul's Epistles by an author now known as 'Ambrosiaster' or pseudo-Ambrose was then often ascribed to Ambrose. This argued that it was unsuitable ('incongruum') that the image of God should be ascribed to a subject being, and maintained that 'the man is made in the image of God, not the woman' ('vir enim ad imaginem Dei factus est, non mulier' – 1845, XVII, cols 436, 240). Pseudo-Ambrose was often cited in discussions of sex and subjection, since his commentaries on Paul's writings assumed and endorsed sexual double standards. Ambrose did not actually write this commentary and doubts about its integrity and authorship were often raised; see, for example, Erasmus [1550?], sig. C1v; Catharinus 1551–2, 280; Tapper 1582, 296; Bellarmine 1602–3, III (1603), col. 1288 (I.II); Campanella 1968, 190; Laud 1847–60, VII, 617; Bingham 1855, VIII, 36.

17. Kipping 1664, 203 ('Manifestus error').

18. Donne 1953–62, III, 242; Donne 1953–62, IX, 190, 192; Augustine 1982, 99 [image]; Aquinas 1872–82, III, cols 1066–7 (Suppl. 39.1 'Sed sexus non est in anima'; 'secundum rem in his quae sunt animae mulier non differt a viro'); Caussin 1634, 253 [devout]. See also Bellarmine 1602–3, III (1603), cols 1290–1 (I.II); Melanchthon 1834–60, XXI, 686; Gualther 1570, f. 552v; Hunnius 1604, 65–6; Gouge 1622, 413; White 1656, II, 105; Reyner 1657, 12; Trapp 1662, 12; Menochio 1678, 233, 535; Poole 1696, II, sig. Hhh2v.

19. *Disputatio* 1641, 27 ('Quis sanus unquam docuit in muliere reperiri vivam fidem? imo contrarium clamat Apostolus, et omnem ei adimit fidem scribens: mulierem salvari non per fidem, sed per generationem'); Cardwell 1841, 599–600.

20. Campanella 1968, 76; Bellarmine 1602–3, III (1603), col. 1291 (II.II 'quis enim credat Paulum velle salutem matris pendere a filiorum perseverantia'?); Lotichius 1630, 53 ('magis benedictae'); Lessius 1621, 148 [procreation]; *Disputatio* 1641, 93–4; Martin 1676, 147, 150 [interpretation]; Willet 1605, 14 [faith]. For the same exegesis, see, for example, Fisher 1597, 268; Carter 1627, 64; Trapp 1647, 305; Dickson 1659, 162; Maldonado 1677, 426–7.

21. Milton 1959, 589, 591; Parsons 1606, 77; Bunyan 1860, 664. Even apart from the obvious thrust of Paul's argument, the Greek was '*aner*' (man as male) rather than '*anthropos*' (man nonsexually, i.e. a human being) so his meaning could not be explained away linguistically.

22. Lapide 1854–9, IX, 239 ('improprie et analogice'); Willet 1605, 14; Swan 1635, 500 [made].

23. Herbert 1648a, sig. E3v–4r [immortal]; Dod and Hinde 1614, 61–2 [modesty]; Fulke 1601, 679; Batty 1581, f. 76r; Lessius 1621, 170 [books]. Compare Bale 1849, 156. It was commonplace for Protestants to accuse Catholics of forbidding women religious education, and for Catholics to accuse Protestants of excluding women from the religious life. Both accusations should be viewed in the context of bitter polemic.

24. Taylor 1828, V, 252 [usefulness]; Rollock 1603, 303 [miserable]. For similar views, see Tasso 1599, sig. C4v; Hunnius 1604, 102; Whately 1617, 40; Salkeld 1617, 103–5; Mayer 1631, II, 499; Wemyss 1633, II, 16; Rogers 1642, 162; Abbot 1653, 17–18, 45; Heydon 1658, 85; Charnock 1866, 398–9; Dickson 1659, 57;

Swinnock 1868, 479–80; Pettus 1674, 65. I Corinthians 7:11 had induced similar ambivalence in the Church Fathers and medieval theologians; see Augustine 1982, 99; Hugh of Saint Victor 1951, 352; Abelard 1855b, 760–1.

25. Willet 1612, 20 [special]; Calvin 1948, 129, 95–6 [different]; Perkins 1628, 368 [honoured]; Barlow 1632, 77 [rejoice].

26. E.,T. *The lawes resolution* 1632, 5 [reasonable]; Aristotle 1984, II, 1999, Politics, (1260a); Xenophon 1923, 423; Tuvil 1616, 138 [grounded]; Plutarch 1603, 483 [compare Tuvil 1616, sig. A5v]; Seneca 1614, 721. See Okin 1979, for a forceful account of classical views of women's rationality.

27. Willet 1608, 278–9 [conceit]; Swinnock 1868, 504 [essential]; Elyot 1980, sig. Bbv and C3–4 [cankered].

28. Whately 1647, 4 [licentiousness]; *Disputatio* 1641, 182 ('Anima rationalis'); Coste 1647, sig. eiiv ('que la distinction du sexe ne distinguoit point la vertu'); Carter 1627, 65 [dwell]; Brenz 1584, sig. D7v [honest]; Geree 1639, 12 [fear]; Taylor 1625, 9 [straightly].

29. Vitoria 1991, 132 [madmen]; Fulke 1601, 549 [womankind]; Wilkinson 1607, 18 [spindle]. Some of the radical civil war sects, in particular the Quakers, provided a possible exception to this rule; they did theoretically give women equality in ecclesiastical matters. See Schlatter 1940, 240 and Mack 1992, 88–120. In her informative and stimulating study, Mack makes it clear that allowing women a role as 'prophets' did not entail abandoning belief in their natural inferiority (106). Moreover, the Quakers rapidly retreated from their eccentric stance (123). For other protestant groups, see Greaves 1985, 4–5; Thomas 1958, 53; Harrison 1992, 54; Potter 1986.

30. Sanderson 1671, 188 [domestical]; Dillingham 1609, f. 3r [many]; Tuvil 1635, 278–9 [outward]; White 1656, III, 208 [estate]. See also Blackwood 1659, 663; Brydall 1680, 9. As Maclean (1980, 27) has stated: 'In theological terms woman is, therefore, the inferior of the male by nature, his equal by grace'.

31. Hayward 1603, 39.

32. Shakespeare, *Henry V*, I.ii.88–95; Bodin 1962, 753 [shifts]; Coke 1662, 101–4; Aylmer 1559, sig. F1v. On Knolles' translation, see McRae (in Bodin 1962, A41). For other attacks on Salic law see Ridley 1607, (especially 98) and Whetstone 1586, 137. One of the few tracts at Anne's accession to mention the problem of women's rule (Michel 1702, 12–13) likewise condemned Salic law 'as in nature unjust, so in religion heretical' (i.e. since contrary to Isaiah 49:23). Algernon Sidney (1990, 60) argued that women 'by nature are unable to perform the duties' of ruling and was one of the few Englishmen clearly to approve of Salic law. However, his fundamental argument was that there was no universal law on this matter 'but every people is by God and nature left to the liberty of regulating these matters relating to themselves according to their own prudence or convenience' (ibid., 61 and see also 505).

33. Tiraqueau 1574, 'De Iure Primigeniorum', f. 40v–41v. See Lemaire 1975, for a fuller discussion of French defences of Salic law.

34. Harpsfield 1878, 171. Hardly surprisingly, the ultra-Catholic Harpsfield thought Edward VI's reign was such that (284) 'it had been, I say, much better for the realm that the king should have had no male child at all'. [*A Glasse of the truth* was possibly written by Henry VIII himself: see Pollard and Redgrave 1976–91, I, 525, No. 11918–9].

35. On Henry Tudor's claim, see Levine 1973, 139, 141. Three males (Edward of

Warwick, John de la Pole, and John II of Portugal) whose titles were better than Henry's were also ousted from the succession by him (ibid, 33–4). At the Exclusion Crisis, supporters of indefeasible hereditary right tried to rewrite the past and maintain that only Henry's marriage to Elizabeth 'turned his usurpation to a lawful sovereignty' (Rider 1680, 30) but this was a distortion of constitutional history. Fortescue did discuss women's rule in the context of dismissing the Yorkist claim to the throne; see Chrimes 1966, 62–4.

36. See Levine 1973 for a fuller account of the dynastic complications surrounding Tudor queens' accession. Interestingly, at 81–2 he adduces evidence that Northumberland's plan in marrying his son Guildford Dudley to Lady Jane Grey (the most plausible claimant after Mary and Elizabeth) was to place any male issue of their union on the throne by persuading the English Parliament to accept a version of Salic law, thus excluding both Mary and Elizabeth from the succession. Edward VI's premature, childless demise scotched such plans, leaving Northumberland in turn with nobody to back but a woman.

37. One modern edition of Knox's *First blast* is Knox 1985, 37–78; Strype 1824, I, 1, 184–5 [Goodman]; Jewel 1850, IV, 665 [one]. Catholics delighted in attacking Knox as part of their polemical campaign to insist that all Protestants were potentially seditious rebels. As Matthew Kellison acidly commented 'In Queen Mary's time because that princess was not for them, then women could not govern: but in Queen Elizabeth's time, because they had insinuated themselves into her protection, women might govern as well as men: and so they are the best temporizers in the world' (1605, 258). Hilarion de Coste (1647, II, 516) advanced the delightfully implausible thesis that it was Knox's tract that provoked Henry Darnley to rebel against his wife Mary, Queen of Scots. In the dreadful doggerel of Frarinus (1566, 'The Table'):

No queen in her kingdom can or ought to sit fast
If Knox's or Goodman's books blow any true blast.

For surveys of the controversy over women's rule that the *First blast* provoked, and bibliography, see Phillips 1941–2; Scalingi, 1978; Jordan 1987; Lee 1990.

38. Aylmer 1559, sig. B2v [will]; Saravia 1593, 54 ('citra ullum discrimen sexus'); Bullinger 1572, f. 44r [ordinance].

39. Becon 1844, 227 [empire]; Allen 1600, 122–3 [queens]; Coke 1662, 101 [content].

40. Kornmann 1610, 115–16 ('dignitatibus hereditariis'); Babington 1615, II, 162 ['grossly'. Jerome 1614, 8, used exactly the same words]; Ness 1696, I, 107 [diametrically]; Whetstone 1586, 138 [busy].

41. Craig 1703, 84, 99. The position that (in default of male heirs) women by the laws of nature and nations should inherit, should be distinguished from that which John Knox adopted when desperately trying to repair the damage done to his relations with Elizabeth by the *First blast*. He then described Elizabeth's accession as 'a miraculous work of God' which he would obey 'albeit that both nature and God's most perfect ordinance repugn to such regiment'. If Elizabeth would admit her rule to be by 'the extraordinary dispensation of God's great mercy ... [making] that lawful unto her which both nature and God's law do deny to all women' he would accept her authority (Knox 1949, 285–6). Knox's position that women can only rule by a special miracle of God, differs from that generally held, viz., that the normal rules of inheritance should be followed in the area of

succession to the crown. Strangely enough, Knox's generous concession did not restore him to Elizabeth I's favour; at Elizabeth's court: 'Of all others, Knox's name, if it be not Goodman's, is most odious here' (Sadler 1809, 532).

42. Ainsworth 1627, 173 [general]; Chambers 1579, 7–8, f. 16v ('Quant au regime des Royaumes et autres inferieurs gouvernemens hereditaires, c'est un regle commune par la loy des gens, qu'en deffault des fils, les filles succedent'); *Zurich letters* 1845, 35 [see ibid., 131 for similar disavowals of Knox and Goodman by Beza].

43. Pascale 1618, 52; Benincasa 1562, f. 72r; Nevizanus 1570, 255; Domas 1705, 87; Lagus 1566, 19 ('videantur iniquiae' 'naturalis aequitas, atque ius commune'), 33 ('non omnino haereditatibus excludi'). See Azor 1600, II, 128; Willet 1608, 278; Jerome 1614, 7; Suarez 1872, II, 7 (V.II.7) for similar arguments.

44. Dillingham 1609, f. 12v [Roman]; Attersoll 1618, 1258 [world]; Dod and Cleaver 1629, 186 [break].

45. Cioni 1985, 101 [Chancery]; Erickson 1993, 19 [comparable]; on medieval practice see Hanawalt 1986, 121, 142–3; Howell 1983, 258. See Warnicke (1989, 67) and Pollock (1993, 13) for the case of Lady Mildmay and her daughter, Grace, where attempts to establish an entail in the male line were overthrown by the courts because proper provision had not been made for female family members. On the occasional circumvention of a daughter's right to inherit, see Hogrefe 1972, 98–9. On family and inheritance, see Baker 1971, 146–61.

46. E.,T. *The lawes resolution* 1632, 6. Hufton (1984, 355) describes eighteenth-century England and France as 'societies in which marriage and motherhood were regarded as the norm, spinsterhood and infertility as a blight'.

47. Nevizanus 1570, 240 ('uxore ducta, gravitatem induit'); Davis 1975, 69 [Frenchwomen]; Griffith 1633, 232 [vocation]. The proportion of spinsters is diffi-cult to calculate accurately, but in England it seems to have varied between 10 and 20 percent during the sixteenth and seventeenth centuries; see Stone 1977, 42–6; Hufton 1984, 357; Macfarlane 1986, 23–8. For an introduction to the legal status of women in Rome and the progress of their rights in civil law, see Corbett, 1930; Maine, 1875.

48. Cajetan 1581, 414 ('aliud absolute'); Poole 1696, II, sig. Nnnn2r [circumscribe].
49. Aylmer 1559, sig. C4v [woman]; Craig 1703, 83 [law]; Rollock 1603, 342 [kind]; Bullinger 1572, f. 43v; Bridges 1573, 859 [simply]; Dillingham 1609, f. 12r–v [family].

50. Byfield 1637, 580 [cottager]; *A Glasse of the truth* [1532?], sig. A3r–v [a 'dote' is a natural virtue or endowment]; Stubbs 1968, 174 [pre-eminence].

51. Aylmer 1559, sig. L4v–M1r [meddle]; Craig 1703, 83 [Deborah]. Deborah was a favourite of those who wanted to maintain the legitimacy of woman's rule (see, for example, Torshell 1650, 86), as she afforded a clear Biblical example of a divinely approved female ruler. As early as the thirteenth century, Bonaventure (1889, 650, XXV.II.I) had accepted that Deborah's case showed that women could hold secular power. The stirring tale of Deborah's part in ensuring that 'all the host of Sisera fell by the edge of the sword' is told in Judges 4:4–19.

52. Tuvil 1635, 141 'though in a civil and politic respect, a public magistrate be more honourable than a private man, yet as he is a son he is to count himself inferior unto him from whom his being is derived'.

53. Aylmer 1559, sig. C4v [inferior]; Leslie 1569, f. 135v [misdemeanours]; Rutherford 1843, 115 [death]. Mary, Queen of Scots' husband (Henry Stewart,

Earl of Darnley) was said by her detractors to have met his death on her orders –
hence the epithet 'the parricide and bloodsucking Medea' (*Zurich letters* 1845,
168, 172). However, his murder omitted the judicial process that would have rendered the deed respectable.

54. Suarez 1872, 188–90, (III.IX.7–10 'propriae personae'; 'superior in ordine ad
politicam gubernationem'); Besold 1623, 14–15 ('nec imperii, nec in administra-
tione'); Pufendorf 1934, 854–5 (VI.I.9).

55. Fulke 1601, 550.

56. Jones 1988, 15 [supporters]; Ferguson 1695, 12, 42 [A 'Queen of Clouts' is a
doll in the form of a queen]. Despite Mary's refusal to make any claim without
William, a number of her supporters insisted in the parliamentary debates on the
matter that William could not accede except jointly with her; see Jones 1988,
149–52, 154, 169, 171–4.

57. Sharp 1702, 16 [trust]; Sherlock 1714, 20. For similar sentiments, see Michel
1702 and Atterbury 1735, 301, 308.

58. Tuvil 1635, 153.

59. Babington 1615, III, 52 [things]; Towerson 1676, 243 [effect].

60. Abbot 1653, sig. A4r [boy]; Lawson 1659, 187 [neglected]; Rutherford 1843, 71.

61. Barker 1624, 200 [exceptions]; Bastingius 1593, f. 210v [equal honour]. On
these points, see Elton 1637, 560; Baynes 1866, 357; *The court of good counsell* 1607,
sig. G2; Willet 1608, 369; Dod and Hinde 1614, 4; Dod and Cleaver 1629, 171;
Wemyss 1633, II, 24; Durham 1675, 313–14; Bernard 1684, 194; Kidder 1694, 109;
Poole 1696, I, sig. Aa4r.

62. Locke 1988, 186 (I.61); Allen 1600, 123 [putteth]; Poole 1696, I, sig. P1v [first].

63. Ken 1838, 283 [comprehended]; Boughen 1646, 55 [settled]; Sedgwick 1624,
72–3 [eminence]; Coke 1662, 101 [impious].

64. Aristotle, 1984, I, Generation of animals, 1127–32 (726a–9a); Rueff 1637, 1–2
[field]; Highmore 1651, 88–9 [vegetables].

65. Galen in Lefkowitz and Fant 1977, 153–6. The scholastics also joined in the
debate. Healy (1956, 11–12): 'Albert the Great, Thomas Aquinas, Peter of
Tarantasia, Egidius Colonna, and most Thomists follow Aristotle and deny any
active contribution of woman in human generation. Alexander of Hales, Saint
Bonaventure and his disciples, Duns Scotus and his followers, according to the
Greek physicians and Avicenna hold that woman contributes actively in human
generation'.

66. Highmore 1651, 84–5 [contributes]; *Aristotle's masterpiece* 1694, 15 [erroneous];
Riolan 1671, 80 [somewhat]. See also the discussions in Albertus Magnus 1596,
sig. B5–6; Manzini 1587, 277 ff. (III.ii); Mercuriale 1591, 158 ff; Lamzweerde 1686,
70 ff. As Noonan 1986, 337, has stated in his brilliant work *Contraception*: 'Exactly
what physiological process the theologians have in mind when they use the
phrase "female seed" is hard to determine'. Medical writers of the period – if not
all theologians – do seem to have discharge from Bartholin glands in mind.

67. Rueff 1637, 9.

68. Poole 1696, I, sig. E1v [principal]; Kipping 1664, 316, ('coniunctim propagent
animam in filios'); Lotichius 1630, 84 ('formatum nutriunt'); Gouge 1622, 484–5
[bond]; Campanella 1968, 196 ('comprincipia naturaliter ad propagandum');
Towerson 1676, 246 [means]; Coke 1662, 75, 101 [authority]. Samuel Clarke (1659,
263) repeated Gouge's view almost word for word. Pricke 1609, sig. B3r; Ayrault
1614, 28–9; Davenant 1627, 437; Bastingius 1593, f. 210v; Stockwood 1589, 18;

Bernard 1684, 194, were among those who shared the same assumptions. Lorenzo da Brindisi (1935, 331) cited Genesis 4; and Wemyss (1633, II, 54–5) and Trapp (1662, 275) cited Leviticus 12:2 to prove that women also contribute 'seed' in reproduction.

69. Gentili 1606, 7 ('testatur clare Aristoteles'). For the same view, see, for example Gentili 1614, 289–90; Sordi 1602, 88; Lagus 1566, 78–9.

70. Gentili 1606, 2–4; Sordi 1602, f. 208r ('non enim, ut saepe dixi, ad imparia iudicantur, nec magis obstricta mater, quam pater, et in matre eadem omnino versatur ratio, quae in patre, proinde idem ius esse debet').

71. Thorndike 1844–56, IV, 358. For unfavourable comment on the Roman law's treatment of mothers, see, for example, Beust 1597, 190–2; Mauser in ibid., 349.

72. Bentley 1582, 35. On the need for both parents' consent, see, for example Gibbon 1591, 5; Carter 1627, 194–6; Dod and Cleaver 1629, 174; Gaule 1630, 97–8; on I Corinthians 7:36–8, see, for example, Campanella 1968, 118; Chemnitz 1978, 761; Stockwood 1589, 46–7. See Chapter 7, Parental consent, for a fuller discussion of the need for parental consent in the marriage of children.

73. Hobart 1671, 10 (*Grisley v. Lother*) [consideration]; Keck 1697, I, 142 [grandmother]; Carlson 1990, 450 [knees].

74. The use nowadays of the abusive epithet 'motherfucker' suggests that a son's sexual relations with his mother are still widely regarded as obviously immoral.

75. Aquinas 1872–82, III (1882), cols 1137–8 (Suppl. 54.3.); Duns Scotus 1639, 832, (XLII.I.3); Hay 1967, 203; Beza 1591, 28; Grotius 1689, 244 (II.V.XII); Vaughan 1706, 243; Taylor 1828, XII, 308–9 [II.II.iii matrimonio]. This argument against a son marrying his mother was standard in theological discussions of the bans on incestuous marriages; it was often extended to nephew–aunt marriage. See, for example, Durandus 1571, II, f. 386v; Campanella 1968, 108; Harpsfield 1878, 166–7; Turner 1686a, 6; Bellarmine 1602–3, III (1603), col. 1412 (I.XXVIII); Menochio 1678, 67.

76. Willet 1608, 369–70 [nature]; Towerson 1676, 246 [head]; Gouge 1622, 485 [doubt]. See Elton 1637, 561; Keble 1685, 482; Fleetwood 1716, 46–7, for the same argument.

77. Gouge 1622, 485 [interposed]; Topsell 1597, 37 [same]; Coke 1662, 101 [nowhere]; Fleetwood 1716, 47 [regarded]; Tyrrell 1681, 52. Compare Rogers 1642, 269; Locke 1988, 310, (II.65.).

78. Dod and Hinde 1614, 3 [special]; Tuvil 1616, 52 [chiefest]. On the mother's right and duty to correct and rule her children, see, for example, Raulin 1512, ff. 45–51; Batty 1581, f. 31r; Allen 1600, 143; Vaughan 1600, sig. N5v; *The court of good counsell* 1607, sig. G2r–v; Beard 1612, 222–5; Dod and Hinde 1614, 61, 68–9; R., R. *The house-holders helpe* 1615, sig. A2v, 2; *Christian parents* 1616, 1–4; Leighton 1844, 153; Shower 1694, 95–7; Ken 1838, 210.

See Todd (1987, 105–13) for the important role given by puritans and humanists to mothers in household piety; Kelso (1956, 129) on a widow's role in educating her children; see Harris's fine articles 1990a and 1990b, especially 261–2 and 280–2 for the importance of widows in family affairs in Tudor England; and Erickson (1990, especially 25, 36) for the legal powers generally held by widows. Compare Archer in Goldberg 1992, and Hanawalt (1986, especially 153) on medieval England.

79. In modern parlance 'mistress' is no longer used in its early-modern English sense as the female equivalent of 'master', just as 'lady' is no longer used as the

female of 'lord'. My text follows early-modern usage because there is no longer
any general term meaning a woman in a position of authority.
80. See Chapter 4, The extent of subjection.
81. Swinnock 1868, 352, 350; Tuvil 1635, 103–4 [ordinance]; Pritchard 1579, 87;
Gouge 1622, 21; Wilkinson 1607, 8, 11 [ship]; Cleaver 1598, 176, 384 [chastise,
compare Smith 1866, I, 34]; Dod and Hinde 1614, 63 [inquire].
82. Coke 1662, 79 [definition]; Baxter 1830, 142 [governors]; Luther 1958, 137–8
[master]; Trapp 1662, 110 [half-wife]; Covarruvias 1597, 140, (I.I.4.9 'dispensatio
familiae'); Jansen 1641, 119 ('fere famulabantur uxori primariae'). On the lesser
status of concubines, see also Peleus 1602, 31; Richardson 1655, sig. D2v; Poole
1696, I, sig. F2r; Butler 1697, 21; Sordi 1602, 7r; Bellarmine 1602–3, III (1603), col.
1330 (I.XI); Calvin 1948, 573; Felden 1653, 141–2.
83. Gouge 1622, 161 [noted]; Stock 1865, 175; Tuvil 1635, 291–2 [distinction];
Menochio 1678, 14, ('familiae fuit domina', 'absolute princeps, et domina');
Turner 1685, 79; Poole 1696, I, sig. E1v. On Scripture, see *The house-keepers guide*
1706, 242; Webb 1653, 68, 114; White 1656, 100–1. For the same arguments about
Sarah and Hagar, see Fosset 1613, 18; Gaule 1630, 56, 60; Tuvil 1635, 283–4;
Whately 1647, 148–50.
84. Byfield 1637, 580–1.
85. Bridges 1573, 804–05; Fulke 1601, 549.

4

The extent of subjection

The characterization of subjection

'Subjection' was used in many contexts in early-modern thought: citizens were subject to their prince, children to parents, servants to employers, and slaves to masters or mistresses. Each relationship involved different rights, duties, and status, and the powers held over the subject varied. The wife was also subject, but early-modern theorists always distinguished conjugal from political, filial, and servile subjection. The great majority denied that a wife should be seen or treated as a slave or servant. They also denied that she was in the same position as a child, although husbandly power was thought to resemble paternal power in some respects, and political power still more. The husband was characterized rather as a senior partner. Slaves and servants served others' ends; the husband and wife worked together (under his direction) for a common end.

Early-modern thought often employed the concept of slavery as a paradigm of the most extensive form of subjection. Slaves held only the most basic rights afforded by the laws of God and nature. A master or mistress could not rightly kill a slave for no reason (as one could an animal). However, a slave had no rights to property or to freedom of movement and could be corporally punished. As the owner's possession (s)he could be sold. Masters could use slaves' labour without regard to the slaves' own benefit. Aristotle – so often cited as authority for treating women unequally – was positive that wives were not simply slaves; 'nature has distinguished between the female and the slave'. He asserted that 'barbarians' did not distinguish between them, but this was just because all barbarians were naturally slaves. When proper distinctions were made, 'the kind of rule differs – the freeman rules over the slave after another manner from that in which the male rules over the female'. Later theorists followed Aristotle on this point just as they did on woman's natural debilities. Divines cited Aristotle to support the belief that to treat one's wife as a 'drudge' or a slave was an 'odious' sin. The Bible was also used as an authority. The Creation story did not contain any explicit assertion or denial that the wife was a slave. But Christian commentators did discover an analogical basis for denying that the wife was a slave to her husband. They argued that Eve was made from Adam's rib, close to his heart, to show that he should love, shelter, protect, and treat her as a companion. She was not made from his foot, to be trampled down like a slave; nor from his head, to rule over him. This metaphor was repeated with tedious frequency in countless medieval and early-modern commentaries.[1] Arguments from the nature of the marital relationship, from the classics, and from the rib analogy were all used in the Middle Ages to prove that wives should not be treated like slaves. Monks taught that Eve was not made 'from the feet, to be subject unto slavery'. Thomas Aquinas insisted that a wife was *not* her husband's slave. He pointed to the wife's rule of the household, and to the fact that Mosaic law (Deuteronomy 21:11–14) freed a slave-woman on marriage, and portrayed slavery and marriage as simply incompatible states.[2] Civilians cited the Roman law that 'a slave-woman who was to be taken as wife could be freed for no other cause than that of marriage'. Scholastics insisted that men wanted wives as 'companions' not slaves. Others thought it 'evident' that wives should be, not slaves, but 'companions, sharers in all things, indivisible companions in every joy and sorrow'. Catholic tradition and casuistry agreed that a wife should not be regarded as her husband's slave.[3]

The belief that a wife was not a slave and should not be treated as one was endorsed by English early-modern theorists. Woman was made to help her husband, not to be 'an underling', 'a drudge and a slave'. A wife should 'guide the household' and, although subject,

> the subjection of the wife must not be slavish or servile, as the Turks and Moors at this day use their wives as their slaves, but it must be sociable and amicable.

Early-modern English theorists often commented unfavourably on the treatment that women received in other parts of the world. Fears of cultural imperialism did not inhibit believers in an absolute and immutable standard of right and wrong. Russians were described as 'but of a barbarous condition', because the husbands treated their spouses 'as servants rather than wives'. With Western pride, early-modern theorists boasted that 'our wives are not obliged to the Asiatic slavery, nor is it their duty to submit to all those vile servilities and imperious humours which the women of those nations are forced to comply with'. Husbands in the countries of the Mediterranean basin,

> are so severe to their wives that if people attempted to introduce the like custom into England, it were to be feared that our women would unanimously join together and by that blessed law of self-preservation depose or put away their husbands.

The tone was humorous, but the basic belief that only lesser breeds treated their wives as lackeys was deeply ingrained. Good Englishmen labelled treating wives like slaves as a nasty, foreign habit in which they would take no part.[4]

Moralists were as capable of condemning their fellow countrymen's conduct as that of foreigners. Joseph Hall, Bishop of Exeter and Norwich, was sure that only savages treated their wives like slaves. Husbands' power should be sweet and gentle but 'is degenerated in the practice of too many into a stern tyranny; according to the old barbarian fashion in Aristotle's time which holds even still, Their wives are their slaves'. Such cruelty, he complained, 'cries unto me daily for redress'; such oppression was 'odious and abominable'. Wives should be given their proper status, far superior to that of slaves; they 'are in the law free burgesses of the same city wherein their husbands are free: both participating [in] the same rights, both enjoying the same liberties'. Slaves were not free but wives were. Moralists unfailingly condemned husbands who 'tyrannize over and trample their wives as if they were not their fellows but their foot-stools, not their companions and copes-mates but their slaves and vassals'. A wife's subjection was 'free and ingenuous' and the husband's conduct must reflect this.[5]

Another key characteristic of slaves was that they were property. A slave could be sold without regard to her or his wishes. Early-modern philosophers had no inherent objection to the idea that one person could own another. Many happily described children as 'part of the father's goods', who 'in the old law' could 'sell their children to supply their necessity'. Civil law had allowed a destitute father to sell his children, and early-modern civilians earnestly debated whether a mother had the same right. Protestant preachers told parents they had: 'as great a propriety in thy children as in any of thy possessions'.[6] In contrast, legal and moral theorists insisted that the wife was *not* part of her husband's property. Even 'in his necessity' a husband was not allowed to 'sell his wife to set himself at liberty, *et uxor non est in bonis*, she is not part of his goods'. A wife was flesh of her husband's flesh and bone of his bone; not a piece of property. A wife was 'certainly' not her husband's slave 'for then a man might sell his wife when he pleased'. All divines preached the same view; a husband's powers over his wife were limited because she 'is none of his goods, none of his possessions'. Slavery and marriage were incompatible; slaves could be owned and even children might be sold, but a wife was never her husband's possession.[7]

Another contrast between slavery and marriage was that a wife could not be given tasks out of keeping with the couple's social status, unlike slaves who could be forced to perform menial and demeaning duties. Catholic casuists taught that a wife must obey her husband in matters of domestic government, but denied that a noble wife was obliged to perform humble tasks for her husband (like preparing meals and sweeping the house) for such jobs 'would dishonour her'. Jean Bodin did not adopt the usual tone of outrage about the husband treating his wife like a slave but even he thought it 'beseemeth ... not the husband under the shadow of this power to make a slave of his wife'. English theorists were no less conscious of the social order. Wives must obey their husbands, argued one, but they 'are not obliged by their conjugal duty ... to be put upon any base drudgeries, which are contrary to their birth, dignity, or circumstances'. Divines urged wives to proper obedience but conceded that if a husband's commands were 'unbecoming their age, credit, quality, and condition they may be safely passed by'. The social hierarchy was as important as the sexual one.[8]

Renaissance thinkers' conviction that a wife was subject did not entail the consequence that she was a servant. A wife's subjection was 'not servile and slavish'; her 'subjection is not servitude'. English moralists castigated the notion that wives should be servants to their husbands as 'heathenish and sottish arrogance'. It was standard for those teaching a husband his proper duties to warn him against thinking of or treating his wife in a demeaning way: he 'must esteem her (not as a servant, but) as his yoke-fellow'. Each member of the household had a proper rank, and a wife's was well above that of a servant.[9] All forms of subjection were not the same; 'given that the wife is subject to the husband, it should be as a companion and partner, not as a maid-servant'. It was argued that just as the Bible taught wives to obey their husbands, so it taught husbands to treat their wives as companions. Scripture condemned the 'carnal behaviour of wretched husbands who use their wives as their servants'. The husband should be obeyed but he 'ought not to require base and servile obedience'.[10]

The alliance of husband and wife was portrayed as a partnership in which both worked for the family's good. A wife's conjugal subjection

> differeth far from servile subjection, for he that is servilely subject worketh for another: but the wife worketh not for another but for herself, for she and her husband are one.

Sixteenth- and seventeenth-century thinkers characterized the relationship of husband and wife as more like that of fellow citizens than like that of master and servant, for 'he that is servilely subject worketh for another, he that is politically subject worketh for his own good'. All early-modern writers about marriage stressed the community of interest that should exist between the married couple. Wives should not be 'commanded imperiously as servants, but entreated lovingly as joint tenants'. Ideally husband and wife had no conflicting interests and so could work together for their own and their children's good.[11] Early-modern moralists all placed great weight on the need for common interest and mutual affection in marriage – 'the supreme and most excellent part of all amity and friendship'. Husband and wife should foster between them 'such society and fellowship, yea and greater, than is between the father and his son and not such as between the master and the servant'. Pressure had to be placed on servants to perform their tasks properly, because their employer would reap the profit of the labours, not the servants themselves. A wife did not need to be compelled like a servant, for she was 'a free citizen in thine house'. At the beginning of the eighteenth century, teachers of proper marital conduct were still harping on the same theme. Husbands must note that wives were not 'menial servants' but 'are in decency to be their partners in their fortunes'. Social propriety and stability demanded that a distinct line be drawn between wives and servants.[12]

No early-modern theorist could quite bring himself to say that a wife was her husband's equal, but many argued that she came close to being so. Her subjection was 'no servile subjection or duty, but duty with a kind of equality'. Books of advice on choice of marriage partner constantly reiterated the counsel that spouses should be of similar age, social status, breeding, and education, for 'the way to live quietly is to marry equally in all respects'. Aristotle's advice that a husband should be 20 or more years older than his wife was ignored or denied by most English theorists, despite the fact that such a marriage pattern is clearly conducive to wifely subjection. It was true, divines argued, that wives were not really their husband's equals but 'they have a joint interest and are one flesh' and so 'must have a joint part in government'. A wife must not be commanded 'in the same imperious manner as a child or servant'. Her husband must ensure that children and servants 'use all singular reverence and obedience towards her'. She was subject to her husband, but the gap between them was not nearly as great as that between father and son or master and servant. Rather, 'the wife's subjection' was 'in some degrees approaching to an equality with her husband', and he should aim at 'using her in a manner as his equal'.[13]

Husbandly power was 'authority tempered with equality'; of all the relationships of subjection in society, theirs was the closest. Because the wife shared in family government, her subjection was 'not so great as that which is due from children to parents, much less that of servants to their masters'. A child and a servant must 'stand in awe', but a wife's obedience should proceed from love. It was generally accepted that conjugal power should lead to less inequality than did parental, contractual, or class distinctions. Husband and wife 'come nearer to an equality' than parent and child, and the wife 'may in some cases be upon equal terms' with him. Early-modern theorists wanted to walk a narrow line between allowing a wife complete equality with her husband and letting him treat her as if she were only there to serve him. The puritan minister, William Gouge, was one of many who complained that wives often regarded themselves as their husbands' equals:

> The reason whereof seemeth to be that small inequality which is betwixt the husband and wife: for of all degrees wherein there is any disparity betwixt person and person, there is the least disparity betwixt man and wife.

The husband was his wife's head but not her owner; she was his subject but not his servant. Of all the relationships of subjection, that of husband and wife was 'nearest to equality'. Women were inferior and wives were subject, but there were several different degrees of subjection and the wife stood far higher on the hierarchical heap than many.[14]

The extent of subjection

Early-modern theorists believed that a wife was subject to her husband, but as we have seen this did not mean *absolute* subjection, for it was less than that of the slave, the servant, or the child. What subjection was required within a marriage? What was the extent of a husband's power over his wife? Were there any limits to his authority? Early-modern thinkers dealt with these questions in the context of their attitudes to government in general. They believed that all relationships within state and society, household and family, should be ones of authority as well as of affection. Wives were not being singled out for special treatment when theorists insisted that they must be subject to discipline; it was part and parcel of the early-modern belief that every human association needed a clear hierarchy of power in order to function properly. As it is uncommon nowadays to think of families as a 'societies' needing 'government', it is most important to grasp these concepts in order to understand early-modern attitudes.

The myth of the pre-industrial extended family has now been laid peacefully to rest, but we must still recognize that early-modern families differed in important ways from twentieth-century Western families.[15] The early-modern household was far more often than now a unit of economic production (although this role was being modified). If a child today fails to wash the dishes, little more than irritation results; if a child in the seventeenth century wandered off when (s)he should have been watching the flocks, economic disaster could result. Furthermore, the early-modern family frequently included servants as well as blood-relatives. In the sixteenth and seventeenth centuries, in both England and New England it was common for children to leave their own home during adolescence and act as servants in another household. Even the children of wealthy parents often did this. Perhaps more than half of early-modern English adolescents were servants. One of the foremost social historians of this period describes the presence of servants in the family as the 'one really telling difference between the family in Stuart England, and the family which we know in our time'. Early-modern theorists took it for granted that there would be servants in all but the poorest families. They spoke of

> yet another necessary member of a family which may make the head of the family ache exceedingly, these are servants of both sexes, men and maids.

Servants were indispensable but troublesome. They needed to be regulated even more than children, because of the absence or weakness of bonds of mutual affection and because a servant did not have the same shared interests as family members. Lazy, dishonest, or destructive children threatened their own inheritance; servants suffered from no such curb on their natural inclinations. Even today, the most liberated couples with the most unbridled children expect home helps to follow orders.[16]

The family was a key unit of law and local government in early-modern England. Taxes were paid by the head of household. He was usually responsible for civil torts committed by family members. In English common law, a husband 'must answer for his wife's faults' and 'shall be punished by the payment of the damages and costs' for his wife's trespasses. 'For as we say, who weddeth the woman, weddeth the debts, and is bound for to pay them.' The head of household could not normally be held accountable for family members' criminal actions, but legislation laid down that he could be in certain cases – for example, failure to attend church. Young, single people living alone were regarded as dangerous and in danger because they lacked the discipline provided by family membership. The family was a social, economic, and legal association of persons which needed careful management.[17]

Early-modern theorists believed that the family's head should possess those powers necessary to govern the household. There was virtually unanimous agreement that in order for any society – including domestic society – to function properly it required one governor; divided authority led to chaos. Both Aristotle and the Bible were believed to teach this: 'Every house divided against itself is brought to desolation; and every city or house divided against itself shall not stand' (Matthew 12:25). Moralists advised sons not to marry until they had their own homes, as 'mixing of governors in a household' would lead to trouble.[18] Families were often compared to other societies; sometimes explicitly, sometimes metaphorically. The English Puritan, Robert Cleaver, in a work significantly entitled A godly form of householde government, said that a 'household is, as it were, a little commonwealth'. Theorists described kingdoms as 'only greater families'. Jean Bodin saw a family as 'the true image of a city', and thought household government provided the 'true model' for civil government. Definitions of the family often incorporated the concept of unitary leadership. The family was 'a natural and simple society ... under the private government of one'. It was not only conservative patriarchalists who viewed families as 'little societies'. The Scottish contractarian theorist, Samuel Rutherford, too, accepted that 'domestic society' was 'by nature's instinct' and that domestic government was natural and necessary. The same arguments were employed at the end of the seventeenth century, for 'a family is the epitome of a kingdom and it naturally resolves into a government, which cannot subsist without rules and order'. The family was a society that differed in size from the state but resembled it in its need for government.[19]

The belief that any society requires government pervaded all early-modern political thought, not merely that on the family. God never left any human association without government: 'Order is of God, disorder of sin'. Husbands would have ruled wives and parents governed children even if man had not fallen, 'for mankind is by nature civil and sociable, and therefore requireth government'. Any association of the corrupted and wicked men that have populated the world since the Fall particularly needed government. The preface to Dillingham's *Christian oeconomy or household government* made clear the implication that this general rule included families too; for as 'it is most meet that commonwealths should be well governed, so it is meet that families should be well governed'. The subjection of the wife to the husband was simply another example of the need for rule and hierarchy: without 'orderly subjection' neither natural affairs, nor political societies, nor indeed the world itself would be able to continue.[20]

Marriage constituted the classic case of the need for superiority in one person, because (by definition) it was a society of two. Democracy was not an obvious option where the vote would always be unanimous or tied. When interests and judgements came into conflict, the association would collapse unless an authoritative power of decision lay in one of the couple. Given early-modern attitudes to the relative talents of the sexes, there was little doubt who was going to draw the short straw.

> It is the wife's duty to be subject to the will and direction of her husband. *Reason*. 1. There must be order in every society without which there follows division and thereupon confusion, Mark 3:25.* Now the fittest to govern in the family is the husband as being of the two the more worthy person every way.

No society could be at peace if rule were divided: 'For in differences and dissensions one side must yield, or else great mischief is like to follow: now of the two who should yield but the inferior?' To early-modern theorists, it was patently obvious that the inferior was the wife. The contractarian theorist, James Tyrrell, vigorously opposed giving the husband an unlimited 'despotic' power, but he too thought there had to be a final judge. 'It is true indeed that the wife ought to be subject to the husband in all things tending to the good and preservation of her children and her family, or the family would have two heads'. Divided authority was a recipe for conflict and could not be permitted in a family any more than it could in other societies. The wife must submit, for 'to allow her a will against her husband is to erect a double judicature in one family, and very inconvenient, and an occasion to make them [sic] unchaste'.[21]

---

*'And if a house be divided against itself, that house cannot stand.'

At the end of the seventeenth century, theorists were still arguing along the same lines. Conjugal society simply could not function democratically. Government involved the inequality of ruler and ruled; 'all equality being the natural enemy to societies'. Just as a body with two heads was a freak, unlikely to survive, so was a society with two co-equal rulers. After the Glorious Revolution, England instituted the dual monarchy of William and Mary and for 6 years England was such a monster with two heads, but even in this case theorists tried to adhere to the principle of indivisible sovereignty. They argued that 'though William and Mary were two several and distinct individual persons of the human species, they were but *un roi*, one singular king in their politic station'. No early-modern philosopher felt comfortable with the division of ultimate power between equals. Spontaneously they enquired, 'Who shall be judge'? 'The true foundation of the power of husbands over their wives seems this', reasoned one analyst, 'that in a society of two it is necessary there should be in one the casting voice'. Theorists thought it 'impossible' for any society to survive 'without a subordination of one to the other'. The longer the association was to last, the greater the need for government. This principle was rammed home repeatedly 'to convince wives ... that there is an absolute necessity of government, which supposes subjection somewhere or other'. The married couple, like any society, required government; the husband should be head and governor of the family and household.[22]

There was little disagreement on these points among early-modern legal and political theorists, but they did debate just how much power was necessary. How far did a husband's power over his wife need to extend in order to ensure that the family would function efficiently and well? Their debates and conclusions are best summarized in two categories: first, his physical powers over her person, particularly as regards punishment and correction; second, his powers over property. Each of these will be handled in turn. In these cases of conjugal power, as all others, a universal limitation was applied in early-modern discussions: the husband could not rightly command any action against morality or religion.[23]

Theorists believed that, because a husband's power was given by God, it could not rightly be used for ends of which God would disapprove. This principle had been long established. Medieval canonists had stated that 'the wife should obey her husband in everything lawful and decent': wives need not obey their husbands if they commanded anything unlawful or indecent.[24] God gave the husband his power 'in trust' for the wife's good, and any husband who abused it 'must one day stand before Him to give an account'. In a wedding sermon of 1617, one divine preached the usual tenets of submissive obedience in the wife and divine authority in the husband, but (just in case the groom was rubbing his hands with glee at the prospects for married life these implied) he spelt out the purpose of a husband's power. 'The end of this, as all other government in nations, kingdoms, countries, cities, and towns' was 'not the satisfying of his desires or procuring of his ease, pleasure, [or] credit, ... but the good and benefit of the party governed, to the glory of God, the chief lord and governor of all'. The divine power of husbands must be used in godly ways. Since the wife also had duties to God, the husband had to respect his wife's conscientious interpretation of God's word even where 'misinterpreted and mistaken'.[25]

Wives' duty to their husbands was as nothing compared with their duty to God, so 'they may not be subject in anything to their husbands that cannot stand with their subjection to the Lord'. No wife should obey her husband if he commanded anything that was 'ungodly'; she need only obey his 'lawful commands'. This limitation was extended to the law of the land. Wives should not obey husbands commanding anything contrary to laws imposed by a magistrate superior to them both. Even Jean Bodin, who gave husbands as free rein as any theorist, accepted that a wife should not follow orders to act dishonestly. Wives should obey 'unless when superior powers do forbid it'. In fact, English law did not actually expect a wife to stand up to her criminally inclined husband and refuse to break the law. A woman who joined in a felony with her husband was assumed (unless the contrary could be proved) to be acting under his control and therefore was not responsible for her actions. The indispensable guide-book to the law of marriage, *Baron and feme*, stated the law 'will not bind her by her acts joining with her husband, because they are judged his acts and not hers; she wants free will as minors want judgement'. Admittedly, this did not display a terribly complimentary view of female resolution, but it was probably preferable to 30 years behind bars. Religious theorists admitted that the laws and customs of England were 'very tender to women offending in the company of their husbands' and attributed this to a presumption of constraint. However, they sternly exhorted wives that 'religion has no such consideration but includes them all under sin who commit any sinful actions'. Wives were not bound by their husbands' commands against the law of God or England, and should refuse obedience if so ordered.[26]

Catholic casuists held that the husband sinned gravely if he tried to prevent his wife performing her religious duties, and insisted that she could attend mass without her husband's consent. Protestants insisted that a wife could refuse to go to mass. Other actions were held to be so obviously immoral that no woman could follow her husband's commands. Self-adornment was the outrageous conduct most often mentioned. Priests complained about women who used cosmetics to improve their appearance, and who in confession excused themselves by saying that it was done to please their husbands. Such wives should refuse to dress immodestly. A husband held power over his wife only as God's delegate; he could not rightly order a wife to do something (such as rouging her cheeks) of which God would obviously disapprove. Puritan divines exhorted a wife to disobey if her husband told her to wear attractive clothes or cosmetics, or to visit the theatre. They ranted against 'painting and tincturing' and added,

> to husbands that require this obedience from their wives we oppose the Apostle's rule, who requireth children, servants, and generally all inferiors to obey them to whom they are in subjection, only in the Lord, that is in those things wherein the laws of God and nature may not be violated and infringed.

God's commands overrode a husband's, just as they did those of father, mother, master, mistress, and prince.[27]

Like the admission of spiritual equality, the limitation that a wife need not obey her husband when he commanded her to do anything against divine or natural law, added little to her freedom of action. It merely reiterated her primary obligation to obey God. For these divines, a wife did not have a *right* but a *duty* to refuse to wear cosmetics. It is unclear whether the average seventeenth-century woman thought these limitations as important as did the divines who made them. One historian has described the admission that wives need not obey against God's law as 'a chink in the otherwise all-encompassing authority of the husband', but it was a chink that almost all early-modern theorists – even the most absolutist – allowed to every subject. It is important that early-modern divines allowed wives to refuse obedience to immoral commands, but to understand the nature and extent of a wife's subjection it is more important to define just what powers were ascribed to the husband and how far these powers extended.[28]

Capital and corporal powers

The limitation that no one could command against the laws of God and nature was accepted about all government – royal and paternal as much as conjugal. The great majority of early-modern political and social theorists were clerics, and religious duties headed their scale of priorities. It is far more interesting that virtually all early-modern theorists limited a husband's powers in a further respect; they denied that he could kill his wife. Even if she flagrantly resisted his authority and committed crimes against marriage, his power over her was not one of life and death. This restriction may seem obvious, but in the context of early-modern thought it was far from a foregone conclusion.

Many thinkers believed that fathers originally possessed the power to execute disobedient or criminal children. The Old Testament commanded that disobedient sons should be stoned to death (Deuteronomy

21:18–21). Roman law had originally accorded fathers the power of life and death over their children. Most early-modern theorists denied that fathers' powers should extend to capital punishment and criticized heathens for 'going too far on a good ground'. But political patriarchalists, such as Sir Robert Filmer, contended that it was fathers' natural power of life and death which formed the basis for the state's capital powers. Even patriarchalists thought that most fathers could no longer exercise this power except by the permission of the chief magistrate – the ultimate holder of patriarchal power – who had now monopolized the power of the sword. Indeed, at least after earliest history only the chief father of each society could execute anyone. Nonetheless, in patriarchalist thought, the civil government's coercive authority essentially stemmed from paternal power. Early-modern thinkers readily countenanced the possibility that the head of a family might punish capitally.[29]

Whether a husband had the right to kill his wife was often discussed, by early-modern theorists, in the context of adultery. Adultery was in their eyes a crucial test case, for if anything could give a husband such a right it was this gravest of crimes against him. A husband was more upset by adultery 'than by his own daughter being debauched or his wife killed'. At some times, Roman law had allowed the husband (and/or the wife's father) the right to kill his wife and/or her lover, if they were caught in *flagrante delicto*. Remnants of this right (called *ius occidendi* or *ius necandi* [right of killing]) survived in the laws of many European countries. In English law, for example:

> Where a man is taken in adultery with another's wife, if the husband shall stab the adulterer or beat out his brains, he is guilty but of manslaughter in the eye of the law.

Divines sometimes seemed to sympathize with secular law on this point, commenting on 'what a just judgement befalleth the adulterer when the jealous husband killeth him in a sudden passion'. Adultery was seen as justifying a violent response.[30] Mosaic law had prescribed the death penalty for adulterous women and their paramours, and many divines thought it would do sexual morals no end of good if this punishment were reinstated. The influential German Reformer, Martin Bucer, thought 'hardly anything would contribute more powerfully to the preservation of marriages once contracted than the reintroduction of the law punishing adultery with death'.[31]

Comparative anthropology proved to early-modern theorists that death was the appropriate penalty for adultery by the laws of nature and nations. 'The adulterer is punished with death by the law of every nation.' Roman law's severe penalties for adultery were considered fundamentally just. With somewhat inaccurate historical scholarship, one moralist argued,

Julius Caesar made a law that if the husband or the wife found either in adultery it should be lawful for the husband to kill the wife, or the wife the husband: Death then by the light of nature is a fit punishment for adulterers and adulteresses.

For a short time in England during the Interregnum, the death penalty was introduced, although virtually no one was actually executed under the provisions of the statute. The death penalty was also on the books in parts of puritan New England; however, as one historian has pointed out: 'Although cases of adultery occurred every year, the death penalty is not known to have been applied more than three times'. Nonetheless, official culture encouraged the belief that adultery should be capitally punished. The view that adultery deserved death was not quickly abandoned by theorists. In 1675, an anonymous tract appeared pleading for the restoration of the death penalty for adultery. It favourably compared the Jews, Athenians, and Romans for executing those guilty of adultery, with the English where 'now it is dwindled down into a peccadillo'.[32]

However, despite condemning adultery in the harshest of terms and despite the law's indulgence to crimes of passion, virtually all theorists insisted that the husband had no right to kill his adulterous wife. Theorists argued that husbands' power was corrective only – it did not extend to capital punishment. They insisted that a husband who exercised the *ius occidendi* was sinning mortally even if the secular laws did not punish him for the deed. In the Middle Ages, Thomas Aquinas had set the tone for these discussions with his unambiguous statement that it was never lawful for a husband to kill his wife on his own authority. The powers of a head of household were designed only for amendment, and only 'corrective punishment' could be used. One summary of medieval schoolmen's views concluded that a husband could imprison and moderately beat an adulterous wife but not kill her. Even if civil law did not punish the husband for terminating her life, he could not do it in good conscience.[33] The civil community possessed the power of life and death; the family did not. This opinion was accepted by all medieval and early-modern writers except a few hard-line political patriarchalists. The Spanish casuist, Domingo de Soto, pointed to this general agreement. No one could be capitally punished except by due process of law, and a husband who killed his adulterous wife 'sinned mortally'. Only the civil magistrate possessed this power. Indeed, many casuists argued that even if the husband were a judge, and acting in accordance with the law, he could not execute her. Soto himself did, however, allow the husband legitimately to act as executioner if this were part of his office as a minister of justice. Standard Catholic political theory held that the power of executing criminals lay only with the civil power; a head of household could use punishment to correct an offender, but he could not punish with death even for the most 'abominable fault'. This view traced its roots to the Aristotelian distinction between familial and political power.

The *polis* was a perfect and autonomous community; the family was not and therefore its head's powers did not extend to life and death.[34]

English divines, Anglican and puritan, also maintained that the husband's power did not extend to life and death. They knew that various legal systems had allowed husbands the *ius occidendi*, but rejected the privilege; these laws 'were never allowed by the law of God, and although before men those [husbands] were not punished yet they were guilty before the Lord'. English divines followed the well-established tradition that characterized the husband's power as corrective only; domestic power did not extend to capital punishment. The magistrate 'hath the power of the sword ... but the husband hath no such power'. His power was only 'directory' or 'reprehensive'. Wives were subject to their husbands, but this subjection did not extend as far as a citizen's to the state, for 'the power of life is proper to the magistrate'. Marriages were made for 'mutual help' and therefore 'it is naturally implied that they shall have no power to deprive one another of life: (howsoever some barbarous nations have given men power of the lives of their wives)'. Civilized countries should not allow the husband power to decide whether his wife lived or died. John Locke was doing nothing more than repeating the standard view when he insisted that the husband's power was 'conjugal', 'not a political power of life and death'. A husband's power was not unlimited but 'proportioned to the manner of their union'.[35]

Agreement that the husband did not possess power over his wife's life was not quite universal. Some political patriarchalists insisted that (originally at least) a head of household had the power of life and death over his wife, children, and servants. They saw the family as the ultimate source of civil power and therefore argued that its sovereign must initially have possessed capital powers. Jean Bodin pointed to Roman laws allowing the husband power to judge and execute his wife for certain offences and deduced 'this power of husbands over their wives to have been common unto all people'. A few English theorists also took patriarchal authority to its utmost extreme and argued that 'the husband's power over the wife is without restriction, and by consequence the husband hath power over the life of his wife'. However, these theorists were exceptional. Most thinkers denied that conjugal power was capital. The German natural lawyer, Samuel Pufendorf, was typical in rejecting out of hand such a right on the familiar grounds that 'the end of marriage' required 'no such sovereignty'.[36]

Virtually all early-modern theorists insisted that a husband's powers over his wife need not and did not extend to the power of life and death. Her subjection was limited. Threat of death is, of course, not the only means of inducing obedience; there are many degrees of physical coercion short of execution. In the early-modern period, one of the questions frequently considered was not 'Have you stopped beating your wife'?

but 'Can you (legitimately) beat your wife?' English law certainly allowed a husband to use some physical violence against his wife. To be precise, English law did not give the husband a positive right to beat his wife; it merely stated that a woman had no legal redress unless she was subjected to severe violence. In fact, this applied in all cases of violence within the family. Normally, under English law, a person violently assaulted by another could sue to the magistrate that his or her attacker be bound over to keep the peace. As a matter of policy, magistrates did not allow a wife, a child, or a husband, to 'pray the peace' unless the violence was extreme. Because it was believed that husbands were generally physically stronger than their wives, this translated into a liberty to use moderate violence against wives. Occasionally the tables were turned. In Bowland, Lancashire in 1534, one husband sued for separation in the ecclesiastical courts, alleging 'that his wife is so violent that he dare not cohabit matrimonially with her, being in fear of life and limb'. Then as now it was far more often the case that husbands beat wives. The law did not intervene if the beating was perceived as limited to 'lawful and reasonable correction'. *Baron and feme* stated that the husband 'may not beat her but she may pray the peace, 1 Edward.4.1., but he may give her moderate correction'. What constituted 'moderate' or 'reasonable' constraint was a discretionary decision. There were limits not only on the degree of force that could be used, but also on the circumstances in which it was lawful to use force at all. Legal students were taught that if 'any man beat his wife for any other cause than [that] for which he may be justly severed or divorced from her, he shall for such injury be punished'. In puritan New England, a husband was forbidden by law to strike his wife at all.[37]

Cases where the wife sued her husband for divorce and alimony do suggest that these limits were enforced, at least if the women were wealthy enough to take their husbands to court. In one case, Richard Greenfield, amongst other offences against his wife, 'threw her on the ground, being with child,' and 'made her eye black and blue'; High Commission granted her divorce and alimony. In another case, a wife sued her husband who 'gave her a box on the ear, and spit on her face, and whirled her about, and called her "damned whore"'. The court promptly granted her alimony, and when the husband's lawyer attempted to overturn this by arguing that 'the husband chastised his wife for a reasonable cause, as by the law of the land he might', he was given short shrift; 'certainly the matter alleged is cruelty; for spitting in the face is punishable in Star Chamber'. The court may seem to have given a strange priority to spitting over physical abuse, but that they were on the wife's side is clear. Elizabeth Cary, Viscountess Falkland, converted to Catholicism at a time when this was both illegal and highly unpopular, but when her husband immured and half-starved her to induce a change in her opinions, she took her case to Privy Council and obtained an allowance of £500 a year. English courts were very reluctant to endorse physical punishment of (wealthy) wives. Whatever the practice, the law of the land did allow some 'chastisement' of wives. English secular political thinkers occasionally defended this. James Tyrrell upheld a husband's power 'to compel' his wife 'by correction', 'if she will not do her duty by persuasion and fair means'. He thought it a power 'very rarely to be used' but accepted that 'some coactive empire' was necessary to govern any commonwealth, including the family. English law and some lay theorists permitted a husband to punish his wife corporally.[38]

English reformed divines did not condone such punishment. The commonest line was that the husband was 'never to lay violent hands on her'. Almost to a man, English ministers adopted this position on physical correction, arguing that the husband 'hath no power or liberty granted him in that regard'. A husband could reprove his wife 'in word only'. They employed the same arguments as were martialled in relation to the *ius occidendi*; the magistrate had 'the power of the sword' but the husband had 'no such power'. If the wife needed physical correction, the husband must turn to the magistrate. The 'Homily on Matrimony' – read out in many Anglican churches each year – set out the Church of England's official stance against wife-beating. Husbands were exhorted: 'Let there be none so grievous fault to compel you to beat your wives'. Beating was 'the greatest shame that can be ... to him that doth the deed'; a violent husband was 'a wild beast' and a 'Bedlam man'. The Anglican, William Heale, devoted most of his book, *An apology for women*, to arguing that it was 'unnatural and uncivil' to beat one's wife. Heale thought beating was ineffective, and that both nature and history forbad it. As for the civil and canon law, he insisted that 'in the whole body of it',

> I have not yet found (neither, as I think, hath any man else) set down in these or equivalent terms or otherwise passed by any positive sentence or verdict, *That it is lawful for a husband to beat his wife.*

Heale was unusual in devoting a whole tract to wife-beating, but his verdict was not. Divines' language grew forceful when dealing with the 'mad or desperate' husband 'who doth beat and [p]unch his wife'. A wife-beater was 'a monster among men, and fitter to live in Bedlam than in a civil society'. Many early-modern English theorists refused to believe that there could ever be sufficient reason for corporally punishing a wife, arguing 'that no man ever did this rudeness for a virtuous end'.[39]

The condemnation of wife-beating was general amongst English divines and indeed Continental protestant preachers.[40] There were exceptions. Paul Baynes was a puritan divine who apparently came close to allowing the husband to beat his wife; he preached that 'if a man should go to blows, she must endure with patience, not striking again nor railing'. Richard Baxter also differed from most Protestants by believing that the husband did have the authority to beat his wife. Baxter merely thought it ineffective and incompatible with conjugal love. Most used the argument from conjugal affection to outlaw beating altogether. They held that parents often chastise their children from love, but that experience and conscience showed this could never be the case with a wife. If she were beaten, it was always from feelings 'diametrically contrary to marriage' like anger and hatred. The Bible was used to justify the ban on corporal punishment of wives; 'we read in Scripture of masters that struck their servants, but never of any that struck his wife'. Physical punishment was appropriate to 'bondmen' not to the wife, who 'like a judge ... is joined in commission with her husband to correct others'. Protestant theologians were generally united in their opposition to violence against wives.[41]

Continental Catholic casuists, in contrast, tended to permit 'moderate' beating. This was one of the few areas of marital relations in which a Catholic–Protestant divide can be detected. The Catholic view was indeed the traditional one. Writing just before the Reformation, William Harrington maintained that the husband could 'moderately correct' his wife. He could not beat her severely even for an extremely serious offence, but should turn to the civil magistrate. If he beat her without a cause, the magistrate should punish him. Whether from the weight of tradition or from less sympathy for wives, most Catholic casuists continued to maintain throughout the early-modern period that (with just cause) a husband could beat his wife. The Spanish Catholic theologian, Thomas Sanchez, devoted a section in his vast work on matrimony to considering whether beating was a sufficient cause for a wife to obtain a separation from her husband. He considered the opinions on both sides of the matter and argued along familiar lines that severe beatings could only be imposed by judicial process. A husband's discipline was limited to 'mild and moderate' beating for serious reasons. When a beating was slight and for a serious cause, a wife could not seek a separation, 'because the husband acts justly'. However, even when the husband was acting *un*justly (that is without good cause), Sanchez still did not allow the wife to separate where the beating was not severe. Casuists thought the right of wife-beating could not be denied to husbands and cited, in support, Saint Paul's teaching on husbands' superiority. It was, however, deemed 'disgraceful' and 'wicked' unless prompted 'by the most grave necessity'. Ordinarily Catholic theorists permitted 'very moderate chastisement'. There were a few exceptions to this rule but, in general, Roman Catholic clergymen were far more likely to allow husbands corporally to punish their wives than were protestant ones.[42]

English protestant condemnation of the physical punishment of wives is surprising since (as the leading social historian, Professor Stone, has commented) the early-modern period was 'the great age of the whip'.

Works of advice on domestic discipline insisted on the need to beat children and servants who went astray, but always excepted wives. Almost every possessor of authority was permitted, indeed often exhorted, to use beating to enforce obedience and godly conduct. The biblical dictum, 'spare the rod and spoil the child' was repeated frequently. The modern belief that corporal punishment is psychologically traumatic and probably ineffective was wholly foreign to the early-modern mind. Children and servants would be reformed and redeemed by the whip; they would come to a sad end, temporally and spiritually, if it were neglected. The counsel was 'better the rod than the tree'. Clergymen told their flocks that it was really merciful to treat their children with what 'is commonly reckoned harshness and severity'. Parents would not be nearly so cruel 'in breaking their children's limbs, in tearing their flesh, in pulling out their eyes', as they would be 'in indulging their visions, sinful dispositions and inclinations'. One theorist advocated beating children in place of spouses; he instructed those husbands and wives who 'bestow blows each on other' that they should 'leave off, reserve them for your children; they will do your sons and daughters more good'. Early-modern protestant moralists believed in beating but insisted that wives should not be beaten.[43]

Moral theorists were making a singular exception when they pressed husbands to refrain from physically coercing their wives. Almost all believed that corporal punishment was salutary, and yet almost all of them denied that this was true in the case of wives. One explanation for this exception might be puritan asceticism. 'The puritan hated bear-baiting, not because it gave pain to the bear, but because it gave pleasure to the spectators.' Early-modern observers were well aware that sadism could be a source of erotic pleasure:

> there is a delight in correction that tickles some men extremely. Else the presbyterian parson would never have taken so much pleasure as he did in whipping his maid. Pedagogues delight in lashing, and are glad when a boy commits a fault that they may be at their beloved sport. And were it the fashion for schoolmasters to teach female scholars, you should find more whipping than there is.

Whether to prevent the husband's pleasure or the wife's pain, Protestants in general condemned any physical violence to wives. In no other area of human relations was the obligation to refrain from physical violence so frequently and so forcefully stressed.[44]

In the early-modern period – in theory at least – the husband's powers of physical coercion were strictly limited, particularly by protestant divines. Theory is, of course, very different from practice, and there is no reason to suppose that moral exhortations against wife-beating were more effective in the sixteenth and seventeenth centuries than they are in the twentieth. Modern studies do suggest that life-threatening violence towards wives met with profound social disapproval and occasional intervention by neighbours. In extreme cases, the husband was criminally punished.[45] The head of household's power extended as far as it needed to in order to rule the family properly. Very few believed that such rule required a power to inflict capital punishment. Most English protestant theorists allowed to family governors the power to punish their children and servants corporally, but denied that a husband should apply it to his wife. Even those theorists who did allow that a husband's conjugal power included a right to administer physical correction set many restrictions upon its exercise. However, physical force is a very crude means of control. Did the theorists allow the husband such extensive powers over property that they could happily dispense with physical punishment without materially undermining the husband's domination? What was regarded as the proper relationship between husband and wife in regard to property?

Powers over property

In England, the legal position in regard to a wife's property was complicated by the fact that (until 1875) two bodies of rules operated side by side: common law and equity. Common law was based on custom and precedent as amended and supplemented by statutory law. Equity was based on the decisions made and implemented by the Court of Chancery. Chancery enforced many duties which the common law did not, and offered legal remedies where the common law could provide none. In order to understand anyone's legally enforceable rights, it is necessary to consider both.[46]

In common law, a wife was not *sui iuris*, but legally subsumed in the person of her husband. This condition was technically known as coverture, and the wife as a *feme covert*. 'A woman as soon as she is married is called covert, in latin *nupta*, that is veiled, as it were clouded and overshadowed, ... she is continually *sub potestate viri* [under the power of the husband].' Husband and wife were – for many legal purposes – one person, just as in civil law, father and son were 'considered one person'. In common law, a wife could not contract separately from her husband nor bring a case alone in court. A wife could testify against her husband or a husband against his wife only in exceptional cases such as treason. A wife could testify where she was the victim of her husband's criminal actions. In the notorious case of Lord Audley, Earl of Castlehaven, who held down his wife while one of his servants raped her, the judges stated:

> That in civil cases the wife may not; but in a criminal cause of this nature – where the wife is the party grieved, and on whom the crime is committed – she is to be admitted a witness against her husband.

In capital cases, also, husband and wife ceased to be regarded as one person, and a husband was not usually held responsible for his wife's criminal actions as he was for her civil torts.[47]

According to English common law, a wife had no proprietary rights separately from her husband. Almost everything she owned became his upon marriage, and almost anything she afterwards earned or was given or bequeathed also became his. In contrast, the husband's property remained his own. Coverture and other legal disabilities would effectively have rendered a married woman without any personally enforceable rights to property had not equity intervened. During the sixteenth century, the Court of Chancery established married women's equitable rights over property. The legal device of the 'trust' was used to allow women to control estates independently of their husbands. Rules on duress were established which permitted a woman to escape contracts made when a *feme covert*. If appointed as personal representatives of estates (and many were), wives could act independently of their husbands in Chancery. In one legal historian's words, 'Chancery began to consider and recognize the feme covert in her own right'. A wife's common-law rights to property were minimal, but common law was only one part of the law of England; for a married woman, equity and ecclesiastical law were more important. The rules enforced by equity by no means gave women equal property rights with their husbands, but they did mitigate the severity of the common law towards wives (or at least to those wives who knew how to sue in Chancery and could afford to do so).[48]

Despite equity and ecclesiastical law, married women's property rights were stringently restricted. In 1680, William Lawrence sketched out 'the manifold mischiefs' a woman suffered by marriage. He commenced his outline of their woes by noting that 'they lose all right of propriety in their own goods to the husband as perfectly as if they had been bought by him for slaves at a market'. A wife's position was worse than a prostitute's, for by law no 'harlot loseth the propriety of her goods and rights to him who lies with her'. Lawrence was not an early and enthusiastic defender of woman's rights; he wrote as a slightly dotty contributor to the Exclusion Crisis debate. Nevertheless, his tract included a stirring condemnation of the legal position a woman was placed in by marriage: she was

> made a villein, a pagan, an alien, an excommunicate, a nun of Lisbon, an attainted in a *Praemunire*, a *non habens personam standi in judicio* against her husband; a person convicted, forfeited, and confiscated to his own use and (what is worse of all) when he hath robbed and rifled her of all she hath, he may exercise his Tartarian authority and beat her because she hath no more.

Lawrence's frenzied rhetoric somewhat overstated the case, but more moderate legal manuals presented a similar picture. 'The husband is the wife's head' noted one 'and therefore all she hath is her husband's'. The law gave a married woman few property rights against her husband. Political and religious theorists generally accepted this as legitimate and argued that control over marital property should be in the husband's hands. However, they did place limitations on this doctrine. First, they allowed the wife to retain ownership and control of some possessions in her own right. Second, they insisted upon the wife's interest in common marital property and upon the husband's duty to use it for his wife's benefit as well as his own.[49]

Marriage was thought to establish a community of goods between husband and wife. Husband and wife were 'both owners and possessors of such goods as they have ... and therefore the temporal goods should not be divided, but it [sic] should be common betwixt them – at least as far as need doth require'. Legal theorists accepted that marriage did not necessarily 'infer a communion of goods', rather it 'depends upon the custom of every country'. However, common property between spouses was thought 'one of the natural foundations of civil society'. The Scottish protestant divine, John Wemyss, referred with approval to Plato's *Republic* where '*meum et tuum* [mine and yours] should not be heard betwixt the man and his wife'. However, Wemyss did not want to place the two on an equal footing with regard to their common property; he leapt directly to the conclusion that 'all should be called the husband's'. It was irrelevant if the wife had contributed more; just as wine mixed with water was still called wine even when there was more water, so 'all should be called the husband's'. This metaphor was taken directly from Plutarch, who had used it 1500 years before to make just the same point. Since theorists regarded the husband as the dominant and really important member of the married couple, community property effectively meant that the husband controlled it all. The wife's rights only appeared against outsiders or re-emerged after her husband's death.[50]

Community of goods did not mean that the wife had the same power over them as her husband. By common law, 'marriage is a gift of all the goods and chattels personal of the wife to her husband so that no kind of property in the same remaineth in her'. This was equitable because 'it cannot be that persons so united as to be accounted *one flesh*, should have divided interests and separate estates'. By marrying, a woman accepted that her husband should care for herself and her goods. Since a wife 'committed her own person' to her husband's charge, 'certainly the disposing of her estate, the managing of her business of right belongs unto him'. As there was to be 'no distinction of mine and thine', it followed that 'the man hath the dispensation of all'. Husband and wife pooled their resources for the children's support and the family's benefit.[51] The husband should decide what was beneficial because he was better equipped by nature and education to handle such matters. Weak little wives should be happy with this arrangement, gaining 'the ease to walk only in the way chalked out, neither cumbered with the managing of the

business if it be difficult, nor blamed for the success of it if it be disastrous'. A wife's divinely ordained subjection was expressed in 'forbear-
ing to dispose of the affairs of the family against his mind or without his
consent'. Moralists warned wives that any money spent without their
husband's consent or wasted irresponsibly was 'but stolen'. 'The husband is the true and only proprietor of all: and though the wife hath a
right to all yet it is only a right of use and not of dominion.' A wife was
'joint-proprietor' with her husband, but in the case of property, as in others, 'no law exempteth her from his government'. Bound by God's law to
obey her husband, the wife was 'not to give anything in a way of disobedience'. Of course, the wife need not obey the husband against her
divine duties. He could not forbid her from giving to the poor if their
need was desperate, and she could save if her husband was improvident.
Thomas Aquinas had argued that a wife could give to charity against her
husband's will 'in a case of necessity'. Priests conceded that a wife
who saved secretly when she saw her husband squandering the
family's goods was not stealing but showing 'great prudence and providence'.[52]

The classical platitude that marital property should be common and
controlled by the husband was interminably repeated, as was Scriptural
teaching that the husband should direct the couple's affairs. But six-
teenth- and seventeenth-century theorists made important exceptions to
these received rules. In Roman law, any property separately owned by
the wife (that is, not given to the husband as dowry) was termed *para-
pherna* or *paraphernalia bona*.[53] She could administer these goods at her
own discretion. Medieval scholastics taught that a wife could own personal possessions apart from her dowry and that her husband's consent
was not requisite to their disposal. Early-modern theorists of civil law
were perfectly familiar with the concept of *paraphernalia* and regularly
discussed the wife's rights and duties in respect to her own, separately
held property. English lawyers accepted that wives held individual
rights over their jointures.[54] Wives' independent rights were sustained in
English courts of law. The term *paraphernalia* came to be applied to personal possessions (particularly jewellery) that the wife owned, and husbands' attempts to dispose of these for their own advantage were
thwarted. Furthermore, the equitable device of the jointure was established in order to prevent a husband from exercising control over his
wife's separate property. Chancery upheld the aim as well as the legal
technicalities of the trust.

> For since Queen Elizabeth's time, it hath been the constant course of this
> court to set aside and frustrate all encumbrances and acts of the husband
> upon the trust in the wife's term, and that he shall neither charge nor grant it
> away. And it is the common way of proceeding for the jointures of women,
> to convey a term in trust for them upon marriage, that it may be out of the
> power and reach of the husband . . .

The common-law fiction that a wife lost all independent property-rights on marriage was treated in Chancery as just that – a fiction.[55]

Both English and Continental divines of the early-modern period also thought it morally legitimate that wives could own and control property separately from their husbands. Divines recommended parents to provide for their daughters, and approved 'the commendable assurance of jointures and dowers in lands or money' as 'a thing allowable by the word of God'. Like their medieval predecessors, theologians often dealt with this question in the context of giving alms, arguing that a wife 'may give of those goods that she hath excepted from marriage'. Excepted goods were those 'reserved upon the match made between them or else ... peculiar unto her by their mutual consent; and of them she may lawfully give without her husband's allowance'. These were distinguished clearly from 'goods which are common to them both; ... these she may not bestow without his allowance'. John Locke was unoriginal in arguing that the husband should rule in family affairs, but added that 'this reaching but to the things of their common interest and property, leaves the wife in the full and free possession of what by the contract is her peculiar right'. The common-law fiction that a wife could have no proprietary rights independently of her husband was not regarded by the political and religious theorists of the period as morally binding.[56] Catholic casuists employed the same distinction between the common marital property and a woman's *paraphernalia*. Hermann Busenbaum – whose *Medulla theologiae moralis* was a standard text for confessors (it ran to 200 editions between 1645 and 1776) – like English divines, followed medieval teaching. He stated that a woman sinned gravely if she spent significant amounts of money against her husband's will or more than was customary for someone of her social status, but not if she took it from her *paraphernalia*. Other casuists likewise argued that a wife had 'absolute liberty' to vow to give alms from her *paraphernalia* and other similar goods, the administration of which belonged to her personally.[57]

Another significant limitation was placed on the husband's control of marital property; this was his obligation to support his wife. Both the law of the land, and moral and political theory insisted on this duty; 'by the Common Law the husband is bound to maintain the wife he takes'. When a woman married, 'the law taketh all away from her' and gave it to the husband, and thereafter 'she cannot work but only to the advantage of the husband'. The gift of her goods and earnings was seen as a form of payment for her support – 'a kind of consideration for her thus to charge him'. Her actions rendered her husband chargeable; 'if she takes bread and eat it, trover lieth against him'. A married woman was normally 'disabled to contract', but 'the disability of the wife is no hindrance to contract for necessaries'; 'the word "necessary" here' being taken to mean 'convenient'. It was difficult for the husband to overturn this legal presumption. In one case of the early 1660s, the wife had deserted her husband and he explicitly forbad a clothing merchant to supply her with goods on his behalf, but only a lengthy legal case allowed him to escape contractual liability for her actions. Even when a husband divorced his wife, the obligation of support remained, although no longer 'according to the degree of her husband'. A husband must support his wife 'so long as she is his wife; though she commit felony or be attainted, he is bound to keep and maintain her'. Whilst a wife lived with her husband, she could contract on his behalf; if separated, he must pay her alimony at a rate decided by the ecclesiastical courts.[58]

No early-modern theorist urged a wife to provide for herself in the sweat of her face; this was Adam's punishment, where Eve's was subjection (Genesis 3:19). 'God hath turned many punishments of sin to be bounden duties; as subjection of the wife to husband, and man's eating bread in the sweat of his brow.' All descriptions of a husband's duties insisted on that of support. The canonist Antoninus's standard late-medieval text on the law of marriage stated that a husband who failed to provide for his wife, when he could do so, was guilty of mortal sin. French Catholics insisted on the duty of support in the fifteenth century just as much as English Puritans did in the seventeenth; 'for an honest man careth for his wife and children as well as for himself'. Marital property was the husband's to control, but he was not free to use it as the whim took him. Aristotle was called on alongside Genesis to prove the husband's obligation of support. Aristotle taught that by nature servants were formed for hard physical exertion, but women were formed for house-keeping and child-care, not heavy labour, and they should not be sent outside to work. 'God hath appointed' women to 'less wearisome' works, because 'of the weakness of their sex and of their pains in child-bearing'. Husband and wife each had 'his and her office, the man to toil without in weary labour and travail, and the woman within doors'; the husband must 'spare her weak body from all toil and labour of worldly employment exceeding her ability'. The responsibility for supporting the family lay squarely with the husband.[59] The husband's obligation of maintenance was as much a theme of early-modern homilies as the wife's duty of obedience. 'If he must be bountiful, surely she must be dutiful; if he must not spare to supply her, she must not fail to obey him'; 'he must allow her cheerfully out of his estate, she must conform herself contentedly to his estate'. The wife was obliged to subjection; the husband had to support her to the best of his abilities. 'The first duty of a husband is to provide for his wife' wrote one theorist. 'The very light of nature teacheth that *in familia prima cura uxoris habenda sit*: in a family the first and chief care should be of the wife.' Few disagreed with him. Books of advice counselled males to marry only when God had given 'convenient means to maintain a wife and family, and not before'. Once married, the husband must allow the wife 'sufficient maintenance' with 'a liberal hand'.[60]

Roman Catholic theorists in no way differed from their protestant counterparts on this point. A husband must make every effort to provide for his wife, and do so in accordance with their position in society. Confessors were to instruct husbands that they 'sinned gravely' if they failed to support their wives. The most basic duty of the husband was that 'he should support his wife and family'. Casuists seriously discussed the question of whether a husband could ever order his wife to go out to work, and decided that generally he could not. The foremost casuist of marriage, Thomas Sanchez, discussed the question of the husband's obligation of support at length; it ceased if the wife left without good reason, but if she left because of 'too much chastisement' the husband must still maintain her. Similarly, if a promised dowry were not paid, the husband did not have to assume responsibility for her maintenance, but if he had knowingly taken her without dowry, the obligation of support was binding. Like the many scholastics whose discussions he cited, Sanchez never disputed the husband's fundamental obligation of support.[61]

Just as the husband was deemed to gain control of all common property even if the wife had contributed more, so he had just as much obligation to support her even if she had contributed nothing. Martin Luther maintained: 'Whatever the husband has, this the wife has and possesses in its entirety'. Just as a common woman gained noble status by marriage, 'so is a woman (though she were never so poor before she was married) as rich as her husband – for all that he hath is hers'. The wife gained a right to use her husband's property whether or not he undertook this explicitly, for

> the communion of their rights and interests is by all nations esteemed so natural and necessary, that the wife – whether she brought any dowry or no – has a right to maintenance out of the husband's estate, though there preceded no Deed of Settlement.

The wedding service in the Anglican Book of Common Prayer required the groom to vow: 'with all my worldly goods, I thee endow'. The groom's words were not treated as empty ones. His commitment was thought to 'signify his consent to make her a joint-proprietor'. In the wedding ceremony, 'a right of participation' was given to the wife by the husband 'in all things which were his'. The wife's rights were definitely subordinate to those of her husband for 'she hath no title but by her husband'. Only her husband had 'discretion to dispose' but 'all is hers in participation to use'. The husband controlled the family property, but the wife had rights to use both his and her contribution to their joint holding.[62]

Roman Catholic canon lawyers and casuists worked out the implications of this assertion in painstaking detail. One argued that if an extravagant husband lost all his money to a third party, that person had an

obligation to support the prodigal's wife and children, for their rights in
these things did 'not disappear'. A wife's property rights were sub-
merged, not annihilated. If a husband were granted legal privileges
(such as immunity from taxation) these must be extended to the wife's
property also, for the two were 'common' and 'indivisible'. In *Tractatus
de alimentis* [Tract about maintenance], the Italian Catholic lawyer,
Giovanni Sordi, discussed such erudite points as whether a husband
must provide his wife with silk clothes, horses to ride, vintage wine, and
expensive medicines. His conclusions all stemmed from one principle – if
the couple's social status were such that these provisions were customary, the husband must supply them. Sordi's view was standard.
Catholics and Protestants, Englishmen and Europeans, had a strong
sense of social rank; noble wives should not be treated like commoners,
nor rich ones like paupers.[63] All early-modern theorists joined in this
chorus. Keeping the wife on short commons while the husband lived in
luxury was just not on: 'for a husband ought to maintain his wife in as
good an estate and fashion as himself; by marriage she is advanced to as
high an estate and dignity in relation to others as he is'. By taking a
woman as his wife, the husband was deemed to agree to support her as
well as he did himself; 'yea a man must *caeteris paribus* [other things
being equal], allow her more liberally in matters of comfort than himself'. A wife assumed her husband's social and economic status, so he
must maintain her as himself, 'allowing her that diet and attire his condition requires, religious modesty permits, and his means can afford'.
Common property gave the husband absolute control; it gave the wife
rights of use. *The Batchelor's directory* accepted that husbands were 'masters' of marital property, but stressed that both spouses 'should share' in
its profits or losses. Alberico Gentili, Regius professor of Civil Law at
Oxford, noted the community of property established between husband
and wife at marriage and maintained that in some sense it made her
'mistress of her husband's goods.' The wife was entitled to an equal standard of living with the husband because she was the joint (though
junior) owner of marital property.[64]

Both the basic rule that marital property should be common and under
the control of the husband, and the assertion that wives had an equal
right of use with their husbands were theoretical exhortations. In practice, many wives – common, as well as noble and gentle – were provided
with separate estates under marriage settlements.[65] Similarly, admoni-
tions that husbands provide for their wives as well as they did for themselves – just like the maxim that wives should be sweetly submissive –
may well have been honoured more in the breach than the observance.
Preachers would hardly have wasted their breath in giving such advice if
practice already accorded exactly with theory. However, there was an
high level of agreement on these precepts. One analyst of social contract
theory has asserted: 'Wives are housewives and housewives, like slaves,

receive only subsistence (protection) in return for their labours'. No Renaissance moral theorist in fact believed this should be the case. Indeed, every early-modern theorist who discussed the issue maintained exactly the opposite. A husband who gave his wife 'only subsistence' was universally condemned.

> Because he hath endowed her with all his goods, ... she hath right to them all as himself .... Beasts have fodder, servants meat and drink for their labour and care; she hath the right of all for his endeavors.

The belief in common property and common interest were essentially related to the tenet that husband and wife should both share the benefits of merging their possessions. Even the husband's right to control expenditure was based on the grounds that he was usually better equipped to manage their common resources prudently. Wives were entitled not merely to bare subsistence, they were common owners.[66]

Early-modern theorists believed that under normal circumstances the husband should control the couple's property. The wife should show obedience by accepting her husband's decisions on its management and expenditure. She must not spend it without his permission or against his will (except in extraordinary circumstances when God's law imposed a higher obligation). However, the husband's powers of administration did not extend to property that the wife owned in her own right. Moreover, the husband's right to control the marital property was counterbalanced by his duty to support his wife. His powers over property were not given for 'procuring his ease', but in the interests of the family as a whole. William Gouge, in the preface to his best-selling guide to the godly family, *Of domesticall duties*, provided an account of his beliefs that can well stand as a summary of this section. The wife was rightly restrained from spending against her husband's will or without his consent, but he was clear that

> the aforesaid restraint be not extended to the proper goods of a wife, no, nor over-strictly to such goods as are set apart for the use of the family, nor to extraordinary cases, nor always to an express consent, nor to the consent of such husbands as are impotent or far and long absent.

Early-modern theorists did not break completely with classical and Scriptural tradition; they reiterated that, as a rule, the husband should control the couple's property. But a number of 'cautions and limitations' blunted the application of this rule. The head of the household should control its common property, but this did not preclude control by the wife of her private property. The husband's right to manage the common property was asserted in tandem with the wife's right to use and profit from it. Common marital property could only legitimately be used for the benefit of the whole family.[67]

Notes

1. Aristotle 1984, II, Politics, 1987, 1999 (1252b, 1260a); Sclater 1650, 107. A very small sample of uses of the rib analogy includes Antoninus 1474, sig. M3r; Raulin 1512, f. 26r; Catharinus 1551–2, 164; Delfino 1553, 45; Lorenzo da Brindisi 1935, 243; Bullinger 1541, sig. A4v; Massie 1586, sig. B1v; Bañez 1594, 28; Campanella 1968, 164; Mauser in Beust 1597, 332; Cleaver 1598, 201; Wemyss 1633, I, 266; Willet 1605, 37; Abbot 1608, 57–8; *The Court of good counsell* 1607, sig. C4r; King 1614, 26; Tuvil 1616, 149; Salkeld 1617, 173; Perkins 1618, 691; Griffith 1633, 243–4; Jansen 1641, 36; Whately 1647, 3–4; Sclater 1650, 106; Needler 1655, 54; Reyner 1657, 11; Swinnock 1868, 489; Archdekin 1679, (Apparatus), 62; Ness 1696, 34; Poole 1696, I, sig. B3r. Moses à Vauts (1650, 68) was one of the very few to object to this analogy, complaining that 'Scripture allows no such terms of relation in this case as body and rib; but confines you to head and body'.
2. Hugh of Saint Victor 1951, 117 (I.6.XXXV); Aquinas 1872–82, III(1882), col. 1201 (Suppl. 65.5). Andrew Willet (1634, 774) used the same logic as Aquinas. For the history of this reasoning, see Chrysostom 1888–9, XII (1889), 150; Hincmar of Rheims 1852, 658; Gerson 1606, IV, 358.
3. Tiraqueau 1555, 5; Settala 1626, 15 ('ut sociam'); Olizarovius 1651, 8 ('perspicuum', 'sed coniuges, rerum omnium consortes, individuas laetorum omnium atque tristium socias'). In civil law, intention to marry a female slave was 'a legitimate motive of manumission' under the *Lex Aelia Sentia* (Justinian 1913, 119–20, I.vi.4). For canon law, see Covarruvias 1597, 171 (I.II.3–7.1).
4. Willet 1612, 6 [Moors]; Nicholls 1701, 94 [Asiatic]; Fletcher in Berry and Crummey 1968, 231 [Russians]; Rider 1680, 24 [self-preservation]. French writers made very similar boasts; see, for example, *La liberté des dames* 1685, 4–9 and 29–30.
5. Hall 1837, V, 464 [barbarian]; Heale 1609, 49–50 [burgesses]; Trapp 1656, 80–1 [foot-stools].
6. Abbot 1653, 33; Gibbon 1591, 7; Stock 1865, 40; Swinnock 1868, 423. On children as part of their parents' goods, see also Stockwood 1589, 21; Pricke 1609, sig. H4r–v; Gaule 1630, 97–8; Wemyss 1633, III, 84; Tuvil 1635, 173; Trapp 1662, 104; Ness 1695, II, 94. Saravia (1593, 8) stated that the Laws of the Twelve Tables gave 'to parents' (*parentibus*) the right to sell their children, but in fact only fathers had this right. For debates on whether a mother could sell her children, see Benincasa 1562, f. 141v–142r; Sordi 1602, f. 257r; Pascale 1618, 4.
7. Wemyss 1633, III, 83–4 [goods]; Tyrrell 1691/2, 34 [sell]; Taylor 1828, V, 265 [none]. Compare Stock 1865, 40. Jordan (1990, 33) argues that in Renaissance thought wives were seen 'as a kind ... of property' (Kelso 1956, 25, speaks in similar terms), but theorists' explicit discussions of the matter stated the opposite.
8. Busenbaum 1663, 145 (III.III.V); Liguori 1954, I, 614 (III.III.II.V.n.352 'ipsam dedecent'); Bodin 1962, 19; Nicholls 1701, 94 [drudgeries]; Fleetwood 1716, 140 [passed]. The insistence that well-born wives could not be forced to perform menial tasks was standard amongst Catholics; see, for example, Arelat (Ripa) 1536, f. 18v; Agrippa 1540, sig. C6v–7r; Tiraqueau 1574, f. 58r; Sordi 1602, f. 152r; Barbaro 1677, 106–7.
9. Dillingham 1609, f. 10r [servile]; Gouge 1622, 269, 358 [arrogance]; Cawdry

1656, 114 [yoke-fellow]. For the same view, see Comber (1679, 77) 'the wife is to be treated as a friend, not ruled rigorously like a servant'; King 1614, 9–10; Davenant 1627, 435; Bernard 1845, 74; Barlow 1632, 85; Byfield 1637, 582–3; Abbot 1653, 38; Webb 1653, 79; Dickson 1659, 293; Clarke 1659, 236; Hopkins 1701, 166. Johnson (1971, 111): 'it is expressly pointed out in virtually every treatise on marriage duties that the wife, far from being rated with the household servants, is second only to the husband in authority over the entire household'.
10. Bellarmine 1602–3, III (1603), col. 1328 (I.X 'Siquidem uxor, licet sit viro subiecta, non tamen ut famula, sed ut socia, et collateralis'); Topsell 1597, 39 [carnal]; Leighton 1844, 146 [base]. See Morgan (1944, 10–12): 'When Daniel Ela told his wife Elizabeth in the presence of neighbors that "she was none of his wife, she was but his servant," neighbors reported the incident to the authorities, and in spite of the abject Elizabeth's protest "that I have nothing against my husband to charge him with," the Essex County Court fined him forty shillings'. Just the same attitudes were displayed by English lawyers; see Gardiner 1886, 267.
11. Wemyss 1633, II, 17 [another]; Dillingham 1609, f. 4r [politic]; Willet 1612, 7 [imperiously]. B.A. Hanawalt (1986, 219) has argued that amongst peasant families in medieval England, marriages were not regarded as 'companionate' or centred on 'domesticity', but as an 'economic and emotional' partnership.
12. Cleaver 1598, 210–12 [The phrase 'a free citizen in thine own house' is also to be found in Henry Smith 1866, I, 27, 'Preparative to marriage']; Fleetwood 1716, 260. See also Whately 1617, 28; Perkins 1618, 700; Tuvil 1635, 124–5; Pettus 1674, 58.
13. B.,Ste. 1608, 78 [equality]; *The court of good counsell* 1607, sig. B1v [quietly]; Baxter 1830, 138–44 [imperious]; Dod and Hinde 1614, 76 [reverence]; Nicholls 1701, 93 [approaching]; Whately 1647; 3–4 [using]. See Houlbrooke 1984, 74–5, 'The ideal of parity between marriage partners was widely emphasized in the literature of counsel'; Wall convincingly details one Elizabethan family where exhortations to 'amicable partnerships' (1990, 36) were honoured in practice; I disagree only with her suggestion that Elizabethan moralists advocated anything else.
14. Wilkinson 1607, 37 [tempered]; Lawson 1659, 191 [great]; Carter 1627, 20 [awe]; Hopkins 1701, 162; Tyrrell 1691/2, 44 [terms]; Gouge 1622, 271; Cawdry 1656, 114 [nearest]. For the same view that husband and wife are more equal than parent and child, see Dod and Cleaver 1629, 199; Gataker 1623, 3, 5; Elton 1637, 549; also Bañez 1594, 28, and Burgersdijck 1631, 289–90 where the Aristotelian roots of this idea are also clear.
15. Laslett 1977, 60ff.; Anderson 1980, 23–35; Mitterauer and Sieder 1982, 27ff. and 68–9; Houlbrooke 1984, 18.
16. Macfarlane (1986, 85–6) has estimated: 'About 60 per cent of the population aged 15 to 24 in the period 1574–1821 were servants'; Laslett 1977, 61; Whately 1624, 54 [head-ache]. See also Morgan 1944, 62–77; Anderson 1980, 23ff.; Houlbrooke 1984, 173; McIntosh 1984; and also Smith in Goldberg 1992, for the medieval background.
17. Chamberlayne 1673, 332 [faults]; Brownlow 1651, 94 [payment]; Cardwell 1841, 491 [weds]; the statutes 7.Jac.I.c.6.s.28; 3.Jac.I.c.4.s.32 made a head of household responsible for its members' church attendance; on the young and single, see Morgan 1944, 84–6 [New England]; Bernard 1845, 41; Lathom 1683, 4.

Compare Vauts 1650, 47; Keble 1685, 362. For the problems of holding husbands responsible for their wives' church attendance, see Kent 1973, 55; Rowlands in Prior 1985, 155–6.

18. For Aristotle, see Camerarius 1581, 45; Whately 1624, sig. A6v [mixing].

19. Cleaver 1598, 1; Dugard 1695, 78 [greater]; Nicholls 1701, 71–2 [every]; Bodin 1962, 8; Perkins 1618, 669; Rutherford 1843, 1, 71; *Marriage promoted* 1690, 46 [subsist]. See also Pollock 1993, 47; B.,Ste. 1608, 40, 63; Olizarovius 1651, sig. h1v; Lowde 1694, 158.

20. Hughes 1672, 11 [order]; Dillingham 1609, f. 10r and sig. H3v [well-governed]; Davenant 1627, 426, ('ordinata subiectione'). 'Economics' was indeed in its original and Aristotelian sense, the science of household management. For similar arguments and analogies, see Albertus Magnus 1890–9, 30, 196 and 269 (XXVIII.4); Aquinas 1872–82, III (1882), col. 1126 (Suppl. 52.3.); Hemmingsen 1566, sig. I3r; Smith 1866, I, 26; Kellison 1605, 269; Niccholes 1809, 281; Fosset 1613, 51; Goodman 1629, 252–3; Sclater 1650, 23; Abbot 1653, sig. A4r; Baxter 1830, 91; Durham 1675, 308; *Treason, popery* 1680, 8; Lathom 1683, 20.

21. White 1656, III, 207 [order]; Gouge 1622, 339 [yield]; Tyrrell 1681, 110; Keble 1685, 208 [judicature]. Tyrrell (1691/2, 41–3) did put forward some arguments to suggest that there was no need for a final judge between husband and wife, but the main thrust of his reasoning was for the traditional view. As Woodbridge (1984, 131) has stated so lucidly:

> Just as the Renaissance could not conceive of democracy as other than mob rule, so the idea of marital equality was foreign, strange, hardly capable of entering the mind.

22. *Marriage asserted* 1674, 22; Ferguson 1695, 33 [*roi*]; Cumberland 1727, 350 [casting]; Fleetwood 1716, 132–3. Compare Locke 1988, 321 (II.82); Comber 1679, 76.

23. The husband's sexual rights are described in Chapter 5, Marital duties.

24. Antoninus 1474, sig. M3v ('Debet ergo uxor viro obedire in omnibus licitis et honestis'). See Biller in Goldberg 1992, especially 89–90 for a discussion of these themes in the medieval period.

25. Elton 1637, 554 [account]; Whately 1617, 21, 34 [procuring].

26. Gouge 1622, 28; Bodin 1962, 16; Baxter 1830, 155 [superior]; *Baron and feme* 1738, 7; Fleetwood 1716, 139–40 [tender]. On the limitation that only a husband's lawful commands need be obeyed, see also Bullinger 1541, f. 54r; Ridley 1548, sig. M6v; Aylmer 1559, sig. G3r; Saravia 1593, 11; B., Ste. 1608, 65; Dod and Hinde 1614, 28; Carter 1627, 48–9; Taylor 1633, 205; Tuvil 1635, 55–6; Byfield 1637, 582–3; Whately 1647, 144; Clarke 1659, 235; Leighton 1844, 146; Baxter 1830, 152–4; Allestree 1683, 300; Tyrrell 1691/2, 41; Hopkins 1701, 168; Nicholls 1701, 93; *The house-keepers guide* 1706, 126. On English law, see also Chamberlayne 1673, 331–2; Keble 1685, 443. It seems that the early-modern legal system in practice was lenient with women offenders in general; Cockburn (in Baker 1978, 75) has reported that 'only about one-fifth of all females convicted at assizes suffered the full legal penalty for their misdeeds'.

27. Trapp 1647, 80 [mass]; Raulin 1512, f. 23v–24r [cosmetics]; Gouge 1630, 34 [theatre]; Tuke 1616, 42–3 [tincturing]. On Roman Catholics' insistence that wives be allowed to worship at mass, see, for example, Gerson 1606, IV, 356; Busenbaum 1663, 146, (III.III.V); Liguori 1954, I, 614 (III.III.II.v.n.352).

28. Lake 1987, 152. Willen (1992, 580) has argued along similar lines about the role of women in puritan devotion that 'godliness tempered patriarchy'.

29. Swinnock 1868, 453 [far]; Filmer 1991, 18–19. On paternal power of life and death in Roman law, see, for example, Azor 1600, II, 111 (II.II.xix); Kraze 1619, 5–6; Pascale 1618, 330; Olizarovius 1651, 127–8. For the belief that all government was originally familial, see Sanderson 1851, 220, (VII.xvi); Shower 1694, 38; Swinnock 1868, 329–30; Lathom 1683, 27. Contractarian theorists were eager to point out that 'Paternal authority in the manner as God had established it under the Law, could not inflict death upon a son but in the presence of judges and upon the hearing of witnesses' (*Reflections upon the opinions* 1689, 24). Compare *Christian parents* 1616, 120. However, patriarchalists considered that Abraham's willingness to sacrifice Isaac showed that paternal power normally included that of life and death (see, for example, Ayrault 1614, 7).

30. Nevizanus 1570, 38 ('quam si filia sibi stupretur, vel mulier occidatur'); *Baron and feme* 1738, 323 [stab]; Wemyss 1633, II, 189 [judgement]. Brundage 1987, gives details of the changing civil law relating to the *ius occidendi*.

31. Bucer 1972, 411. For other theorists arguing that adultery merited death, see Bullinger 1541, 40v–41r; Joye [1541?] sig. d7r; Latimer 1844–5, I (1844), 244; Ames 1639, 221 [=211] third pagination; *Disputatio* 1641, 117–18; Beust 1597, 136; Piscator (n.d.) *Appendix, passim*; Dove 1601; *Christian parents* 1616, 128; Blackwood 1659, 249; Trapp 1656, 81, 461–2; Durham 1675, 353–5; *An essay towards* 1697, 111. Thomas More (1989, 83) in Utopia punished only a second offence with death; first-time offenders got off lightly with 'the strictest form of slavery'.

32. Drexel 1642, 238 ('Adulter iure gentium omnium morte mulctatur'); Dillingham 1609, f. 13r [Roman law never gave the wife any right to respond violently to her husband's infidelity]; Firth and Rait 1911, II, 387–9; Durston 1989, 158 [Interregnum]; Morgan 1942, 602 [New England]; *A letter to a member* 1675, 6. See also Allestree 1673, 174; Wood 1712, 302. The treatment of adultery in the seventeenth century resembles that of drunken driving in the twentieth; condemned in the very strongest terms in theory, but punished quite mildly in practice. However, theorists insisted that 'adultery, though there be small ado about it nowadays, yet in its own nature it is always foul and dishonest' (Ayrault 1614, 20).

33. Aquinas 1872–82, III (1882), col. 1175 (Suppl. 60.1); Nevizanus 1570, 40, 60.

34. Soto 1619, 362 (V.I.iii 'mortaliter peccet'); Laymann 1625, 413–14 (X.III.II.7–9 'delictum atrox'). For the standard Catholic view, see Durandus 1571, II, f. 382r; Bañez 1594, 327ff.; Covarruvias 1597, 237–40 (I.II.7–7); Lessius 1606, 79–80; Campanella 1968, 198–200; Diana 1660, 24; Busenbaum 1663, 152 (III.IV.I.II); Huygens 1697, 250–1; Liguori 1954, 629 (III.III.II.V.n.376); For Aristotelian roots, see Burgersdijck 1631, 282; Lathom 1683, 7. Despite the Spanish husband's reputation in England as a willing executioner of an adulterous wife, Fitzmaurice-Kelly has stated that Spanish moralists taught 'whatever the extent of her sin, he is never justified in killing her. Francesch Eximenic devotes several chapters to an exhortation on this point' (1927, 602).

35. Wemyss 1633, II, 6–7 [Lord]; Elton 1625, 145–6 [sword]; Gouge 1622, 659 [directory]; Baxter 1830, 165 [barbarous]; Locke 1988, 174 (I.48); Domas 1705, 17 [proportioned]. Compare Smith 1982, 130; Taylor 1828, V, 264, XII, 478 (II.III.iii); Keble 1685, 366; Tyrrell (1691/2, 16–17, 34).

36. Bodin 1962, 16–17; Coke 1662, 77 [restriction]; Pufendorf 1934, 863. Bossuet also rejected Bodin's assertion that husbands have a power of life and death over their wives: see Hulliung 1974, 412.

37. Keble 1685, 637 (*Bradley v. his wife*, 1664–5); cited in Hair 1972, 54 [limb]; E.,T. *The lawes resolution* 1632, 128; *Baron and feme* 1738, 9; Ridley 1607, 56; Morgan 1944, 10 [New England]. Amussen (1988, 61) has stated that husbands were on occasion bound over to keep the peace against their wives.

38. Gardiner 1886, 265 [black and blue]; Godolphin 1680, 511–12 [The court did allow that the case might have been altered if the wife had been shown to have sufficiently provoked her husband]; cited in Fischer in Hannay 1985, 225–7; Tyrrell 1681, 110–11.

39. Cleaver 1598, 168 [hands]; Perkins 1618, III, 691–2; *Certain sermons or homilies* 1756, 454–5; Heale 1609, 28; Niccholes 1809, 281 [mad]; Elton 1637, 555 [punch]; Taylor 1828, V, 267–8 [virtuous]. For the same position see Ames 1639, 207–8; Gentili 1614, 492; Carter 1627, 24; Tuvil 1635, 122–9; Byfield 1637, 583; Abbot 1653, 44; Webb 1653, 87–8; Trapp 1656, 256–7; Hopkins 1701, 163.

40. Ozment 1983, 51–5; Sprunger in Greaves 1985, 58. Kelso (1956, 86) has stated that in Renaissance theory '[m]ost advisers are against wife-beating'.

41. Baynes 1866, 337; Baxter 1830, 169–70; Davenant 1627, 436 ('matrimonio ex diametro contraria'); Smith 1866, 26. As Moses à Vauts (1650, 55) pointed out, there was indeed no Scriptural precept for wife-beating, but 'we see as slender grounds for women's partaking of the holy supper'. Vauts wrote a lengthy tract in defence of husbands' right to beat their wives, but his views on this matter – like his confused theology – were beyond the pale.

42. Harrington 1528, sig. D2v; Sanchez 1654, III, 401 (XVIII.14–16 'levi et modera-ta'; 'Quia vir iuste agit'); Soto 1619, 384 (V.I.3 'dedecus est, atque adeo nefas, nisi gravissima ingruente necessitate'); Laymann 1625, 413 (V.X.III.II.3 'castigatio moderata magis'). See, for example, Gerson 1606, IV, 357; Sordi 1602, f. 212r–v; Raulin 1512, 27v; Pascale 1618, 332; Campanella 1968, 198; Busenbaum 1663, 146 (III.III.V); Liguori 1954, IV, 151 (VI.II.II.III.n.972). For Catholic theorists who con-demned wife-beating, see Settala 1626, 103ff.; Olizarovius 1651, 54.

43. Stone 1977, 217, 163–6 and 176–8; Hopkins 1701, 155 [rod]; Howe 1822, VII, 521–2 [limbs]; Barker 1624, 202 [bestow]. For recommendations to beat children and servants, see Bullinger 1541, 70v–71r; Hooper 1843, 360; Batty 1581, f. 15v, 20v; Allen 1600, 143; *The Court of good counsell* 1607, sig. G2r; Beard 1612, 222–5; Barker 1624, 25; Carter 1627, 125; Gaule 1630, 63; Tuvil 1635, 425; Abbot 1653, 55; Trapp 1662, 82. See also Houlbrooke 1984, 140–5.

44. Macaulay 1913–15, 142, vol. 1, chapter 2 [baiting]; *Fifteen real comforts* 1683, 53 [tickles].

45. See Amussen 1988, 61, 129; Gillis 1985, 81; and on the medieval background, Hanawalt 1986, 208.

46. For a description of the relative spheres of common law and equity, see Baker 1971, 38–49. A woman's property rights were also affected by the canon law administered in the church courts. Many of canon law's proceedings had impli-cations for property; for example, wives sued for separation and alimony in the ecclesiastical courts. For a stimulating and scholarly study of canon law's opera-tion in early-modern England, see R.H. Helmholz, 1990.

Regrettably, I was able to obtain Erickson 1993, only at a very late stage in the preparation of my book, but am certain that this thorough and impressive study is indispensable to any student of women's legal rights over property in early-modern England.

47. E.,T. *The lawes resolution* 1632, 125 [covert]; Benincasa 1562, f. 165r ('Et pater et filii eadem persona censentur'); Raymond 1696, 1; Howell 1809–28, III, 414 [Castlehaven had little chance of a sympathetic hearing as he was also being tried on charges of homosexual acts with his servants, one of whom he had encouraged to have carnal knowledge of his twelve-year-old daughter]. On testifying against a spouse, see E.,T. *The lawes resolution* 1632, 206; Hobart 1671, 97; Grey 1730, 141–2.

48. E.,T. *The lawes resolution* 1632, 129–30; Cioni 1985, 281 [Chancery]; Erickson 1990, 24 ('it would not have occurred to seventeenth century Englishwomen and men that inheritance and marriage were governed exclusively by common law. They were well aware of the alternatives available in equity, ecclesiastical, and manorial law'). Anyone wishing to understand women's property rights should consult Cioni and Erickson's scholarly studies. Chancery also upheld the rights of legally separated women to sue in their own right (Keck 1697, 35). English law also allowed wives the condition of *feme sole*, in which they could make legally binding contracts, sue in court, etc., independently of their husbands; wealthier women did take advantage of this, see Barnes in Baker 1978, 8–9. Women held extensive property rights from a fairly early period in English history, see Macfarlane 1978, 81, 91. See also Stone 1977, 331.

49. Lawrence 1680, 67–9; Finch 1627, 41 [=40]. Compare Smith 1982, 130.

50. Harrington 1528, sig. D4v [both]; Wood 1712, 46 [custom]; Wemyss 1633, 268–9; Plutarch 1603, 319 [Compare Barbaro 1677, 39]. For the expression of similar views on the Continent, see Grenaille 1640, 221–2; Bullinger 1541, f. 57r; Beust 1597, 231.

51. Gouge 1622, 299–300 [gift]; Goodman 1629, 257 [committed person]; Taylor 1828, V, 261 [dispensation]. See also Dillingham 1609, f. 20r; Abbot 1653, 37; Clarke 1659, 235.

52. Thomas 1661, 64–5 [cumbered]; Cawdry 1656, 107; Hopkins 1701, 169 [stolen]; Baxter 1830, 150; Gouge 1622, 291–3; Herbert 1648a, sig. W2r; Aquinas 1872–82, II (1882), col. 284 (2.2.32.8); Raulin 1512, f. 25r ('prudentiae et providentiae magnae'). A biblical example often used to justify disobedience in a wife, was I Samuel 25. Abigail helped David against her husband Nabal's will and 'Jehovah smote Nabal, so that he died'; David promptly added Abigail to his growing collection of wives. Poole (1696, I, sig. Iii4r) defended Abigail's disobedience because she was acting in 'a case of apparent necessity'; Herbert (1648a, sig. W2r) used this example to prove that wives could give to charity without their husbands' consent; see also Abbot 1653, 42.

53. The English word 'paraphernalia' is no longer generally used in this sense; but the Oxford English Dictionary still gives a primary definition as 'those articles of personal property which the law allows a married woman to keep and, to a certain extent, deal with as her own'.

54. On Roman law, Corbett 1930, 202–3; for medieval views, Aquinas 1872–82, II (1882), col. 284 (2.2.32.9); early-modern descriptions of wives' separate property rights, see, for example, Hotman 1585, 17–18; E.,T. *The lawes resolution* 1632, 91; Wiseman 1657, 87. For discussions of whether a wife had to support an indigent husband from her paraphernalia, see Sordi 1602, f. 26r [=E4r]; Pascale 1618, 333; whether she needs to give her husband money she has earned by work outside the home, Sordi 1602, f. 152r; Nevizanus 1570, 310–12. On jointures, E.,T. *The*

lawes resolution 1632, 182–3.

55. Keck 1697, 240, 225. *The lawes resolution of women's rights* displayed, as Prest (1991, 178) has noted, 'a broad sympathy throughout the text with the lot of women or at least those propertied women whose concerns ... [it] chiefly addresses'. See also Ames 1639, 207; Clarke 1659, 234.

56. Topsell 1597, 216 [commendable]; Perkins 1628, 352 [peculiar]; Locke 1988, 321 (II.82). Cahn (1987, 131) has written of 'the insistence by ideologues that wives should not have their own property and should rely on their husbands for all comforts and necessaries', and Woodbridge (1984, 132) has stated that 'property ownership for women ... was opposed by preachers on moral grounds'. These statements cannot be accepted without modification. Divines did support the common ownership of marital property under the husband's control. However, few or none objected to wives owning some property independently, and I have encountered no theorist who objected to adult spinsters' and widows' ownership of property.

57. Busenbaum 1663, 146 (III.III.V); Laymann 1625, 215 (IV.IV.VII.12 'absoluta libertate') See also Azor 1600, II, 711 (II.XII.x); Liguori 1954, I, 614 (III.III.II.V.n.351).

58. Keble 1685, 363, 70–1, 83, 447, 365; the case was *Manby v. Scot*; ['trover' is an action at common law to recover the value of property wrongfully seized]. See also, Godolphin 1680, 509; *Baron and feme* 1738, 276–7.

59. Gouge 1622, 593 [bounden]; Antoninus 1474, sig. R5v; Raulin 1512, f. 24v, 44r; Bernard 1845, 6 [honest]; Settala 1626, 25 [Aristotle]; Dod and Hinde 1614, 45–6 [weakness]; Rogers 1642, 187, 219 [spare]. See also Luther 1958, 205; Topsell 1597, 244; Carter 1627, 12; Swinnock 1868, 475.

60. Hieron 1613, 24, 18; Dillingham 1609, f. 18r-v [first]; Whately 1624, sig. A4v [convenient]; Whately 1617, 32–3 [liberal]. Alice Clark (1919, 12) has argued that 'In the seventeenth century the idea is seldom encountered that a man supports his wife; husband and wife were then mutually dependent and together supported their children'; but this is clearly untrue. The obligation of the husband to support his wife was accepted throughout society; as Wiener (1975, 41) points out, 'magistrates believed that wives could and should be economically dependent on husbands'. See also Stone 1977, 201–2.

61. Busenbaum 1663, 146 (III.III.V. 'Peccat graviter'); Laymann 1625, 411–12 (X.III.II.2 'Quod maritus uxorem et familiam alere debet'); Sanchez 1654, III, 179–84 (IX.IV.19–IX.V), 180 ('nimiam castigationem').

62. Luther 1958, 137; Pritchard 1579, 80 [hath]; Turner 1698, 59 [Deed]; Baxter 1830, 151 [joint-proprietor]; Hooker 1977, 405–6 (V.73.7); Wilkinson 1607, 20–1 [title].

63. Bañez 1594, 72 ('non cesserunt'); Suarez 1872, II, 379–80 (VIII.XI.2 'communia, et quasi indivisa'); Sordi 1602, f. 121vff. See Benincasa 1562, 73vff. for the same assertion that 'alimenta' extend to luxury goods.

64. Elton 1625, 143–4; Gouge 1622, 402 [fashion]; Baynes 1866, 339 [liberally]; Herbert 1648a, sig. V6v; *The batchelor's directory* 1694, 245–6; Gentili 1614, 184 ('bonorum mariti eam dominam facit'). See also Cardwell 1841, 491; Pritchard 1579, 77; Dod and Hinde 1614, 57; Gouge 1622, 300–3; Dod and Cleaver 1629, 204; Davenant 1627, 432, 435; Bernard 1845, 15; Barlow 1632, 85; Tuvil 1635, 87–8; Byfield 1637, 634; Rogers 1642, 250; Abbot 1653, 44; Taylor 1828, V, 261; *The housekeepers guide* 1706, 71; Fleetwood 1716, 261.

65. Erickson (1990, 37): 'For most people marrying in early-modern England, the

principal purpose of a marriage settlement was the protection of the wife's property. And it was *this* type of settlement which was regularly employed by ordinary people with ordinary wealth'.

66. Pateman 1988, 124; Stock 1865, 176.
67. Gouge 1622, sig. IT3v.

5

Sexual equality

Sex and subjection

The previous chapter dealt with the extent of a husband's power over his wife, but one area was passed by in silence – that of sexual relations. In the view of most early-modern writers the principal end of marriage was reproduction. It was in large part because a couple had sexual congress that marriage differed from any other association. Sexual relations were almost essential to the definition of marriage, and marriage was a relationship of subjection. What then was the relationship between sexual activity and subjection? Was a wife subject to her husband in sexual matters, and if so to what extent? What sexual rights and powers did husband and wife each have?

Early-modern writers had a great deal to say about sex. Those who wrote in Latin – veiling their pronouncements in the decent obscurity of a learned language – were usually more explicit and detailed than those who wrote in the vulgar tongue. All were agreed that sexual relations were not simply to be conducted in accordance with the inclinations and desires of the parties concerned. Strict regulations on what conduct was, and was not, permissible were prescribed by the laws of God and nature. The prescriptions of Renaissance morality were based on three axioms, underpinned by their concept of nature and their interpretation of the Bible. First, sex was instituted for the procreation of children and this (directly or indirectly) was its only legitimate object. Second, sexual contact in virtually any form was only permissible between husband and wife; only marriage gave a couple rights to use one another's bodies in sexual acts. Third, the rights (and duties) of the husband and wife over one another's bodies in relation to sexual acts were absolutely equal.

God was believed to have created two different sexes so that mankind could reproduce.* Saint Augustine reasoned in this way in the fifth century. Aristotelian thought also placed sexual differentiation and the female sex in a primarily reproductive context. Theorists moved with

---

*'So God created man in His own image, in the image of God created He him; male and female created He them. And God blessed them: and God said unto them, Be fruitful, and multiply, and replenish the earth, and subdue it' (Genesis 1:27–8).

complacent ease from Aristotle to Genesis. It was necessary to distinguish male and female, wrote one Italian theorist, so that the male could
generate and the female incubate 'as Aristotle teaches'; the object of this
distinction was the procreation of children as 'holy Genesis' showed.
English writers took the same position as their Italian counterparts. The
distinction of male and female sex was by nature for they could 'achieve
nothing' singly, but 'together' could generate children. Sexual differentiation was essentially connected with sexual reproduction.[1]

The existence not only of the two sexes, but the female sex specifically,
was attributed to reproduction. Woman was made 'as a help to generation'. Medieval and early-modern theorists could not think of another
reason why God would have produced the inferior half of mankind.
Augustine had established this view in his commentary on the account
of Eve's creation given in the Book of Genesis. The thirteenth-century
monk, Vincent of Beauvais, in his enormously comprehensive encyclope-
dia quoted Augustine word for word. Thomas Aquinas took up
Augustine's arguments and argued that God made woman to help by
having children,

> not indeed as a help in any other tasks (for as everyone says, in any other
> task you can think of a man can be more suitably assisted by another man
> than by a woman) but as a help in generation.

A male was a better assistant than a female in every area except that of
reproduction.[2] The opinion that God created the female sex with the aim
of enabling man to reproduce was accepted in the sixteenth century as it
had been in the Middle Ages. Even the humanist divine, John Colet, who
was clearly disgusted by the carnal aspects of marriage, accepted that
Eve was made 'fit for bodily propagation'. The Catholic cardinals,
Cajetan and Bellarmine reached the same conclusion as Augustine and
Aquinas: woman was necessary for reproduction and for no other purpose.[3] The Protestant Reformers disagreed with the scholastics and their
Counter-Reformation opponents on many points of theology, but were
in wholehearted accord on this. For Martin Luther, woman was 'needed
to bring about the increase of the human race'; God created Eve because
'this sex was to be useful for procreation'. A wife fulfilled additional
roles, such as housekeeping, but reproduction was her original and primary function. Luther rejoiced that 'the entire female body was created
for the purpose of nurturing children'. Theorists repeated Augustine's
arguments unchanged, asserting that reproduction was God's intention
or otherwise they would 'have been both created males'. The male is
after all 'more perfect than the woman' so 'why should they not both
have been created male; if it had not been for their multiplication by the
ordinary course of generation?' Males were better in virtually every
respect than females but women were needed to incubate offspring.
'Man might have helped man in labour and conference, but the woman

hath a womb and breasts and is a meet help for the conceiving and conserving of children.' John White was a puritan minister who helped in
the planting of the Massachusetts colony, but on this point he was content to repeat Catholic orthodoxy unchanged. White wrote that God created woman with reproduction in mind,

> otherwise man might have had as much comfort in society, and help and
> assistance in his employment by the creating another man; only he could
> have no issue but by a woman.

A woman supplied companionship and ran the home, but what God was
aiming at in her creation 'more especially' was the propagation of children.[4]

Richard Hooker, in his immensely influential *Of the laws of ecclesiastical
polity*, followed in the long tradition that God had created women 'for
man's sake to be his helper in regard of . . . the having and the bringing
up of children'. Conformist Anglicans and radical Calvinists alike agreed
that woman 'was meet for man, necessary for the procreation and education of children, and profitable for the disposing of household affairs'.
Woman was useful for doing housework; but absolutely indispensable to
having children. God made the male 'fit to beget' and woman 'fit to conceive, bear, bring forth, and nurse children'. Unless discussing housework, virtually no one suggested that women had anything of
importance to offer mankind other than their capability to produce and
raise a family.[5] To the end of the seventeenth century and beyond, theorists portrayed woman's primary function, purpose, and use, as provid-
ing children. Woman was given as a helper 'and such a helper she is that
man could not have been capable of that blessing, *Increase and multiply*,
without her'. Discussing Eve's creation the German natural lawyer,
Samuel Pufendorf, repeated Augustine's argument virtually unchanged:
were it not for sexual urges 'men would rather gather by themselves and
could find of their own number a more practical helpmeet'. The only
thing a woman could do that a male could not was bear children, so this
must be the reason for her existence.[6]

Both Catholics and Protestants also agreed that just as sexual differentiation was divinely aimed at reproduction so was man's 'vehement
desire' for sexual relations. Both sexes possessed a 'mutual appetite for
conjunction' so that they would fulfil God's order to be fruitful and multiply. The belief that sex is valuable as a source of pleasure, as a way of
expressing affection, or as a means of bonding a couple emotionally was
virtually never voiced by early-modern theorists.[7] It was not only theologians who argued that God made mankind with sexual drives so that
they would be motivated to have children. English laymen argued that
God intended mankind to be fruitful 'and for that purpose He instilled
into mankind an appetite for procreation, which instinct is nature in
them, and the appetite for the same is natural'.[8] Civil lawyers saw the

instinct to have children as so clearly natural that they used it to describe just what the law of nature was. Early-modern civilians (summarizing earlier views) defined the law of nature as

> that which nature hath taught every living creature, as the care and defence of every creature's life, desire of liberty, the conjunction of male and female for procreation sake.

Nature taught not life, liberty, and the pursuit of happiness, but life, liberty, and the production of children.[9]

God made sexual intercourse enjoyable so that 'reproduction should be perpetually served'. A seventeenth-century doctor put the same idea in seventeenth-century medical terminology. The two sexes were made to reproduce:

> For that reason each is subject to a certain desire and pleasure by which fire, as it were a certain burning, they fall into mutual embrace. In the male it is aimed utterly at the ejaculation of semen, in the woman's womb at receiving, retaining, and nourishing it.

Sexual desire was naturally directed towards impregnation. Were it not for sexual pleasure, men would 'never want' to approach women and therefore God implanted a 'vehement desire of the other sex' in order to perpetuate the species. Sex was instituted with the serious object of having children, *not* because it was enjoyable. 'Why doth nature give to love so great pleasure? *Answer*; For preservation of mankind.' Sexual relations were naturally aimed at 'reproducing someone similar to oneself'. Wherever possible, early-modern theorists tied the urge to copulate to the desire to have children. Once again the book of Genesis was portrayed as describing not only divine laws but human nature. Man did not desire sexual relations merely to enjoy a peculiarly pleasant physical sensation; rather, man desired 'to leave after him a posterity conformable to this order of the Creator, "Increase and multiply", which is the principal end of marriage'. The modern notion that sexual pleasure is an end in itself and children merely a side-effect was entirely absent from the early-modern official morality.[10]

From these attitudes stemmed the total ban imposed in Christian writing, from the primitive church almost to the present day, on any non-reproductive sexual activity. Bestiality, homosexuality, masturbation, sodomy, and the marriage of the incurably impotent were all forbidden because they produced venereal pleasure but not children. The Ten Commandments were held to forbid not only adultery but also 'all other kinds of filthiness' – in particular, bestiality, sodomy, and 'pollutions of the flesh'. Sexual pleasure was only permissible as a by-product of reproduction. In the words of the Anglican *Homilies*, God forbad 'all unlawful use of those parts which be ordained for generation'. Nature's purpose in sexual relations was not its participants' 'gratification', but the production of children. Anyone who copulated merely for the fun of it was acting unnaturally: 'they offend against this purpose who find in it nothing other than a means to alleviate their desires'. To be legitimate, sex must be aimed at producing children not pleasure.[11]

It is important to grasp the absolutely basic character of the tenet that sex and reproduction were necessarily connected by God and nature, because it was implicit in virtually everything written during the sixteenth and seventeenth centuries about marital rights and duties. Women had been created to bear children. Males' desire for females and females' desire for males should be aimed at begetting children. Sexual relations were not a singularly enjoyable activity instituted by God to cheer up mankind during long winter nights. All sex should be ultimately directed at the production of children since it was for this reason alone that God had made mankind a sexually differentiated species.

Sex and marriage

Heterosexual relations were morally permissible because they produced children, but even these relations were restricted: they must be within marriage. Sexual relations between unmarried couples were also potentially fertile, but this did not render them acceptable to early-modern theorists. The arguments used against fornication illustrate the second major constraint on sexual relations; that is, that only a married couple could licitly copulate.[12] Other sexual relations were 'illegitimate', like the children who resulted from them. Virtually all medieval and early-modern theorists believed that marriage was natural, reasonable, and necessary. The object of sexual relations was children and it was believed that only marriage guaranteed their proper nurture. Children might be born from casual sexual relations between an unmarried couple, but they would not be well reared. Early-modern theorists argued for marriage and against fornication as follows: children need a male to help support and educate them. No male will incur such effort and expense unless he is sure the offspring are his own. Where children are born from promiscuous fornication no male can be certain of this. Fornication probably results either in children with no male to support and guide them, or else in abortion or infanticide. In such cases individuals suffer and, if the practice were general, society could not function well. Therefore, fornication is unreasonable and immoral, while, in contrast, marriage is natural and moral.[13]

All theorists stressed the need for fathers in child support. In doing this they were describing the facts of early-modern social life. Europe was not then made up of welfare states. It had no effective social welfare system, only the most rudimentary provisions for the completely destitute. The state's role was to provide defence, law, and order; the family was regarded as the proper unit of social support. No early-modern theorist expected the state to provide unemployment benefits, old age pensions, free education, maternity-leave, crèches, or nursery schools. Apart from (notoriously negligent) wet-nurses, there were no private commercial child-care services, and virtually no well-paid employment opportunities for women that would have enabled them to pay for these if they had existed. Pregnant and nursing mothers were expected to be dependent upon male support. Marriage was necessary because 'women would not be able to undergo the pains and care of educating their children without the assistance of the fathers'. No one doubted 'the weakness of a woman and her need of support in pregnancy.' Women were assumed to be dependent upon their husbands while they bore and nursed their children.[14]

Supporting his family was seen as a father's natural responsibility. Civil lawyers insisted that the husband must always maintain a pregnant wife, whether or not it was certain she was pregnant and whether or not she was wealthy enough to support the child herself. A father was so obliged to support his children 'that he is said to kill them' if he neglected this duty. The father's duty of support was thought to stem from natural law. English law and theorists upheld the same principle. A husband 'must be careful to provide for his house, to feed and clothe his family, to instruct his wife and children'. To the vast majority of theorists, it was clearly 'a law proceeding from nature' 'that parents should educate their children and supply them with maintenance'. The duty was thought incumbent on both parents but particularly the father, for 'the care of the instruction and institution of children is a duty required rather of the father, who is better able, than of the mother, who is every way the weaker vessel'.[15] Early-modern thinkers thought it natural for fathers to feel affection for their own children. They believed that males would be instinctively willing to support and maintain their own offspring, for 'nature itself doth teach the very brute beasts to nourish, cherish, and comfort their young'. They readily adopted Cicero's belief that nature 'engendereth in especial a singular love' to one's own children; for this reason males 'study to prepare things necessary for clothing and food, not for themselves only but for their wives, children and other whom they love and ought to defend and keep'. Men, like all animals, 'company together because of generation and a diligent oversight of their posterity'. The laws of nature 'enjoin a peculiar benevolence of parents towards their children'. Fathers love their children more than themselves, argued one author. Another described the 'natural affection' that led man to desire children 'for the conservation of his name and family' and so 'that he may outlive his own death in his posterity'. Males desire 'to bring to perfection' their child 'and to leave it alive after him that begot it. And for this purpose nature has instilled in the sire a love for the thing begotten, which urges him to take care of the education and nurture of it'. Fathers want children in order to inherit 'their name and their goods', but they would be reluctant to raise another man's children to this end.[16]

The belief that people were disinclined to support others' children was fundamental to medieval and early-modern thinking. Parental affection was thought to grow from the fact that the child was like oneself, a part of oneself. Every man 'coveteth to preserve his being in his posterity'. 'It is nature's instinct to generate her like; her ambition to live in her image and set up her name in her succeeding offspring.' Barring gross incompetence on the part of the maternity ward, a woman can always be sure her child is her own. Usually so generous to the male, nature has given him no such assurance. The counsel of perfection – that every child should be equally loved whether or not it is a blood-relation – was not believed to have sufficient motive force to persuade someone to cherish another's children. It was a tag of Roman law that the mother was always certain, even if she conceived by prostitution, while only marriage indicated who was the father. In the early-modern period, just as in antiquity and the Middle Ages, this was not a legal fiction but an unavoidable fact of life. The state of scientific knowledge was such that no reliable proof of paternity could be provided where a woman had sexual relations with more than one male. Early-modern authors argued that mothers loved their children more than did fathers, because the former were 'more certain' that they were their own. Whatever the trust between wife and husband, the children 'are always more surely her children than his'. Before blood, tissue, and bone-marrow typing, it was a wise father who knew his own son or daughter.[17] The distrust of step-parents, found in official theory and popular culture, embodied the belief that it was natural for parents to love their own children, not other people's. Remarriage was condemned, because it was thought the children would fare badly under a step-mother or step-father. Mothers had no problem recognizing their children, but the institution of marriage was needed to provide fathers with sufficient certainty of issue. 'Common sense' showed that mothers were certain, but fathers uncertain and merely 'presumed so in law' not shown 'by true and certain proof'. Pregnant and nursing women needed assistance from males: 'Yet what man would offer his support unless he were sure he was the father? – a matter of uncertainty apart from marriage'. Sexual promiscuity led to radical uncertainty about the paternity of children.[18]

The early-modern solution to the problem of uncertain paternity was restricting sexual relations to within marriage. Each woman must bind herself to one male only. The Scots canonist, William Hay, lecturing in the 1530s, insisted that 'generation can never lawfully take place among men except in marriage'. The prohibition stemmed primarily from social necessity: 'A fleeting union is repugnant and contrary to the welfare of the child, family, and state, to natural reason and Sacred Scripture'. It must be against the child's interest, because in the case of one born from

casual sex, 'no one would know who was the father, who ought to educate the offspring, look after the mother, and succour her in want'. For
many years Charles Blount carried on an adulterous liaison with
Penelope, wife of Lord Rich, and he was therefore singularly well qualified (practically if not morally) to comment on the uncertainties that
resulted from sexual promiscuity. Without marriage, Blount thought 'no
man should know his own children' and so there would be 'no care of
their education, wherein consisteth their well-being'. The Dutch political
theorist, Hugo Grotius insisted that the institution of marriage was the
best way of establishing 'who was probably the father of each child'.
James Tyrrell was swayed by the same logic: fathers would only be motivated to rear legitimate children,

> for as to children got out of marriage it is uncertain who is their father: who
> can only be known by the declaration of the mother; and she sometimes can-
> not certainly tell herself. So that no man is obligated to take care of or breed
> up a bastard.

Parents love their own children, and children need the care of both parents; the exclusive, permanent tie of marriage was thought to give the
father assurance of his children's identity and therefore the motivation
and ability to nurture them. Each woman must agree to restrict her
embraces to one male and all other males must recognize her as out of
bounds.[19]

Virtually all medieval and early-modern theorists maintained that fornication was against the law of nature, as well as against God's law.
They argued that anyone – Christian or pagan – who thought about the
matter seriously would conclude that sexual relations were licit only
within the permanent relationship of marriage. Some pagans had not
properly appreciated that fornication was naturally immoral, but careful
reasoning would convince any unbiased thinker. Fornication was as
unnatural as theft; men instinctively desired both the opposite sex and
their neighbours' goods, but reason would tell them that neither of these
could simply be taken at will. The seventh commandment forbad adultery and was believed to prohibit simple fornication also. Theorists
admitted that 'the law of nature affords not such evident proofs of the
unlawfulness' of fornication as it did of adultery; but insisted that 'the
light of nature is not without some proof of its unlawfulness'. Early-
modern theorists used arguments from nature and reason to prove why
the commandment must be construed so broadly.[20] It was believed that
sexual relations outside marriage deviated from their natural object of
reproduction: 'the whore desireth not ... the procreation of children'.
Fornicators were sinfully indifferent to their potential offspring's welfare: 'inasmuch as that due care which ought to be taken for the issue of
the begotten is not attended unto'. They were only concerned with 'the
satisfaction of a present lust'. Whatever he did in practice, Henry VIII

accepted in theory that every act of generation outside marriage was 'damnable'. Protestant theorists echoed the orthodox view. Male and female were naturally inclined to coitus but 'every conjunction of them except in marriage is unlawful and can only produce an illegitimate off-spring'. Throughout the sixteenth and seventeenth centuries, theorists insisted that 'vagrant lust' was naturally evil because of the probable neglect of the fruits of such unions.[21]

Early-modern writers saw their analysis of the evils of sex outside marriage substantiated by the fate of bastards. The begetters of bastards 'are not so careful of the good education of such children; . . . such fathers care not'. The folly and sinfulness of fornication were apparent:

> For the bastards begot by such vagrant lust are wholly neglected in point of education, wanting the care of a father and the cohabitation of parents, and so both an accursed posterity is begotten and beggary increased.

Marriage was essential to children's welfare and to the proper function-ing of society. Moralists bemoaned how often 'the son of fornication falls an early sacrifice to conceal his mother's shame', and described how the little bastard would probably grow to be a thief and a beggar or, if female, a prostitute like her mother. Another tract of the same period saw bastards as the cause of a 'charge, which is too common, of parishes being put to the expense of maintaining children left at doors', for 'noth-ing fills our streets more with poor than the issue of unlawful beds'. Early-modern social observers thought that common experience proved fathers were not inclined to care for bastard offspring, and that bastards were less likely to grow into responsible and honest citizens.[22] In early-modern England, what social welfare system did exist was extremely local – generally based on the village. The costs of bastardy were not sub-merged in an extra penny of income tax. The villagers who paid the 'Poor Rate' to support bastards were all too certain that ordinary citizens paid the price for others' fornication. As one modern historian has observed, moral and financial concerns were interrelated: 'The objections to bastardy were expressed very largely in terms of its contribution to the problem of poverty'.[23]

It was not only illegitimate children and the local community who would suffer the consequences of fornication; all of society was threat-ened. All 'histories sacred and profane' abounded 'with instances of fam-ilies and nations being ruined by concubinage and other sorts of whoredom'. Theorists recognized that some societies had not pro-nounced fornication illegal, but 'notwithstanding any toleration of this wicked practice among the heathens, yet they were not insensible of the turpitude and certain immorality of it'. Even pagans sensed the immoral-ity of promiscuous fornication because – despite their ignorance of God's law – reason informed them that families were necessary to civil society. Cities and states were regarded as made up of families rather than

individuals, because it was families that socialized children into responsible modes of conduct and formed the bonds which made co-operative
association possible. And families stemmed 'from marriage'.[24] Marriage
and the family were basic to distinguishing humans from animals.
Without them, society would fracture into egocentric, competing individuals. Promiscuous reproduction would make men 'like the ostriches
in the wilderness, who lay their eggs in the sand and leave them exposed
to be crushed by the feet of every accidental passenger'. To leave children ignorant of their own parentage 'would render us like brute beasts'.
Early-modern theorists portrayed the married couple as the basis of all
orderly society, for 'all kindred and affinity in the world take their birth
from this root'. Without marriage, 'men would live dispersed like savage
beasts and irrational creatures, without distinction or separation of tribe
or family, which are the first parts of the commonwealth'. Marriage was
essential to the existence of the family and the family essential to the
existence of the state. *Marriage promoted*, developed this theme:

> nor can there be a true propagation of human nature but by matrimonial
> contract; whereas were there a promiscuous use of women, the world must
> be peopled with mankind in the same confusion that the wilderness
> abounds with beasts; herds not communities would inhabit the Earth, and
> natural affection would hardly continue with such brutes to rear up their
> increase till they had strength to provide for themselves; and then what
> inhuman violence these savages would mutually exercise upon each other is
> not easy to imagine.

Early-modern commentators thought that no society could function
peacefully without clear family ties. Amicable human relations arose naturally among men, because they were not '(as the ancients fabled the
first men to have been) born of some big-bellied oak, or ... like Mr
Hobbes' sticklers allied to no one in the world'. In such conditions there
might 'be some show of reason why they should not be solicitous for
any'. It was no accident that the Hobbesian state of nature – where all
men war with one another – was one of autonomous individuals lacking
any familial ties. In many early-modern theorists' view, it was marriage
and the family which formed the affective bonds that prevented such
anarchy arising.[25]

   Renaissance theory held that people would readily perceive that children would not be well reared unless a permanent tie bound the parents.
Marriage would follow from reproductive desires, because the law of
nature taught 'every animal' to join in rearing their offspring. From
mankind's 'strong desire of propagating their species' resulted
'covenants relating to the maintenance and government of their families'.
Dogs mate promiscuously, but humans marry. Sexual desire was naturally aimed at rearing children and this involved marriage. Like so many
of the beliefs held by early-modern theorists, these had deep roots.
Aristotle is famous for having described man as a 'political animal' – an

animal naturally inclined to live in a *'polis'* or city. For Aristotle, marriage was still more basic to human nature: 'for man is naturally inclined to form couples – even more than to form cities'. It was natural for male and female to procreate and live together. Children were a 'bond of union' and 'parents love their children as being a part of themselves'.[26] Aristotle's arguments were regularly cited by all subsequent theorists to prove that, even without the Bible's guidance, pagans had recognized the necessity of marriage. Although Aristotle did not stress the need for the permanence of the marriage bond, as did later Christian writers, his arguments seemed entirely compatible with Christian teaching.[27]

Thomas Aquinas cited Aristotle to show that the children's good was the principal end of matrimony. Children require 'existence, nourishment, and education' from their parents and only the marriage bond guaranteed their supply. Aquinas looked to the animal kingdom as proof of these assertions. Animals whose young needed the support of both parents, paired for as long as was necessary. Human children required a long period of support by both parents, therefore there was the 'greatest bond' between male and female. In the early sixteenth century, one theorist combined Aristotle's and Aquinas's authority and added a little sociological support: parents' whole lifetime was scarcely enough to ensure the proper rearing of children, who could not properly support and maintain themselves until they were 20 or 25 years old. Given the average age of marriage, the parents would then be 'in old age'; supporting children involved a lifetime's joint commitment. John Locke and other Restoration theorists repeated Aquinas's arguments and examples (with a little seventeenth-century scientific jargon added) in their accounts of why human marriages should last longer than the casual couplings of animals. Many other animals rapidly separated after a short period of care but 'in human kind' parental union was 'of much greater continuance; children standing in need of tender usage and incessant provisions much more and much longer than the young of any other creatures in the world do'. Human children could not survive or prosper without a long period of parental care and they certainly would not grow up to be good Christians and responsible citizens if they received the level of affection and attention given to the average bastard. To be well cared for, 'the father's care is all along as requisite as the mother's indulgence'. It was both rational and natural for men to marry in order to nurture and protect their offspring.[28]

Early-modern historical and anthropological knowledge was also martialled to support the contention that marriage was *the* natural basis for society and the family. Theorists were aware that promiscuous sexual unions had occurred throughout history, but they believed that the only successful *institutional* arrangement for reproduction was that which bound one woman (or more) to one male in a long-term exclusive tie. They observed, around the world, cultures which endorsed monogamous and polygamous heterosexual unions, but they saw no evidence of societies based on unrestricted sexual freedom, group marriage, polyandry, or homosexual communes, and did not consider these as feasible arrangements. In early-modern thought, the family was prior to society both historically and logically. Families were the building-blocks, the basic units of society and they could arise only by marriage. From the society between male and female, preached one divine, 'spring all other societies which God hath ordained for the behoof of the life of man'. Family life was essential to human society. There was never a race 'so barbarous, and so deviant from law and good practice, so foreign to all humanity' as not to esteem marriage. Sixteenth- and seventeenth-century writers regarded the virtually universal acceptance of marriage in all societies at all times as proof that it was natural. Marriage 'hath been continued from the beginning of the world, in every law, in every manner of people, in every city, and in every time'. Marriage was not simply a Christian or a European institution: 'not only do the Scripture of God commend it, but all laws, all policies and commonwealths, yea, that never knew or heard of Christ do evince it'. Early-modern theorists found very few exceptions to this statement. *Marriage asserted* turned to comparative anthropology, and argued that in the Cape of Good Hope:

> there they never marry, but couple promiscuously (with baboons sometimes when the spirit moves them) without being scared with the damned bulbeggars of adultery, incest, and other yet more horrid crimes; and for relation of wife, brother, son, uncle, and so forth they have no occasion for the names.

The author recognized that other arrangements for reproduction than marriage were possible, but clearly did not deem them worthy of serious consideration. More soberly, one seventeenth-century historian argued that a few barbarous nations practised community of wives or promiscuous adultery but 'the main current of the heathen laws were [sic] against such practices'. For the great majority of early-modern theorists, marriage was the only rational, natural, practical basis for the organization of a civilized society.[29]

Early-modern moral and political theory envisaged necessary connections between sex and reproduction. Sexual differentiation, the female sex, sexual desire, and sexual relations were all tied to the procreation of children. They also asserted that children and marriage were necessarily connected. Only married couples should have sexual relations, for only then could the father be sure that the children were his own and be motivated to cherish and rear them. Marriage was as natural as reproduction and a necessary prerequisite of civilized society.

Marital duties

Sexual relations must be aimed at reproduction and must take place within marriage. These dogmas formed the basis for early-modern theorists' discussion of the sexual rights and duties of husband and wife. A husband had the right to sexual relations with his wife in order to procreate children; she had exactly the same right. Another (and indeed far more common) way of stating the same rule was that – if asked by the other – each had a duty to enter into reproductive sexual relations. However, there was one key rider to the doctrine that sex must always be aimed at reproduction. This was the belief that marital relations were also a means of averting fornication. Marriage was primarily instituted for the sake of procreation, but its secondary purpose was helping sinful men to control their sexual passions. This was not God's original intention, for when Adam and Eve were spotlessly innocent in the Garden of Eden, they were in no need of help. Every sexual act would have been aimed at begetting a child and would have achieved its object. Unmoved by sexual libido, sinless men would have copulated simply in order to have children and for no other reason. This view was inherited from medieval theology. Augustine had described the idyllic conditions that would have obtained had not our first parents fallen: 'The man then would have sown the seed and the woman received it as need required, the generative organs being moved by the will not excited by lust'. Albertus Magnus took the same line; sex would have been lots of fun in Paradise, but it would have been 'under the control of reason'. In the seventeenth century theorists still argued that sex in the Garden of Eden was purely and exclusively reproductive, 'not in the manner of reigning and raging lust'.[30]

However, mankind had fallen and afterwards sexual urges were 'moved and restrained not at our will but by a certain independent authority so to speak', stated Augustine, whose *Confessions* showed that he spoke from experience. Since 'the fall of our first parents', mankind had lost 'command of itself'; sexual desire was 'tainted and mixed generally with much filth' and 'grown so violent, impetuous, and headstrong with the most', that 'grievous inconveniences' would result were it not for the panacea provided by marriage. Saint Paul was interpreted as having endorsed this medicinal function of marriage when (after making clear how superior he considered his own virginity) he had stated, 'But if they cannot contain, let them marry: for it is better to marry than to burn' (I Corinthians 7:9). Since the Fall, sexual desire was not controlled by the will; marriage provided a legitimate outlet for man's naturally uncontrollable lust. Saint Paul's words were the subject of almost endless debate from an early date; the supporters and the opponents of marriage in general, and of clerical celibacy in particular, both quoting this passage in their support. Whatever their attitude to clerical marriage, most theologians accepted Paul's reluctant concession. Since the Fall, marriage had taken on a role in addition to procreation, that of 'a remedy against carnal desires'.[31]

Saint Paul's unwilling concession that marriage was better than burning led some theorists to consider the possibility that, as a reason for marriage, remedy might be more important than reproduction, or at least as important. John Colet warned readers of I Corinthians 7,

> Let all here beware of thinking that Paul allows marriage for any other rea-son than one's inability to restrain oneself ... because marriage has no good-ness in it except in so far as it is a necessary remedy for evil.

Colet was going further than most theologians in arguing in this way; even the ascetic Augustine had conceded that the desire to conceive new worshippers of God was a legitimate motive for marriage. Much of Antoine Hotman's *Traicte de la dissolution du mariage par l'impuissance & froideur* [Tract on the dissolution of marriage for impotence and frigidity] debated whether the role of remedy had replaced that of reproduction for Christians. This was plausible, he thought, because of Paul's and the Fathers' assertion that virginity was superior to matrimony. Reproduction only populated the earth, virginity peopled heaven. Nevertheless, Hotman rejected the contention and insisted on the prima-cy of reproduction. For Hotman, as for generations of earlier theologians, venting lust must be 'not the primary, but only the secondary end'. Like most early-modern theorists, Hotman had a profound distaste for non-reproductive sexual pleasure. He did not wish to permit marriage to eunuchs capable of giving and receiving sexual pleasure but incapable of procreation.[32]

Hotman was unusual in debating this view.[33] Most divines accepted that reproduction was the primary end of marriage and believed that under ideal circumstances sexual activity would always be directed at this goal.[34] Theorists conceded with little enthusiasm that in a fallen world 'there is a double end of marriage; first to beget children; secondly the remedy against lust'. Husband and wife were permitted to have sexual relations with one another to prevent either partner from 'burning' and falling into sin. Even in these circumstances, any steps taken to *prevent* the conception or birth of children would be sinful. However, this proviso did allow marital relations to take place even when there was no reasonable expectation that they would be fertile.[35] The commonest examples of permitted infertile sexual relations involved pregnant women and aging couples. Casuists taught that sexual relations could, without sin, take place during pregnancy to reduce the temptation to adultery. Moreover, marital congress 'serveth not only for the necessity of generation ... but for the relief of such as are past it'. Not all approved of such indulgence. One puritan divine argued that women over 60 years old should not marry, for they could no longer have children, and the end of avoiding fornication

> they should be far from; seeing the body is dead, the heart should not grow rank with filthy lust: the lecherous old person is hated of God.

However, orthodox Catholic and protestant theology did allow those too old to be fertile to marry, although even this minor concession to non-reproductive sexual pleasure cost the more austere theorists some painful heart-searching.[36]

Some theologians argued that sexual relations 'as remedy' could only licitly take place where the object was to prevent the *other* spouse from falling into sin. In other words, if (say) a wife believed her husband would go and obtain sexual satisfaction elsewhere if she did not provide it, she should do so even if reproduction were not the object. 'It may be a fault in him at sometime to require it; yet is then thy yielding no fault in thee, if otherwise he will not be satisfied.' They doubted whether a partner could, without sin, request sexual relations to prevent himself or herself from straying. The belief that non-reproductive marital relations were at least a venial sin was based on Augustine's authority.[37] More frequently, however, divines conceded that it *was* licit to initiate sexual relations to ease one's own temptation. The canonist, Antoninus, argued that it was not sinful to ask a spouse for sexual relations to prevent oneself from 'burning' if there were no other way of achieving this end; it was sinful only if this means were chosen because it was a particularly enjoyable way of dealing with the problem. Many casuists adopted the laxer view and taught that sexual relations to prevent one's own incontinence were not sinful 'if done in due circumstances'. Some were quite explicit that conjugal relations were licit in order to avert incontinence, 'not only in a partner but even in oneself'. But as fashions in casuistry moved between austerity and indulgence, it was always 'greatly disputed' whether, without sin, one could request sexual relations to prevent one's own incontinence.[38]

A few theologians went so far as to suggest that there was nothing intrinsically sinful about having sexual relations with one's spouse simply from affectionate desire. The Italian Catholic theologian, Giovanni Delfino, argued in the mid-sixteenth century that, provided a person were not overcome by passion and acted moderately, and provided that only the person's spouse was, or would be asked, 'the matrimonial act can be aimed at pleasure and enjoyment without sin'. Delfino was very unusual amongst both Catholics and Protestants in arguing for such indulgence of the pleasures of the flesh; most insisted that marital relations must be either procreative in intent, or a refuge of last resort when lust was gaining the upper hand.[39] In the later seventeenth century, the English Protestant, John Corbet, was one of the few theorists to give this issue detailed consideration in the English language. His position was similar to Delfino's, though less generous to the lustful. He argued that there were two good reasons for conjugal relations: reproduction was the 'more noble' and rendering the debt to a craving spouse was 'an act of justice'. Corbet conceded that the couple could also use 'the marriage-bed', even when 'respect to procreation or the avoiding of sin doth not urge it'. However, he still insisted that if 'pleasure' was 'ultimately intended' it was 'a mortal sin'. The admission that sexual relations within marriage might be aimed at averting sexual sins was not meant as a charter for libertines. Reproduction was the proper end of sex; the avoiding of incontinence a distasteful (though necessary) concession to fallen man. Almost everyone agreed 'that we should not mutually embrace for the sake of pleasure' and condemned those 'insatiable lechers' who placed 'obscene purposes and beastly pleasure' before the generation of children. However, early-modern theorists thought sexual desire was so strong, and the capacity for chaste celibacy so rare, that without the remedy supplied by marriage the devil would fill the whole world with promiscuous fornication, prostitution, and sinfulness.[40]

Procreation and averting fornication both placed sexuality at the centre of marriage. When seventeenth-century theorists discussed the purpose of marriage, they always stressed reproduction; they often added either the avoidance of fornication or companionship, or both of these. The Catholic catechist, Saint Peter Canisius, was typical of those who listed all three ends:

the propagation of mankind ... a familiar and faithful living together of man and wife: And finally the avoiding of fornication in this imbecility of a corrupted nature.[41]

Others concentrated on reproduction and remedy.[42] Many moralists listed some of these ends on one occasion and others on another. The Catholic priest, Nicholas Harpsfield, for example, stated the ends of marriage were reproduction and the 'communion of mutual life', then (later in the same tract) reproduction and 'to staunch and remedy the heat and fervour of concupiscence'. There was almost no dispute between Catholics and Protestants, nor amongst Protestants themselves, about the relative importance of these three ends of marriage or the order in which the three ends were listed. The stress laid on each of the three varied in accordance with the particular controversy or didactic aim. What *is* clear is the almost universal consensus that reproductive sexual relations lay at the very heart of marriage. John Milton's contention in his divorce tracts that marriage was primarily an affective and only secondarily a physical union was completely heterodox. It is difficult to find anyone other than Martin Bucer who shared his strange opinion.[43]

This is not, of course, to deny that early-modern theorists deemed companionship and co-operation an important part of marriage. Aristotle and other ancient authorities indicated that (even apart from the obvious purpose of having children) males should marry in order to make their lives more comfortable by having someone to run the household. A husband could confidently deputize his wife to manage household affairs, while he dealt with public business, because of the mutual goodwill and affection that were deemed a normal part of marriage. The Aristotelian view readily merged with Saint Augustine's teaching that – even apart from sexual intercourse – there would have been between the sexes 'a kind and friendly union, of the one ruling and the other obeying'. So great was Aristotle's and Augustine's influence that the married couple's co-operation and companionship were often placed at the heart of marriage. God had created Eve to be a 'trusty keeper' of her husband's 'life and goods'. Wives should help their husbands by 'housewifery and industry and skill to manage household affairs' because 'to that end they were made at first'.[44]

All Christian commentators – protestant and Catholic, puritan and Anglican, medieval and early-modern – agreed that husband and wife should be loving companions. Scripture commanded husbands to love their wives (Colossians 3:19). Saint Paul's words could not be ignored. Pagans could rest happy with the physical union and the subjection of wife to husband that the law of nature dictated, but Christians must go further; this was why Paul taught that the merely natural desire of husband for wife was insufficient, and required 'spiritual and gratuitous love'. At the opening of the sixteenth century, the French Catholic pastor, Jean Raulin, preached that conjugal love should be 'enduring and persevering'; it should be 'love uniting the hearts in a perfect manner'. Husband and wife must face the world together – a case of, 'As it is commonly said, "Love me, love my dog"'. English divines charged that 'love in marriage should exceed all other love, under the love of God'. Marriage was 'made by voluntary love', and 'there is no worldly love greater than the love between man and wife, who be one heart and one mind'. Casuists taught that a husband sins if 'he does not love his wife' or if 'he only loves her carnally'. The love of husband for wife must be 'sincere and cordial', and neither should stoop to 'an inordinate, sensual, or worldly love'.[45] The insistence on the need for conjugal love was preached by members of all religious denominations. Early-modern theorists believed that a married couple would not rear children well, nor be deflected from sexual incontinence, unless they loved one another. Without 'mutual conjugal affection', marriage simply would not function well. John Milton was unusual in detecting tension between the 'companionate' and 'reproductive' ends of marriage. Most early-modern writers thought them complementary rather than conflicting. Between husband and wife there was 'unity, rather than simply connection or affection', because they were 'one body and one flesh'. Sexual relations were stressed because they were the key factor that distinguished marriage from any other society, partnership, or friendship. Companionship could be obtained in many legitimate ways, but only in marriage could it be joined with sexual relations; 'comfort is no sufficient cause for marriage because it may be had without marriage, but children cannot'. Children were marriage's 'greatest blessing', because 'all other blessings may be had without marriage' whereas 'only children must come of ... wedlock'.[46]

In many early-modern accounts of the purposes of marriage, mutual companionship was implicitly connected with sexual relations. The Calvinist divine, Andrew Willet, argued of the Virgin Mary and Joseph that they must originally have intended to have sexual relations. Otherwise,

> they should have married neither for avoiding fornication, nor for procreation, which are the two chief ends of marriage: as for the third, which is mutual comfort, it ariseth of the former.

Roman Catholics were avid enthusiasts about Mary's virginity, but they, too, held that every marriage had reproduction for its end, 'even that of Saint Joseph with the Blessed Virgin'. Males associated with other males for intellectual and social intercourse; they associated with females for sexual intercourse. Archbishop George Abbot's comments upon the notorious Essex divorce case of 1613–14 embodied the same attitudes. Frances, Countess of Essex, sued for a nullity on the grounds of her husband, Robert's impotence. Abbot reasoned that 'marriage in young couples is for carnal copulation and procreation thereupon, and that it is the intendment of those which contract matrimony to receive satisfaction in that kind'. If Robert were in fact impotent, he was doing Frances 'a very great injustice to retain her as his wife'. Remarking on the same case, James VI and I was still more explicit, for he asserted: 'that the essential point of matrimony cannot be accomplished *sine copula* [without copulation] is warranted by express Scripture'. Almost every early-modern theorist agreed that a marriage was null and void if either party were permanently incapable of sexual relations. The only question at issue in the Essex divorce case was whether the Earl really was incapable of sexual relations with Frances.[47]

Discussions of sexual relations within marriage generally used the terminology of duties rather than that of rights. Indeed the term 'marriage debt' was a common synonym for coitus. Marriage entailed consent to sexual relations and that consent could not later be withdrawn without good reason. One partner could refuse to grant the other's request for coitus only if it were made in unreasonable circumstances. Saint Paul had ordered that husband and wife should always render the other spouse his or her due, 'except it be with consent' (I Corinthians 7:3, 5). Neither should deny 'the duty unto which marriage is destined as a remedy of incontinency'. Both husband and wife were obliged to 'giving each the seasonable, moderate use of the other's body'. The right to

sexual relations was the logical correlative of the duty to avert fornication. Divines lectured husbands on the limits to their authority in the
area of sexual relations. It was mortally sinful to use their wives merely
as a source of sexual gratification, in unnatural ways, or even to copulate
too frequently. And if husbands replied,

> 'She is my wife. I can use my own in accordance with my own will'. The
> response is that this is not true. For one should neither abuse one's own, nor
> use it except for the appointed end ...

God did not give anyone the right to use another person as a source of
sexual pleasure. Married couples could copulate in order to have children and in order to avoid sins of sexual immorality; He granted them
no other rights.[48]

Moralists always insisted that husband and wife had the right only to
natural sexual relations; 'natural' meant those aimed at reproduction.
Neither spouse was obliged to perform any sexual favour for the other
that was not reproductive in its object. Roman Catholic theologians and
canon lawyers were generally very reluctant to allow a wife to disobey
her husband, but they all insisted that a wife should absolutely refuse
her husband's orders to co-operate in 'unnatural' sexual acts, such as fel-
latio or sodomy. Canon lawyers argued that she could obtain a separation if he attempted to compel her to participate in such wicked
practices. English Protestants continued to teach basically the same doctrines. By the standards of many early-modern divines, John Corbet was
lax on sexual morals, but he wholly accepted that 'preternatural ways by
which human nature cannot be propagated ... are justly to be abhorred
by all who have not lost the sense of humanity'. God had commanded
that 'everyone direct the motions arising from that natural instinct, ...
for the propagation of their kinds, unto the lawful use of marriage'.
Protestant divines restricted sexual activity to procreative heterosexual
relations, and forbad 'all mixtures' but those 'hallowed by marriage and
the order of nature'. Neither husband nor wife need satisfy 'unnatural
lusts'.[49]

With almost complete unanimity, early-modern moralists maintained
that in the area of sexual relations husband and wife had equal and reciprocal duties and rights. This equality was based primarily on the Epistle
of Saint Paul to the Corinthians. The Apostle had little sympathy for
indulging the urgings of the flesh, but had conceded that 'to avoid fornication let every man have his own wife, and let every woman have her
own husband'. He then spelt out that:

> The wife hath not power of her own body, but the husband: and likewise
> also the husband hath not power of his own body, but the wife. (I
> Corinthians 7:2-4)

Under Saint Augustine's influence, throughout the following centuries
this text was interpreted as meaning that in one area – sex – husband and

wife were equal. The husband was head and the wife was subject, but in this one matter of the 'matrimonial debt' they were on an equal footing. Husband and wife were equally bound exclusively to fulfil one another's reproductive and sexual desires. Theologians granted the husband power of household government, but qualified his authority, stating 'that in those things which relate to the matrimonial debt, they are equal'. Equality in sexual duties was one of the rare areas of sexual equality.[50]

Canon lawyers and theologians described marriage in quasi-legal terms as 'a mutual donation and translation' of the spouses' bodies. Marriage was a contract of exchange. Each spouse gave and received 'a right in the body of the other for the procreation of children'. The definition of marriage given in the mid-sixteenth-century *Reformatio Legum Ecclesiasticarum* [Reformation of the ecclesiastical laws] reiterated the orthodox Catholic idea of marriage as a reciprocal contract.

> Marriage is a legal contract, inducing and effecting a mutual and perpetual union ... in which each surrenders to the other power over his body, for the purpose of begetting offspring, or of avoiding harlotry, or of controlling life by means of reciprocal obligations.

Husband and wife each subjected themselves sexually to the other by marriage and gained exactly corresponding rights and duties in sexual matters. Spouses were 'reciprocally obliged' to one another, principally 'to the conjugal act', with rights over one another's bodies and 'corresponding dominion and subjection'. A couple were 'one flesh' because each spouse was 'ruler of the partner's body'. In sexual matters, the wife was subject to the husband, but he was also subject to her. A married male 'no longer belongs to himself'. Just as theologians condemned the marriage of a son with his mother because it placed him in a position of authority over her, so they condemned a daughter's marriage to her father because (whereas naturally she should be subject) marriage rendered her 'his companion and partly his mistress, having power over her father's body'. The one area where the wife ruled the husband as he did her was the carnal.[51]

The same stance was adopted by English protestant theorists; they believed 'the woman hath the same liberty that the man hath; because that – howsoever the man be the head – yet in respect of the body of the one or the other, they have an equal power'. The normal rule of subjection of wife to husband did not obtain in sexual matters as in others. In sexual activities,

> the Lord hath granted as great power to the woman over the man as he hath granted to the man over the woman, as in the mutual use of their bodies: and in this case, he is as well subject to his wife as he is her lord: but in other things the man hath the superiority over the woman.

When sex was the point at issue, early-modern English divines supported sexual equality. The husband possessed greater power in family government, but no greater sexual freedom. The husband controlled the common property of the household, but control of the property they had in one another's bodies was mutual and reciprocal. Husband and wife were 'mutual proprietors one of another: by self resignation of each to other, they give themselves mutual power one over another'. It was entirely conventional to maintain that husband and wife 'are so equal in the matter of wedlock that both of them are both superior and inferior in asking and rendering the due'. In most areas of the marriage contract the wife gave and the husband took. She made a gift of her goods and chattels to him; she subjected herself to his rule; she agreed to live where he wanted; she took second place to him in guiding the children. In one area only – sexual relations – did early-modern theorists describe the couple's rights in the same terms without distinction of sex. For a change, the inclusive 'he' was used of both spouses. 'Every married person hath *ipso facto* surrendered up the right and interest he had in and over his own body, and put it out of his own into the power of another'.[52] One historian of Renaissance theory has argued that in the sixteenth century, 'Duties between man and wife ... are mutual but never reciprocal'. In virtually every area of marriage this was true; the one significant exception was marital relations themselves. Another historian of political theory has asserted that in the case of husband and wife in early-modern social contract theory, 'the right is not to one another's bodies; the right is that of masculine sex-right'. In fact, the one area where masculine rights were proclaimed equal (*not* superior) to feminine rights, was sex. Early-modern theorists proclaimed a husband's masculine sex-right to his wife's body; they also insisted on the wife's feminine sex-right to her husband's.[53]

Virtually all early-modern theorists afforded husband and wife mutual, equal, reciprocal sexual rights over one another's bodies. This assertion meant little until explained and applied. How was their belief construed when it came to the practical problems of sexual relations? Did theorists insist on equality in the abstract and then actually grant the husband greater freedom than the wife? Was there a 'double standard' that endorsed males obtaining sexual pleasure where they wished, whilst restricting women to that provided by their husbands? Could a husband have many wives, while a woman was limited to one husband? The answers to these questions provide the acid test of sexual equality in sexual affairs, and I shall now turn to them.

Notes

1. Augustine 1977, 278 (XIV.22); Settala 1626, 21–2 ('ut docuit Arist[oteles] ... sacra Genesi'); Primrose 1655, 1–2 ('nihil possit efficere' 'mutuus'). See also Poullain de la Barre 1989, 140–3.

2. Hugh of Saint Victor 1951, 117 (I.6.XXXV); Augustine 1982, II, 75 (IX.5); Vincent of Beauvais 1624, 2217 (I.30.8); Aquinas 1872–82, III (1882), col. 1091 (Suppl. 44.2), I (1872), col. 1376 (I.XCII.1 'non quidem in adiutorium alicuius alterius operis, ut quidam dixerunt, cum ad quodlibet aliud opus convenientius iuvari possit vir per alium virum quam per mulierem, sed in adiutorium generationis'). Compare Albertus Magnus 1890–9, 30, 103 (XXVI.7) See Brundage 1987, 85–6 for the patristic origins of this view.

3. Colet 1867, 53 ('ad carnem propagandum'); Cajetan 1581, 413–14; Bellarmine 1602–3, III (1603), col. 1286 (I.II). Compare Catharinus 1551–2, 154–5. See also Lorenzo da Brindisi 1935, 237; Jansen 1641, 35–6.

4. Luther 1958, 115–19, 202; Salkeld 1617, 181 [perfect]; Tuvil 1635, 94 [womb]; White 1656, II, 79.

5. Hooker 1977, 402 (V.73.2); Willet 1605, 36. See also Pricke 1609, sig. K7r; Babington 1615, 14; Ainsworth 1627, 12; Lawson 1659, 205; Maynard 1668, 176; *Marriage promoted* 1690, 1. Leites (1986, 84) has argued that 'The Puritans do not share the outlook of Augustine, who thought that she [Eve] would have been of no use to Adam had she not been the bearer of his children. For Puritans she was also his companion'. Tavard (1973, 176) has asserted the same about Calvin. In fact, the Puritans did not differ from Anglicans or Catholics on this issue; all were agreed that woman was created primarily for reproductive purposes and secondarily for housework – her value as a companion stemmed from these two. The commonest description of the wife in both Catholic and protestant, both medieval and early-modern tracts (most of which were written in Latin), was *'socia'* – which means companion. The notion that Protestants in general, or Puritans in particular, were original in regarding the wife as her husband's companion cannot survive even a cursory reading of medieval and Counter-Reformation works.

6. Grantham 1641, 5 [blessing]; Pufendorf 1934, II, 883–4 (VI.I.24).

7. Campanella 1968, 60 ('impetuoso desiderio'); Melanchthon 1834–60, XXIII, 676 ('mutuam coniunctionis appetitionem'). For a few (rare) exceptions to the belief that sexual relations should have no aim but reproduction, see Noonan 1986, 324–6.

8. Plowden 1816, I, f. 303r.

9. Ridley 1607, 2. The basis of this definition was *Decreti* c.7.d.1; for very similar definitions, see, for example, Dawson 1694, 17; Freig 1591, 5; Raulin 1512, f. 29v; Campanella 1968, 60–2; Suarez 1872, 149 (II.XVII.3); Turturetti 1629, 290–1; Burgersdijck 1631, 284; King 1702, 73.

10. Nifo 1641, De amore liber, 109 ('ut perpetuo servetur generatio,'); Primrose 1655, 2 ('Ideo libidinem utrique et voluptatem quandam inservit, qua igne quasi quodam flagrantes, in mutuos amplexus. Viro penem dedit ad semen eiiciendum, mulieri uterum ad illud excipiendum, retinendum et fovendum'); Hotman 1595, f. 43r ('ne voudrions jamais approcher d'une femme'); Le Grand 1694, 395; Chartier 1596, 5 [pleasure]; Settala 1626, 15 ('ad generandum sibi simile'); *The batchelor's directory* 1694, 23 [vehement].

11. Poole 1696, I, sig. P1v [filthiness]; Herbert 1648a, sig. W8r–v [pollutions]; *Certain sermons or homilies* 1756, 114; Pufendorf 1934, 842 (VI.I.4) [alleviate].

12. The word 'fornication' was often used in early-modern English to mean any illicit sexual activity. In this text (except when quoting others), it will be used for what was strictly called 'simple' fornication, that is sexual relations between an unmarried male and an unmarried woman.

13. As Laslett (1977, 104) has pointed out, early-modern European views are not unusual:

> for the rules which forbid the begetting of children outside marriage have to a large extent succeeded in ensuring to every person born a mature male protector and provider. This appears to be true of all societies at all times.

14. Wood 1712, 304; Pufendorf 1934, II, 845 (VI.I.5).

15. Sordi 1602, f. 3v; Benincasa 1562, f. 70r ('filios necare dicitur'); Keble 1685, 443–4; 18 Elizabeth cap. 3; Massie 1586, sig. A8v [provide]; Wiseman 1657, 8–9 [proceeding]; Attersoll 1618, 133 [vessel]. For other lawyers and casuists insisting that the father's instinct and duty to support mother and child was natural, see for example, Bartolus 1570, f. 6r; Nevizanus 1570, 38, 253–5; Lagus 1566, 7–8; Bastingius 1593, sig. A5r; Bañez 1594, 191; Gentili 1606, 'De secundis nuptiis', 22–3; Pascale 1618, f. 168ff., 311.

16. Carter 1627, 99 [brute]; Cicero 1534, sig. A7r; Cumberland 1727, 32–3; Nevizanus 1570, 240; Lessius 1621, 131–2 [outlive]; Plowden 1816, I, f. 303r; Venette 1688, 202, ('pour heritiers de leur nom et de leur bien'). Compare Brydall 1699, 2.

17. Blount 1911, 324; Gaule 1630, 89 [generate]; cited in Roby 1902, 169; Turturetti 1629, 299 ('certius agnoscit'); Nevizanus 1570, 45 ('certiores'); *A discourse of the married and single life* 1621, 120.

18. Kornmann 1610, 195 ('Communis ratio' 'iuris praesumtione' 'non vera et certa probatione'); Pufendorf 1934, II, 845 (VI.I.5). For suspicion of step-parents, see, for example, Arelat 1536 (de Garronibus), f. 24v; Bellarmine 1602–3, III (1603), col. 1351 (I.XVI); Vaughan 1600, sig. N6v–r; *Christian parents* 1616, 129–31; Houlbrooke 1984, 211, 217–18.

19. Hay 1967, 35–7; Blount 1911, 324; Tyrrell 1681, 13–14; Grotius 1689, 282, (II.VII.VIII 'probabiliter constaret qui esset partus cuiusque pater').

20. Jansen 1641, 296; Towerson 1676, 413. On the unnaturalness of fornication, see, for example, Alexander 1948, 548; Chemnitz 1978, 718; Hemmingsen 1566, sig. I5r–v; Vitoria 1991, 160; Calvin 1948, 98; Covarruvias 1597, 151 (I.II.1.6); Suarez 1872, 105 (II.VII.4); Jansen 1641, 156; Gott 1670, 427; *The batchelor's directory* 1694, 67; Huygens 1698, 374–6; Morgan 1942, 593. Durandus 1571, II, f. 377v and Locke 1987, 195, were unusual in seeing it as against divine positive law only.

21. Wemyss 1633, II, 178 [whore]; Barlow 1690, 15 [due care]; Henry VIII 1521, sig. q1r, ('damnabilis'); Mauser in Beust 1597, 336; Domas 1705, 17; Davenant 1627, 355, ('ex vago concubitu'). See also Arelat (Bertrandi) 1536, f. 90r; Delfino 1553, 39; Bernard 1845, 20; Burgersdijck 1631, 197; *The batchelor's directory* 1694, 98, 74 [second so numbered].

22. Barker 1624, 267–9; Rogers 1642, 333 [lust]; Turner 1698, 43 [sacrifice]; Beverland 1698, 56 ('quoque prostibulo') *Marriage promoted* 1690, 39, 61 [doors]. See also Bullinger 1541, f. 35r; Nevizanus 1570, 245; Butler 1697, 30; Barlow 1690, 15; *Reasons for passing* 1699, 14.

23. Kent 1973, 44. For the local, financial consequences of bastardy, see Amussen 1988, 115–17; Ingram 1987, 276–91; Quaife 1979, 90ff.; Willen 1988, 561.

24. *Concubinage and poligamy* 1698, 73 [histories]; Turner 1698, 24 [turpitude]; Tunstall 1518, sig. B2r ('ex matrimonio'). See, for similar arguments, Lagus 1566, 11; Gualther 1570, f. 412r; Ayrault 1614, 39; Whately 1624, 23; Cawdry 1656, sig.

A5v; Clarke 1659, 19; Leighton 1844, 122; Lawrence 1680, sig. a2r; Le Grand 1694, 395; Huygens 1698, 376–7.

25. Comber 1679, 34 [the ostrich's reputation as a paradigm of parental neglect stemmed from Job 39:13–16]; *Marriage promoted* 1690, 9; Settala 1626, 42–3 ('nos brutis animantibus reddit'); Niccholes 1809, 257 [affinity]; Dugard 1695, 51 ['sticklers' are sticklebacks – small fish living in shoals]. For similar patriarchalist criticisms, see Dawson 1694, 29; Lowde 1694, 159. The contractarian James Tyrrell (1691/2, 10) tacitly admitted the force of the patriarchal assertion that 'we did not at first spring up out of the earth like mushrooms' [*pace* Hobbes' *De Cive*].

26. Cochlaeus 1535, sig. F2r ('omnia animalia'); Cumberland 1727, 156; Antoninus 1474, sig. l6r ('non canino'); Aristotle 1984, II, Nicomachean Ethics, 1836 (1161b–2a).

27. See, for example, Duns Scotus 1639, XI.2, 787 (XXVII.i.n.9); Cochlaeus 1535, sig. F2r; Albertus Magnus 1890–9, 30, 290; Soto 1619, 389–90 (V.III.3); Gataker 1623, 29, 37; Settala 1626, 40ff.; Grenaille 1640, 28; Kraze 1619, 3; Comber 1679, 32–4; Astell 1730, 15.

28. Aquinas 1872–82, III (1882), col. 1078–9 (Suppl. 41.1 'esse, nutrimentum et disciplinam', 'maxima determinatio'); Subertus 1508, f. 50 ('in antiquitate'); Locke 1988, 319–20, (II.79–80); Dugard 1695, 9. Nicholson (1986, 143) has stated: 'The modern family increasingly came to be seen as a mutual survival unit' as is evidenced by Locke's arguments that 'the major function of the family is to raise children'. Since Locke's arguments were taken wholesale from Aquinas, it is difficult to see how this was distinctively modern.

29. Abbot 1608, 27 [spring]; Olizarovius 1651, 6 ('tam barbara fuit, adeoque extra lege mores proiecta, tam ab omni humanitate aliena'); Harrington 1528, sig. A2v; Cardwell 1841, 76 [Scripture]; *Marriage asserted* 1674, 66–7 ['bulbeggars' are monsters invented to frighten children]; Bingham 1855, 6–8 [current].

The belief that marriage is 'the first and most natural of all societies' (Comber 1679, 1) was endlessly repeated; see, for example, Raulin 1512, f. 4r; Portius 1537, sig. Civ; Agrippa 1540, sig. B1; Hemmingsen 1566, sig. L7v; Massie 1586, sig. A3v; Bastingius 1593, f. 137v; Gataker 1623, 40; Baudius 1638, 361–2; Olizarovius 1651, sig. h1v; *The batchelor's directory* 1694, 98–9; Brydall 1703, sig. A2v. As Davis (1975, 143) has rightly stated: 'Europeans of the fifteenth to eighteenth centuries found it remarkably difficult to conceive of the institution of the family as having a "history", of changing through time'. Indeed, they explicitly insisted that 'only domestic society, being clearly natural, is to remain one and the same throughout all ages and nations' (Ames 1968, 308).

30. Augustine 1977, 280 (XIV. 24); Albertus Magnus 1890–9, 30, 106 (XXVI. 7 'sub imperio rationis'); Salkeld 1617, 181. See also D.,L. 1692, 29, 35.

31. Augustine 1977, 276 (XIV.17); Gataker 1623, A wife indeed, 37; Ames 1968, 319 [carnal].

32. Colet 1985, 191; Hotman 1595; Antoninus 1474, sig. P4r–v ('non tamen primarius: sed secundarius').

33. Hotman was not alone in doing so. Peleus (1602, 97ff.) also discussed at length whether the marriage of eunuchs was valid.

34. See, for example, Bernard 1845, 19; Stock 1865, 183; *The batchelor's directory* 1694, 24; Niccholes 1809, 258; King 1614, 18–19; Ochino 1657, 48; Comber 1679, 32.

35. Wemyss 1633, II, 142. All the theorists were agreed that even when sexual

relations were being employed as a remedy against lust, the couple must use no form of contraception. Contraception, and early-modern attitudes to it, will not be dealt with here, since this task has been brilliantly executed in Noonan, 1986.

36. Campanella 1968, 170; Rogers 1642, 7; Bernard 1845, 19. Henry Smith (1866, 9) also argued that procreation was the proper motive for marriage; those marrying 'where there can be no hope of children for age and other causes' were not acting as lawfully, for they must be motivated by lust or greed. See Topsell 1597, 44 and Ness 1695, II, 173–4 for similar arguments. The tenet that sexual relations during pregnancy were not sinful had been orthodox at least since Wycliffe's opposite view had been condemned by the Council of Constance of 1414–18 (Hotman 1595, f. 47v). See Beverland 1698, 87–8 as an example of one whose distaste for allowing sexual relations during pregnancy was apparent. For more standard views on the permissibility of the marriage of the aged and of sexual relations during pregnancy, see Antoninus 1474, sig. P6v–7r; Agrippa 1540, sig. B3v; Mauser in Beust 1597, 337; Kornmann 1610, 'Linea amoris', 99–101; Gouge 1622, 223–4; Fuller 1845, 18; Huygens 1696, 33.

37. Hacket 1607, 11; Augustine (1955, 17, cap. 7) expressed the view that 'it is no sin to render the conjugal debt, but to exact it beyond the need for generation is a venial sin'. Faithful Augustinians such as Portius (1537, sig. E4v) adopted his view without modification. As John Corbet (1685, 243) noted, this view seems internally contradictory:

> they who hold the use of the marriage bed without respect to procreation to be sin, do hold it no sin in that case to render the due benevolence when required, because it is a point of justice. But if this thing be a sin on the demander's part, the rendering of it cannot be a due.

38. Antoninus 1474, sig. Q3r; Raulin 1512, f. 30v ('si fiat cum debitis circumstanti-is'); Laymann 1625, V, 418 (X.III.IV.2 'non tantum in coniuge, sed etiam in seip-so'); Huygens 1696, 35–6 ('Multum disputatur'). On the medieval background, see Brundage 1987, 447–53.

39. Delfino 1553, 33, 41 ('ut sine peccato ullo matrimonialis actus ad voluptatem et delectationem referatur'). Diana (1660, 193), argued that the belief that indulging in marital relations solely on account of pleasure was at least a venial sin was the 'more probable' ('probabilius').

40. Corbet 1685, 243–5; Barbaro 1677, 98 [embrace]; *Aristotle's masterpiece* 1694, 52 [lechers]; Pareus 1631, 483. For other discussions of this point, see Perkins 1618, 689; Colet 1867, 27; Pufendorf 1934, 843 (VI.I.4) Archdekin 1679, 322–3; Comber 1679, 29–30. For the view that Puritans in general did allow sexual intercourse solely as a source of 'pleasure and comfort', see Leites 1986a, 12 and *passim*, and Leites 1986b, 115.

41. Canisius 1592–6, 252–3. All three ends were listed in, for example, Antoninus 1474, sig. P4r–v; Harrington 1528, sig. A3v–4r; Delfino 1553, 31; Chemnitz 1978, 721; Smith 1866, 9–12; Batty 1581, f. 4v–5r; Covarruvias 1597, 152 (I.II.1–1.6); Cleaver 1598, 156; King 1614, 7; Abbot 1653, 25; Lawson 1659, 205; Swinnock 1868, 464; Corbet 1685, 225–6; *The batchelor's directory* 1694, 24–5.

42. For instance, Guido [1500?], f. 52v; Hooper 1843, 380–1; Lagus 1566, 406; Beust 1597, 96; Mauser in ibid, 337; Fulbecke 1602, f. 25v; Abbot 1608, 45; Biderman 1621, 58; Herbert 1648a, sig. X3v; Olizarovius 1651, 5; *Marriage asserted* 1674, sig. A5v; *An essay towards* 1697, 126.

43. Harpsfield 1878, 239, 244. Thomas Cranmer (Pocock 1870, especially 366–90), in his loyal attempt to liberate Henry VIII from Katherine of Aragon, argued that the bond of marriage arose from natural love rather than from coitus, but his aims were very different from those of Bucer or Milton. Other examples of listing different purposes of marriage at different points in the same treatise include Cardwell 1841, 75–6, 594; Hunnius 1604, 66, 101–2; Watson 1558, f. 166v, 181–2; Bernard 1845, 19, 45.

44. Augustine 1955, 9 (cap. 1); Agrippa 1540, sig. B3r [trusty]; Gataker 1623, *A good wife*, 19. For others placing the commodity of common life at the heart of marriage: see, for example, Delfino 1553, 31; Vettori 1584, 474, 484; Peleus 1602, 168; Grenaille 1640, 23; Maldonado 1677, 437.

45. Tapper 1582, 274 ('spiritualem et gratuitum amorem'); Raulin 1512, f. 21r ('permanens et perseverans', 'amor perfecte cordium unitivus', 'Vulgo dicitur. Qui me amat, amat canem meum'); Harrington 1528, sig. D2r [exceed]; Watson 1558, f. 167–8 [heart]; Huygens 1696, 110 ('si uxorem suam non diligat', 'carnaliter tantum diligendo'); Bernard 1684, 303–4 [sig. Qq1r–v]. On the medieval background, see Biller in Goldberg 1992, 70: 'There is continuous reference to a high ideal of married love in the English texts'. See also Houlbrooke 1984, 97–8.

46. Huygens 1696, 58 ('mutuo coniugum affectu'); Nevizanus 1570, 243 ('potius est unitas, quam coniunctio, vel affectio'); Topsell 1597, 44; 256. John Halkett has rightly pointed out that 'there is no clear distinction between Puritan and Anglican teaching on marriage' (1970, 5). On the companionate versus reproductive ends of marriage, there was no clear distinction between Catholics and Protestants either. The thesis that Puritans stressed the companionate over the physical more than Protestants in general, who in turn stressed it more than Catholics, continues to find acceptance: see, for example, Turner 1987, 99–115. Broad reading of Continental Catholic sources, rather than merely English puritan ones, shows this thesis to be untenable. For persuasively informative descriptions of the similarity of puritan, protestant, and Catholic teaching on married life, see Todd 1987, and Davies in Outhwaite 1981 (although the implication that 'humanists' differed from other Catholics should be treated with scepticism).

47. Willet 1634, 516; Campanella 1968, 202 ('etiam matrimonium S. Ioseph cum Beata Virgine'); Howell 1809–28, II, 846, 800. Both Catholic and protestant jurists allowed annulment of marriage for precedent incurable impotence; see Guido [1500?], f. 53r, 61r; Monner 1561, 145; Campanella 1968, 114; Pareus 1631, 484. English law adopted the same view: E.,T. *The lawes resolution* 1632, 57; Conset 1685, 256; Vaughan 1706, 220–1; *Baron and feme* 1738, 432.

48. Sedgwick 1624, 107–8; Swinnock 1868, 476; Raulin 1512, f. 32r ('Uxor mea est. Possum uti re mea iuxta voluntatem meum. Dicitur ad hoc quod hoc non est verum. Neque enim debet quis abuti re sua: nec ea uti nisi ad finem ad quem ordinata est'). On the duty to allow one's spouse sexual relations, see, for example, Bullinger 1541, f. 24r; Veron 1562, f. 11r; Beust 1597, 231; Mauser in ibid., 320, and Schneidewein 1585, sig. G1r; Covarruvias 1597, 221 (I.II.7–2.1.); Perkins 1618, 689; Biderman 1621, 58; Diana 1660, 188; Archdekin 1679, 475; Swinburne 1686, 137.

49. Corbet 1685, 241; Sedgwick 1624, 98; Taylor 1828, XII, 477. For the wife's duty to refuse her husband's attempts at intercourse except in 'the proper vessel', see Antoninus 1474, sig. P6r; Guido [1500?], f. 91–2; Raulin 1512, f. 31; Pascale 1618, 67–8; Huygens 1696, 80.

50. Augustine 1955, 70; Albertus Magnus 1890–9, 30, 196 (XXVIII. 7 'quod verum est, quod in his quae pertinent ad debitum matrimonii, sunt aequales'). See also Antoninus 1474, sig. M3r, and Biller in Goldberg 1992, 84. In the neat epigram of Brundage (1987, 93), 'Equality of the sexes in marriage meant equality in the marriage bed, but not outside of it'. Not everyone agreed with Saint Augustine's interpretation of Saint Paul's meaning. Erasmus ([1550?], sig. I3r) commented,

> albeit Augustine would have that the woman should have in all things as much power as the man, which thing, like as he doth earnestly and constantly affirm, so can he not find to make his word good.

51. Delfino 1553, 2 ('donatio, ac translatio corporum suorum', 'ius in corpus alterius ad procreandam prolem'); Spalding 1992, 91; Huygens 1696, 27 ('mutuo obligati', 'actum coniugii', 'dominium et servitutem correspondentem'); Menochio 1678, 4 ('dominus corporis'); Luther 1959, 100; Campanella 1968, 142 ('socia et ex parte domina habens potestatem super carnem patris'). For the same view, see, Pocock 1870, 341; Nevizanus 1570, 25; Tapper 1582, 290; Bañez 1594, 28. The repetition of this standard definition by Locke (1988, 319, II.7) has been portrayed by Clark (1977, 720) as a distinctively 'materialistic' definition of marriage appropriate to that arch-liberal, defender of private property. Boxer and Quataert (1987, 20) have spoken of the changes brought about by 'The substitution of Protestant contractual for Catholic sacramental marriage'; however, it should be noted that canonists and scholastics had long viewed marriage as an essentially contractual relationship. See Chapter 7.

52. Mayer 1622, 59 [liberty]; Wemyss 1633, 265–6 [power]; Reyner 1657, 23 [proprietors]; Corbet 1685, 226 [superior]; Sanderson 1671, 262 [facto]. See also Musculus 1548, sig. b7r; Hooker 1977, 405, (V.73.7); Willet 1608, 395; Ames 1639, 198 and 1968, 320; Sedgwick 1624, 102; Whately 1624, 31; Wemyss 1633, I, 268; Sclater 1650, 106; Trapp 1647, 79; Reyner 1657, 23; Dickson 1659, 122; Locke 1987, 199–200.

53. Jordan 1990, 4; Pateman 1988, 168.

6

Double standards?

Adultery

Adultery is often cited as the classic case of a 'double standard' in sexual morals. In numerous societies adulteresses are censured and punished, while married males are not penalized for copulating with unmarried women. Indeed, many cultures have *defined* adultery as sexual relations with a married woman – a married male not being deemed to commit 'adultery' if he has sexual relations with an unmarried woman. In early-modern England, the double standard on adultery was established within the language itself. The word 'adulteress' meant a married woman who had sexual relations with someone other than her husband; her lover was called an 'adulterer'. Husbands who were unfaithful with single or widowed women were more likely to be termed 'fornicators' or 'whoremongers' than 'adulterers'. Similarly, their partners were called 'harlots', 'Misses', or 'concubines', but not 'adulteresses'. As one contemporary commented,

> custom, which is the master of language, [has] in a manner appropriated the title of adultery to the falseness of the wife, . . . and absolved the husband from the imputation of it where he did not defile another's bed . . .

The language itself embodied the presumption that the marital status of the woman, not the male, determined whether a liaison was adulterous.[1]

Early-modern theorists were well aware that in popular culture only wives' adultery was generally condemned as a serious offence. Social disapproval and legal sanctions were far more likely to fall on an adulteress than on an unfaithful husband. A husband's infidelity aroused mild sympathy for the wife, but 'cuckolds' were 'scorned by all men'. Custom, the master of language, has given us no feminine form of 'cuckold'. Satirical pamphlets warned males about committing their honour 'into the hands of a silly woman' and risking being 'dubbed a knight of the forked order, and having their names enrolled in the colony of cuckoldom'. A wife's infidelity undermined her husband's status far more than his did hers. A husband almost automatically obtained a separation for his wife's adultery, but a wife who attempted to sue for a separation *solely* on the grounds of her husband's adultery – without being able to prove cruelty or desertion also – was unlikely to have success. Theorists wrote in the knowledge that the 'double standard' on adultery had a firm place in society at large.[2]

Despite contemporary attitudes, virtually all early-modern theorists insisted that the unfaithful husband also sinned. The 'double standard' was condemned. Moralists defined adultery to include the roving husband as well as the straying wife. There were a few exceptions to this consensus. Moreover, most theorists believed the wife sinned more gravely by committing adultery. Nonetheless, there was a high level of agreement that adultery was also evil in a husband. The seventh commandment – 'Thou shalt not commit adultery' – might seem sufficiently unambiguous for it to be unsurprising that the Christian theorists of the Renaissance felt obliged to condemn adultery in both husband and wife. However, this interpretation should not be taken for granted. Just as generations of theologians have thought war and capital punishment entirely legitimate activities despite the commandment, 'Thou shalt not kill', so too it is quite possible to understand the seventh commandment in a way that allows married males variety in their sex lives. Like all the commandments, the force of the seventh depends on its interpretation and application. Three options allow 'Thou shalt not commit adultery' to be entirely compatible with husbands licitly having sexual relations with many women. First, simply restricting adultery's definition to sex with a married woman, declares open season for all males who can confine themselves to widows, spinsters, and divorcées. Second, permitting polygamy or concubinage effectively allows (at least wealthier) married males multiple sexual partners without moral condemnation.[3] Finally, letting husbands divorce wives at will for any cause including simple dislike (i.e. 'repudiation'), permits a serial monogamy which renders adultery unnecessary for all but the most libidinous. Early-modern theorists were perfectly aware of all three options, for they were amply evidenced – and generally endorsed – in the classical sources. The Old Testament strongly suggested that all three escape routes were morally acceptable. Both Jews and Romans reserved the punishments of adultery for sex with a married woman. Indeed, the problem for Christian theorists was to find any society which classed male infidelity as adultery. In all the history and societies of which they were aware, the double standard was the normal standard.[4]

By the early-modern period, opting for any of these three ways of easing the married man's burden would have involved a complete break with Christian tradition.[5] Saint Paul had only grudgingly conceded to males one woman each. With the same austere attitudes, Saint Augustine wrote in as zealous castigation of a husband's extra-marital sexual activity as of a wife's. Canon law followed their teaching: it was 'generally received' that a married man copulating with a single woman was not an adulterer in civil law but was so by papal law. The medieval scholastic tradition had condemned adultery by either spouse with great fervour. For hundreds of years, priests (who were allowed no wives at all) insisted that each layman must limit himself or herself to one sexual partner.[6] Even the most radical Reformers would have found it difficult to attack the belief that 'both man and woman ought to suffer ... sharp death, rather than they should once consent to commit adultery'. Protestant divines showed no inclination to do so. Whatever accusations their Counter-Reformation detractors made (and the protestant support for clerical marriage provoked many), the Reformers were no more inclined to permit sexual indulgence than Catholics. Protestant theologians no longer included matrimony among the sacraments, but this omission was certainly not intended to ease the prohibition on extra-marital sex by either spouse. Divines of all denominations persistently defined adultery in a culturally idiosyncratic sense as the copulation of a married male with a single woman, and tried to restrain married males as much as married women.[7]

Few theorists maintained that on balance an unfaithful husband was a greater sinner than an adulterous wife; but many did argue that he was worse in some respects. The most common way he was thought to sin more was because he 'is or ought to be more strong'. As the superior half of mankind, males should live to higher standards than women or suffer greater condemnation for falling from common ones. Some asserted that adulterous husbands should suffer more severe punishment 'by how much the more it appertaineth to them to excel in virtue and to govern their wives by example'. The French Catholic, Jean Raulin, echoed Augustine's arguments that husbands, being wiser, should 'be more restrained'. The German protestant jurist, Basil Monner, argued that it was inequitable to punish a wife who copulated with an unmarried male and not to punish a husband sinning with an unmarried woman: because males should show 'greater prudence and strength of mind' than women, whose intellects and judgements were 'weaker'. Early-modern theorists thought women were less self-controlled than males. How could irresolute women be expected to restrict the indulgence of their sexual desires to one person if males could not?[8]

Stiff penalties were recommended for unfaithful husbands. The English *Reformatio Legum Ecclesiasticarum* of 1552 proposed that an offender should restore his wife's dowry, forfeit half his goods to her, and either be imprisoned for life or perpetually banished; this was a punishment no less severe than that imposed on an adulteress. The May 1650 Act against adultery was less even-handed but did dictate the death sentence for a husband who was twice convicted of infidelity with a single woman (an adulteress was to be punished capitally for the first offence). English ministers preached vigorously against adultery by the husband. Theorists often castigated males as the chief offenders and accused them of suspecting their innocent wives because they were fearful the women would follow their own shameful example.[9] Sexual *mores* relaxed in Restoration society, but official theory remained unchanged: adultery was indefensible in either husband or wife. Theologians insisted that Saint Paul made 'the case of the man and of the woman to be equal in the point of infidelity and desertion'. Allestree's popular guide to pious conduct, *The whole duty of man*, recognized that the world seemed to regard husbands' infidelity 'with less abhorrence' but insisted that before God 'the offence will appear no less on the man's side than the woman's'. Moralists defined adultery in a sex-neutral way and heaped denunciations on all perpetrators of 'such filthiness'; 'those who have burnt together in lust shall burn together in unquenchable flames'. Throughout the sixteenth and seventeenth centuries, the majority of theorists disregarded the original Jewish interpretation of the seventh commandment as applicable only to married women and insisted on a single standard of marital fidelity. 'By the divine law', asserted one theorist – calmly ignoring Mosaic law – 'adultery is every violation of the conjugal rites and marriage bed, committed either by the man or woman'.[10]

Most early-modern theorists denounced societies that had given the husband greater freedom and power in the matter of adultery. Theologians recognized that under secular laws husband and wife 'appear unequal' in the case of adultery, but insisted that Christian law made them equal. They complained that when 'the ancient Romans' gave husbands 'more privileges', this did not 'stand with the tenor of God's law'. Roman laws on adultery were rejected as 'unjust and cruel'. The behaviour of the Old Testament Jews was characterized as a case of unfair male oppression: 'Men having the rule did abuse it to the woman's injury'.[11] Theorists were just as willing to condemn the double standard in their own time, though (naturally enough in the case of Englishmen) the main object of condemnation was generally foreigners.

> Among many nations, as the Italians, Spaniards, and Turks, it is counted a capital crime in the wife to tread awry but in the husband it is usual and venial: the Scriptures give them no warrant thus to presume: they are both under the same law of wedlock and must both draw in the same yoke.

Italian and Spanish husbands were attacked for being 'very unchaste themselves' and yet severely punishing the slightest fault in their wives.[12]

Despite the prevalent attitudes in society, virtually all those who discussed adultery treated it as a crime in the husband as well as the wife and – to some extent at least – an equivalent one. Theorists supported their advice of chastity to husbands with comments like 'Wouldst thou expect thy wife a conqueror when thou liest foiled with the same weapon?' and 'he that strikes with the point must be content to be beaten with the pommel'. If a husband wanted a chaste wife, he must 'seek to be virtuous himself first'. Woman was naturally jealous 'and since her husband breaketh with her, she will not stick to break with him and privily borrow a night's lodging with her neighbour'. Such warnings implicitly subverted the double standard. If the adulteress's sin were qualitatively different from that of the unfaithful husband, an argument based on the assumption of equivalence would not have been persuasive. Wives would not adhere to higher standards than husbands;

> it was the great argument which the Spanish lady used to herself that she had not done much amiss to admit her page to her bed, because she knew that her husband was abed with an inn-keeper's daughter of the town at the same time.

Continental Catholic writers used the same argument as English Protestants: it was 'grossly unjust' that the husband should exact chastity from his wife when he did not show it to her. Wives followed their husbands' bad example as they did their good, 'and the adultery of husbands' would produce 'lewdness in wives'. The wife was characterized as retaliating on the same level, not disproportionately.[13]

The radical Puritan, William Ames, argued that 'commission of the same fault on both sides seemeth to take away the right of divorce from either of them; for faults of an equal nature are sometimes abolished by a compensation'. The conformist Anglican, John Dove, maintained that a husband could not separate from an adulterous wife 'if the woman be able to plead compensation against her husband; that is, if he have been incontinent as well as she.' Both were referring to the same legal principle: *'Paria enim delicta mutua compensatione tolluntur'* [For equal offences compensate for one another]. Almost every casuist and canon lawyer who dealt with the problem reiterated this dictum.[14] The medieval church had allowed a wife to avert judicial separation for adultery by a plea of the husband's equal guilt. Roman Catholic casuists argued that either spouse could separate or refuse to have sexual relations if the other committed adultery, but if both had been unfaithful, neither could. Canonists' arguments embodied the tenet that husband's and wife's duty of fidelity was equal and reciprocal. The maxims that husband and wife 'should not be judged unequally' and that they were 'reciprocally related' [*correlativa*] were endlessly repeated. Civil law applied the same reasoning as canon law: a husband who had been unfaithful could not proceed criminally against his wife for her adultery. English law expressed recognition of the (partial) equivalence of the two offences in the legal defence of 'recrimination'. A wife being sued for divorce could debar the action if she could show that her husband had also committed adultery. In the legal systems of Europe, an adulteress was penalized more severely than an unfaithful husband, but he could not err with impunity; his own offence offset hers. Even after ecclesiastical courts had lost control of many matrimonial cases, the canon law's insistence on the equal treatment of husband and wife in cases of sexual misconduct remained influential.[15]

Cuckolded husbands and dishonoured fathers attracted more sympathy than betrayed wives. Preachers fulminated against adulterers who gave their 'unlawful lusts scope to range abroad to the wives or daughters of other men'. Pamphleteers appealed 'to the very whore-mongers and adulterers whether they themselves would resent anything as a higher injury ... than to have their own wives and daughters debauched'. Yet although the belief that male rights were infringed by any unchastity was sometimes implicit in their arguments, virtually all the conventional theorists persistently and self-consciously insisted that husband and wife should be held to the same standard in sexual matters. Their unanimity is unsurprising since Christian tradition was firmly and enthusiastically against sexual pleasure in general, and any double standard in sexual fidelity in particular. Christian theologians were reluctant to set aside Saint Augustine's insistence that husband and wife must be held to the same standard and his assertion that Saint Paul had taught just this.[16]

Secular theorists were far less likely to reach the Christian conclusion. Those attempting to treat adultery solely in terms of abstract reason, without concentrating on the Gospel's teaching, found it difficult to deny that the double standard was the rational one. Their reasoning was straightforward. The systematic secular justification of matrimony was centred on the need for certainty of issue required to motivate males to support their offspring. A woman was tied to one male to give him this assurance. Early-modern theorists believed that without such motivation the family (and society) would fall apart.[17] Adultery by wives ended that certainty as effectively as did promiscuous fornication. Of course, the logical mind might argue that if *all* males refrained from intercourse outside marriage, there would be just as much certainty of issue as if all women did so. Indeed, if all males restricted intercourse to within marriage, no woman (however wanton her inclinations) could possibly conceive a child that was not her husband's.[18] As much as female chastity, universal male continence and fidelity guarantees that no spurious children (or unwanted bastards) will be born. Logical as this approach is, it did not appeal to the seventeenth-century mind, which tenaciously considered the particular case rather than the general rule. In any particular case, adultery by a husband does not mislead his spouse into caring for a child that is not her own. Adultery by a wife might.

Virtually every theorist who maintained that the wife sinned more based his case on this fact. Only a naive husband needed to be reminded that if his wife turned adulteress, she might 'bear for thee strange children (which although thou fatherest yet thou didst not beget) to sit at thy fires and inherit thy goods'. Of course, unfaithful husbands might be persuaded by the 'flatteries of vile dissembling harlots' into attempts to divert property from the 'right heirs' to their spurious offspring. But the law placed obstacles in the path of bids to legitimize bastards where there were legitimate children, and it was difficult to conceal a child's real mother.

> However, the great lady that called all her gallants to her bedside when she lay a-dying and assigned to everyone his share is a convincing argument that a man may toil and moil and cark and care and, when he has done, bestow the sweat of his brows in the wrong Christmas-box.

Early-modern theorists put this 'convincing argument' in more abstractly theoretical terms. They insisted that the wife's greatest duty was sexual fidelity. This was not only what motivated husbands to marry, 'it is what God Almighty did principally design by instituting single marriage'. Wives must be faithful for marriage to fulfil its function of establishing the paternity of children.

> For unless the wife does preserve the bed entire, men would fall into all the difficulties and uncertainties in relation to their offspring as they would find in a vagrant love, which they enter matrimony purposely to avoid.

In Renaissance thought, female adultery undermined the essential purpose of marriage in a way that male infidelity did not. Of all sexual sins, only a wife's unfaithfulness would 'directly fight against the purity of posterity and human society'. Both husband and wife equally infringed their wedding vows by having sexual relations with someone other than their spouse, but the *consequences* of a wife's adultery were commonly regarded as much worse.[19]

In early-modern Europe, the impact of female adultery on the husband was compounded by laws which made it difficult for him to prevent a spurious child from inheriting his name and property. First, English law maintained the fiction that any child born within wedlock was the husband's regardless of the probabilities of the case: 'Albeit the wife were as common as the cart-way'. Common law was not alone in this; civil law also presumed the husband's paternity whatever the wife's conduct, even if the wife were a prostitute. Second, the land of many of England's wealthier families was 'entailed'. An entail was a legal device aimed at conserving familial wealth by severely limiting the owner's power to alienate property outside the family. When land was 'entailed', a cuckolded husband could not bequeath it to someone other than the spurious child. As one guide book to English custom summarized the law:

> If a wife bring forth a child during her husband's long absence – though it be for some years – yet, if he lived all the time within this Island, he must father that child; and if that child be her first-born son, he shall inherit that husband's estate if entailed or left without a will.

Even if an English father were aware that his wife's child was not his own, he would still find himself bound to support and maintain it, and might not be able to prevent the spurious child inheriting his property. Third, English marriage law made it almost impossible for a husband to divorce his wife and marry another woman who might give him a genuine heir.[20]

These laws compounded the problems of cuckolded husbands. In a society where contraceptive techniques are unknown or unreliable, potential cuckoos in the nest are always seen as a threat to the husband and legitimate children. Furthermore, Western European society was (and had been for centuries) one where birth and inheritance were the major determinants of status and wealth. Marriage ensured that the children of the nobility and gentry succeeded to their natural positions of wealth and power, and 'the respect and dignity of blood is preserved nowhere but in the channels of marriage'. The adulterous liaisons of upper-class women were therefore inherently a threat to social order and degree. 'It is of public concern that there should be no supposititious births and that the dignity of families and of the different ranks of men be preserved entire.' Why should anyone defer to the lord of the manor if he were only the gamekeeper's bastard? Only the wife's adultery posed a threat to lineal succession so for centuries her adultery seemed the more dangerous.[21]

Thomas Aquinas and almost all other medieval Catholic discussions of this problem had stressed the injustice and damage caused by spurious issue. Early-modern Catholic casuists and canonists repeated their arguments. Both spouses should be chaste, but 'it more concerns the woman' or a false child might succeed to the husband's estate. Handbooks for confessors pointed to 'the more serious inconveniences' of wifely adultery such as 'damage to the true heirs' and 'uncertainty of issue'. The consequences of a wife's infidelity were simply more disruptive.[22] Protestants saw the matter no differently. The Reformer Henry Bullinger argued that adultery was 'horrible' in either spouse, 'yet in women it is most hurtful and detestable' because 'the adulteress altereth the inheritance' and 'stealeth it from the right heirs'. Only the wife's adultery diverted inheritance from (what early-modern morality deemed) the proper heirs. An unfaithful husband did not deceive his wife into caring for another woman's children, but an adulteress forced her husband 'to bring up those adulterous children which are not his own'. Infidelity by either party was thought to involve 'filthiness and falseness', but only a wife's transferred her husband's 'estate to strangers and other men's children'.[23]

Laymen seem to have been still more conscious than theologians of the dissimilar consequences of infidelity. They insisted that 'everybody desires his own blood should succeed him' and that therefore the husband's offence was not equal, for 'a man by his folly of this kind brings no spurious issue to inherit the lands of his wife'. One response was that unfaithful husbands might infect their wives with venereal disease, 'which sticks longer by them than their children'. Early-modern theorists were well aware that whore-mongering 'frequently brings the lascivious prodigal more than circumcised from the surgeon and sends him nose-less to the grave'. They also knew that venereal disease was no respecter of the guilt or innocence of its victim. However, the force of the argument from spurious issue still held more power in the early-modern mind.[24] Theorists who wished to hold husband and wife to the same standard felt obliged to deny the significance of adultery's consequences. They admitted that greater injury was done by the wife's adultery, but insisted that 'the sin was equal' because both equally violated matrimonial fidelity. William Fleetwood, Bishop of Ely, reasoned that

> if there be some difference betwixt the offence of one party and of the other with respect to reputation and the confusion of families, it is not what concerns the conscience; it falls not properly under our consideration.

Since Fleetwood did not vouchsafe why the consequences of a sin should not concern the conscience, he may well have lost the sympathies of his audience.[25]

Throughout the sixteenth and seventeenth centuries, most theorists continued to insist that husband and wife should be judged equally in cases of matrimonial infidelity. But the friction between traditional Christian dogma and the tendency to judge adultery by its effects grew – not least because the conflict between authoritative Christian tradition and consequentialist rationalism raged within the mind of each theorist. (Perhaps it raged more strongly in married protestant divines potentially able to beget legitimate heirs as their Catholic counterparts never could.) Certainly, as the Church's power in society waned and secular theorists grew more articulate, the lay view seems to have gained ground. The natural lawyer, Samuel Pufendorf, was typical of the newer, less Christian, and more consequentialist approach to ethics. He followed the scholastics on many issues but was clearly influenced by the *mores* of secular society. In his comprehensive discussion of adultery (although supposedly simply summarizing the arguments on both sides), he made it clear that he considered wives' infidelity far more serious than husbands'. The appeal of the double standard grew stronger when the grip of Christian authority slackened.[26] Nevertheless, only slowly – if at all – did arguments from the natural purposes of marriage undermine the Augustinian view. Divines remained eager to condemn extra-marital sexual pleasure in either sex. The Church courts could, and did, punish proven adultery and fornication in both males and females, particularly if a bastard resulted. But both because the woman was left holding the baby, and because of a general acceptance of the double standard, in practice the mother was more likely to suffer (and to suffer more severely) than the father. The double standard persisted (as it had always done) despite, not because of, official Christian teaching, which insisted that

the husband also sinned if he had sexual relations with any woman other
than his wife.[27]

Polygamy

A common way that societies have granted males sexual variety is by
allowing them to have more than one wife at a time whilst forbidding a
woman to have more than one husband. The former arrangement is
termed polygyny and the latter polyandry. Technically 'polygamy' is a
sex-neutral term denoting any arrangement that allows a plurality of
spouses. However, in normal usage both now and in the early-modern
period, polygamy implied a plurality of *wives*. Since very few societies
have permitted polyandry and very many have permitted polygyny, the
term polygamy came to be used of the case most frequently
encountered.[28]

Europe has been predominately monogamous for most of its recorded
history. Roman society was monogamous, and polygamy has 'always
been detested' by the Western Church. Early-modern anthropological
knowledge suggested that there were few or no polyandrous societies.
Lacking empirical examples, both medieval and early-modern discussion
of polyandry often drifted into a debate about Plato's suggestion in *The
Republic* that male and female members of the governing élite should
interbreed freely and the children be raised in common. Knowledge of
Plato's recommendations came largely through Aristotle, whose arguments for rejecting this 'community of wives' were universally accepted.
On the other hand, medieval and early-modern theorists were well
aware that polygamy – a plurality of wives – was not unusual. Old
Testament Israel was polygamous. Christian Europe had for centuries
been in constant (if frequently hostile) contact with Islamic
Mediterranean societies which saw no moral objection to polygamy.
Voyages of exploration to Africa and the New World only added examples of societies wholly ignorant of the notion that the male should be
'cursedly confined' to one woman. Early-modern observers knew well
'that among the Turks and nearly all barbarous nations' polygamous
custom was established.[29]

The possibility that polygamy might not be unnatural had been tentatively broached in the early sixteenth century by the Roman Catholic
Cardinal Cajetan, who held innovative views on many aspects of marriage. His writings found little favour with Catholic theologians, even
when Henry VIII's marital difficulties gave them a real interest in considering such ideas. In October 1530, Henry's agent to the papacy wrote that
the pope had considered dispensing Henry to marry a second wife, but
'now of late the pope shewed me that his council shewed him plainly
that he could not do it'. This decision was in itself a strong indication

that Catholic theologians regarded polygamy as against the laws of God and nature, for the pope's power to dispense from all human laws (civil or ecclesiastical) was generally accepted.[30]

In the heady early days of the Reformation, when many Christian doctrines were considered afresh, a few radicals wanted to throw out the monogamous baby with the popish bath-water. In Munster from 1534 to 1535, John of Leyden briefly instituted a polygamous regime amongst his Anabaptist followers before being defeated and executed by the outraged German nobility. The Protestant Reformers, Martin Luther, Philip Melanchthon, and Martin Bucer, countenanced a polygamous solution to the marital problems of Henry VIII. Bucer's half-hearted, confidential admission that polygamy might be justifiable was (much to his embarrassment) promptly acted upon by Philip of Hesse, who supplemented his drunken wife with a younger and more sober woman. The leading Reformers, however, advocated this particular doctrinal innovation only privately and only to potential patrons powerful enough to forward the protestant cause. In 1563, 'that filthy beast and shameless apostate and back-slider', the radical Italian Protestant, Bernardino Ochino, made the mistake of publishing his support for polygamy in print and was banished from Switzerland and from polite theological circles for his pains. There were other advocates of polygamy for Christians. In the late seventeenth century, Johannes Lyser wrote in defence of polygamy a tract of more than 500 immensely learned, delightfully eccentric pages. His work incongruously combined a practically paranoid dislike of women with the seemingly masochistic assertion that males should marry umpteen of them. Lyser was in all respects an oddity. The champions of polygamy were few and far between, and monogamy was firmly anchored in Western European custom and law. Nonetheless, almost every serious theorist of marriage felt compelled to argue against polygamy.[31]

For the most part, the Christian consensus against polygamy was firm. Most exegetes thought the New Testament was unambiguous on the virtues of monogamy. Quizzed by the Pharisees on divorce, Christ asked rhetorically if they did not know that God made man

> male and female and said, For this cause shall a man leave father and mother, and shall cleave to his wife; and they twain shall be one flesh? Wherefore they are no more twain but one flesh ... And I say unto you, Whosoever shall put away his wife, except it be for fornication, and shall marry another, committeth adultery. (Matthew 19:4–9)

The use of 'wife' in the singular and the word 'twain' [two] were understood as implying Christ's preference for monogamy.[32] In another sense, also, this text seemed to imply disapproval of polygamy, for many theorists – protestant as well as Catholic – thought Christ was here forbidding a husband's remarriage after divorce 'and why he should be forbid after a divorce to marry again, if he might marry two or more wives before, I can't yet conceive'. Similarly, Saint Paul's admonition, 'let every man have his own wife' (I Corinthians 7:2), did not appear to leave open the option of *wives*.[33]

Although there was general agreement that Christ had forbidden polygamy, early-modern theorists still perceived a problem. Was monogamy *naturally* the only permissible system or was it simply commanded by Christ and therefore binding only on Christians? Unassisted by Scripture, pagan societies had instituted marriage, but many had not adopted monogamy. Did this imply that polygamy was a reasonable form of marriage? Did the law of nature permit a husband to have many wives? Or a woman to have many husbands? Moralists were concerned to decide whether by natural reason all men should recognize that monogamy was the only legitimate form of marriage. Early-modern theorists debated polygamy on the basis of three accepted axioms about the natural ends of marriage. Marriage must ensure: (i) the birth and care of children, (ii) remedy for mankind's excessive lust, and (iii) companionship and help within the household. If all three ends could equally well be achieved in polygamous or polyandrous marriages, they accorded with the law of nature; if not, they did not.

The birth of children can clearly result from polygamous and polyandrous unions as from monogamous ones. But to the early-modern theorist, it seemed clear that to achieve fecundity it was much more efficient to pair one male with many women than one woman with many males. Even in these sexually liberated days, a farmer who tried to establish a breeding herd with one cow and a dozen bulls would be thought a little loopy. To the early-modern mind, the biological logic appeared inescapable: the impregnation of many females by one male was *natural* in a way that the opposite was not. From the Middle Ages onwards, virtually all discussions of the relative (un)naturalness of polyandry and polygamy took as their starting point the biology of multiple impregnation.[34] Two further beliefs about human physiology informed their discussion of polyandry. The first was that if a woman copulated too frequently or with too many males, she would become infertile. The second was that copulation during pregnancy (especially with a male other than the father of the foetus) might injure the foetus. Although there was increasing scepticism about the second, these ideas were held by medical experts throughout the classical, medieval, and early-modern periods.

Both these beliefs were inherited from classical medicine and asserted as fact in the influential medieval tract *De secretis mulierum* [About the mysteries of women]. Helkiah Crooke's standard seventeenth-century textbook of human anatomy accepted that 'too frequent copulation, as in harlots ... bringeth barrenness'. Popular works put forward the same theory:

Wherefore do not common harlots conceive: or if they do, it is very seldom?
Answer. The diversity of the seeds doth let [i.e. prevent] conception and causeth that the same cannot be retained.

Medical treatises taught that pregnancy was occasionally terminated 'in superfetation, where after one conception another cometh'. There were growing doubts amongst early-modern physicians about whether it was possible to impregnate an already pregnant woman, particularly as Aristotle had insisted that the wombs of pregnant women were 'closed', but scientific theories take time to spread.[35] In Sussex in 1621, a woman was presented to the ecclesiastical court for being 'delivered of two children'. She contended that 'Henry Smith, now of Horsham, and William Walter, late of Rusper, are fathers and that they lay with her, one one night and the other the next night'. Both the possibility of 'superimpregnation' and the sterility of whores were generally received as scientific truths. The belief that fertility diminished when a woman copulated frequently and/or with many males seems to have been generally accepted into the late seventeenth century. Theorists still argued that 'the womb is quickly vitiated by commixtures and either destroyed by heat or over-cooled by diversities', and that sexual congress with many males was 'destructive to propagation'. Promiscuity 'by a natural cause prevents conception', for the 'beaten paths are always barren and never productive of any fruit'.[36]

Another belief pointed to the 'unnaturalness' of women having many sexual partners. This was the assertion that sexual relations – particularly with a male other than the baby's father – rendered nursing mothers' milk less nutritious. Even a woman's own husband should try and refrain from intercourse with his nursing wife in case she conceived and ceased to lactate. However, they could copulate for 'if the child's father has intercourse with the nursing mother, the milk will not be any less fit and proper food for the child, as it would be if she had intercourse with another man who was not the child's father'. Writers discouraged excessive sexual indulgence during lactation.

Wherefore commeth it that the milk in a woman's breast suddenly decayeth, if she give herself to be immoderate in lust.
Answer. Because the menstrual blood doth not ascend to the breasts to nourish the child.

Motherhood and promiscuity were thought antagonistic, not complementary.[37]

These ideas about the biology of sexual reproduction were directly applied by moralists to the question of whether polyandry or polygamy might be natural. Polyandry was unnatural because 'it is possible for one man to make several women pregnant', but if several males copulated with one woman, 'they will impede each other owing to the different composition of their seed'. Since 'the most important purpose of marriage is the multiplication of offspring', polyandry could never be lawful. Theorists argued against a Platonic community of wives because the natural end of marriage, the procreation of children, was 'impeded' when one woman has many husbands. The same did not hold true for polygamy for one male could impregnate several females just as a farmer could cultivate several fields. Males were 'by their nature' capable of reproduction at any time whereas women were (usually) incapable of conception during pregnancy. The main object of marriage was children. Polygamy was conducive to that end; polyandry was not. Throughout Europe and without regard to religious denomination the same argument was employed. A male can impregnate many women, but a woman could not 'be made fruitful by many males'. The object of sexual congress was reproduction; a woman could usually only give birth once in each year no matter how many husbands she had, so there was no excuse for additional ones. Polygamy did not 'cross' the end of reproduction; polyandry did.[38]

Early-modern theorists also granted that eventually wives can be multiplied to the point where the first end of marriage cannot be satisfied for all of them. King David and King Solomon, they contended, had simply bitten off more than they could chew; no husband could impregnate quite so many women. One Hebraist argued that the maximum number of wives was 'nowhere expressly determined' in the canonical books of Scripture, but reported that 'the Rabbis' put the maximum at eighteen. The reproductive purpose of marriage was coupled with the belief that 'two women will be better impregnated by two men than by one', and this was held to place some natural limit on polygamy.[39] Science and experience were also held to prove that males would be debilitated and rendered less fertile by frequent copulation. The contention that frequent sexual intercourse was bad for the health was endlessly repeated in the sixteenth and seventeenth centuries. The children born from a male who copulated too often would be sickly. The more a couple enjoyed coitus the more it would enfeeble them. Sparrows died young because they copulated continually. Males would outlive females were it not for too-frequent copulation. Immoderate intercourse made hair go grey and fall out. Excessive sexual indulgence had rendered Englishmen 'as lascivious as satyrs and as impotent as pygmies'. Polygamy 'debilitates the fathers' and so 'naturally creates a weak and infirm issue'. Monogamous Europeans reproduced more rapidly than polygamous Asians because Christian males copulated less. A male simply did not have sufficient semen to fertilize hosts of women and this was why Mohammed had been able to beget only three daughters from his fifteen wives. King Solomon 'had but one son by many house-full's of wives, when many a poor man hath an house full of children by one wife'. All these beliefs tended to suggest that polygamy was not the healthiest, most natural, or most reproductive form of marriage.[40]

Nevertheless, the view that woman's fertility was particularly curtailed by promiscuity was more firmly held in early-modern Europe. Polyandry was thought to inhibit reproduction more than polygamy. Polygamous societies did exist and function; polyandrous ones were exceptional oddities. As the legal theorist, Thomas Wood concluded of polygamy: 'Several nations do suffer it in the male, but not in the female because propagation would be prejudiced by such a liberty'. Polyandry was thought more inimical to the principal end of matrimony than polygamy.[41]

Marriage was also aimed at the proper nurture of children and this, all the theorists were agreed, required that parents be certain that the children were their own. Polygamy did not undermine either parent's certainty as polyandry was thought to do. Once again, misconceptions about conception played their part. Medieval and early-modern theorists did not believe that paternity could be established by working back from the date of birth to the date of conception. They believed one male's semen could lie dormant in a woman's womb until 'raised up' by another male's (Genesis 38:8). 'And for this cause the woman bringeth forth oftentimes a child resembling the first husband, if there be no long time between the death of the first and the marriage of the second.' Roman law had forbidden a widow's remarriage for a year after her husband's death and early-modern commentators on civil law argued that this was because nobody could otherwise be certain whether the child was that of the first or second husband. Moreover, they believed that if a woman had sexual relations with a male, his semen could alter the physiology of her womb in such a way as to affect the characteristics of children born from it, even if begotten by coitus with another male. A mother was affected by every male with whom she had sexual relations, for a bitch mounted 'by several kinds of dogs ... brings forth her whelps fashioned and coloured like to all those she coupled with'. A child's paternity could not necessarily be deduced from any physical resemblance to a particular male since a woman's 'imagination' partly determined her child's appearance and 'so powerful is its operation that, though a woman be in unlawful copulation, yet if fear or anything else causes her to fix her mind upon her husband, the child will resemble him although he never [be]got it'. Certainty of issue and numerous husbands were simply incompatible.[42]

Marriage was directed at the child's 'pious and religious education', but in polyandrous marriage 'it would not be known who was the father of the child, so that several would take less care of the instruction and education of the child than if there were only one husband'. The common argument against communism (that where everyone is responsible no individual takes responsibility) found one of its earliest expressions in discussions of polyandry. Copulation of many males with one woman would lead to 'uncertainty of offspring ... and therefore an uncertainty

of children's maintenance and education'. In the late seventeenth century, these scholastic arguments were still repeated: 'The end of formal
marriage – which is certainty as to one's offspring and mutual assistance
– can be secured no less in polygamy than in monogamy'; in stark contrast, polyandry was 'repugnant to natural law'. 'For the natural and
orderly end of marriage is for one to secure his own particular offspring,
and how could a man recognize his own in such promiscuity?' The need
for certainty of issue was seen as clinching the reproductive argument
against a wife having many husbands; it was an outrage against 'nature
itself' for fathers to be left unsure. One sixteenth-century priest, Jean
Raulin, pointed out that even if the husband of a polygamous marriage
knew that all the children were his own, this did not entail him having
the ability to support them all properly. Proper nurture required material
resources as well as the motivation that stemmed from certain paternity.
But Raulin was unusual. Most theorists thought that both the birth and
care of children were far better ensured by polygamy than by
polyandry.[43]

The second (and secondary) end of marriage was remedying lust;
helping mankind 'avoid whoredom'. It might seem that unlimited
spouses would promote this object, for 'surfeiting the appetite might
sicken and so die'. Early-modern theorists took the opposite stance for
two reasons. First, their psychology of sexuality implied that the more
sexual pleasure and variety a person were offered, the more (s)he would
want. Sexual appetites were not sated by being fed but were enhanced;
more spouses would make someone more lustful. Second, they doubted
whether one woman could satisfy many males or one male many
women.

Early-modern theorists explained that many wives would provoke
extreme sensual desire, just as 'many dishes' lead to eating excessively.
Polygamy would not be a 'restraint' but a 'stimulant' to the libido.
Theodore Beza used this reasoning in his response to Bernardino
Ochino's defence of polygamy. God ordained marriage as a remedy, but
polygamy has the opposite effect, inflaming those 'otherwise chaste' to
immoderate lust, 'just as abundance of food causes gluttony and abundance of wine drunkenness'. In women, too, lust was kindled more by
use, 'just as in whores'. God had granted the use of the marriage-bed
with the object of quelling sexual desire, of 'assuaging and destroying'
lust 'not inflaming and stimulating' it. Fallen man was prone to unlimited sexual desire, but it would be 'diminished' by restriction to one partner. Monogamy was like diets that limit their victim to one food only;
faithful adherence suppresses appetite.[44] English divines based their case
on these assumptions. Polygamy crossed the second end of marriage,
'for avoiding of lust; for this diversity of concubines made a way rather
to increase lust than to quench it'. 'God had ordained marriage as a
means to bridle men and restrain them from extravagant lusts.' This was

achieved by limiting each to one partner. If a male were not satisfied with one woman then giving him more would only increase his desires; 'neither variety nor multitude can allay the rage of those imperious desires, which never knew what it was to be checked by modesty and moderation'.[45]

In the Renaissance mind, the other problem related to marriage as a remedy for lust was whether one person could sexually satisfy many spouses. A few theorists discussed the possibility that one woman would be enough for many husbands, although this was generally considered either too ludicrous or too distasteful to warrant lengthy consideration. Early-modern writers were not ignorant of the practicalities of sexual relations, knowing that at a single sitting 'a woman is able to have to do with more men than a man can with women'. If sating sexual desire were the main object of marriage, polyandry might be more practical than polygamy. However, orthodox theology held that the end of averting lust was only secondary; procreation was primary. Throughout the early-modern period, theorists countered the argument from physical capability with the standard scholastic view: even sex as a remedy must be potentially reproductive to be legitimate. In a polyandrous marriage 'since conception may occur but once, such intercourse serves no end but the satisfaction of lust'. Polyandry satisfied the secondary end of marriage only by ignoring its primary one.[46]

Polygamy presented different difficulties for marriage as a remedy for lust. Catholic orthodoxy held that a husband was duty-bound to have coitus with his wife to avert her from incontinence as well as to procreate children. The medieval canon lawyer, Antoninus, cited numerous authorities to support his view that the husband should not forbid his wife from requesting sexual relations without reasonable cause, nor deny them when she did ask, 'on account of the many dangers'. Allowing husbands to ration sexual relations would risk provoking wives into committing mortal sins of sexual incontinence; this was just what marriage was designed to avoid. Standardly, early-modern theorists adhered to this view. If a wife indicated sexual desire 'by word or gesture', a husband should satisfy her 'freely and promptly', if it was within his power. The husband must render his marital due to his wife not 'denying it her unless his health, reason, bodily purity, or Christian mortification, approve his refusal'. Marriage was instituted as a remedy for the sexual desire of both spouses.[47]

The majority of sixteenth- and seventeenth-century theorists thought a polygamist would be unable to keep all his wives happy in this regard. A husband could not pay many wives the 'marriage debt' without exhausting himself and undermining his health. Saint Paul had taught husband and wife that they must not deny one another their marital obligations so 'that Satan tempt you not' (I Corinthians 7:5). Any husband 'that defraudeth his wife of the right of marriage commiteth no small sin'. He owed his wife 'cohabitation', and 'polygamists could not perform this duty to their wives to dwell with them'. European travellers in the Orient ascribed the Turks' 'jealous' confinement of their wives to 'polygamy, which makes a husband guilty of insufficient correspondence, and therein fearful that his wife may seek a further satisfaction'. Polygamy naturally resulted in frustrated and potentially adulterous wives;

> it being impossible for one man to satisfy the desires of many women; they must necessarily burn and lay hold of every opportunity to quench the flame, and many times commit crimes against nature rather than not have it effected.

'Crimes against nature' meant non-heterosexual activity – in this case masturbation and/or lesbianism. Polygamy was satisfactory in regard to the primary end of marriage – reproduction – but inadequate as a remedy for lust.[48]

To sustain the case for polygamy, its few advocates abandoned the orthodox view of husbands' sexual duties. Bernardino Ochino argued that it would not be hard for a polygamist to satisfy all his wives because 'he is not obliged to satisfy all the carnal desires of his wife, but only such as are moderated with reason'. Johannes Lyser argued that a husband's only obligation was to impregnate his wife; he was not required to satisfy 'inordinate carnal inclinations'. Indeed, it would take 'ten or twenty' husbands to satisfy one woman 'who always lusts if not pregnant'. In arguing in this way, Lyser was diverging both from orthodoxy on husbands' obligations, and from the generally accepted view that female sexual desire increased (rather than abated) during pregnancy. Even its champions tacitly accepted that polygamy could not satisfactorily fulfil the second end of marriage – preventing 'burning'. They merely broke with tradition and argued that a husband was under no obligation to satisfy (quite possibly insatiable) wives. In the late seventeenth century, Samuel Pufendorf followed the supporters of polygamy and argued that if co-wives limited their desire for sexual relations to becoming pregnant, one male could satisfy them all. If they wanted more, it was merely 'intemperance'. Unusually for Pufendorf, he deviated on this point from the scholastic view of marriage and sexuality. After all, if sexual desires could be held under the control of reason and limited to the purpose of having children, the second end of marriage – 'remedy' – would be unnecessary. The majority of early-modern moralists continued to adhere to the view that wives' concupiscence must be remedied just like husbands', and that a polygamist would not be able to meet this end by servicing all his wives to their satisfaction.[49]

The final end of marriage was companionship and help within the household. In this case, too, polyandry was perceived as a total failure. The obvious point that a wife is worn out by looking after one husband, let alone more, was not made. Instead, early-modern theorists pointed to the endless dissension that would result from many husbands attempting to share one wife. They believed that the whole animal kingdom showed it was natural for males to compete for the exclusive sexual attention of females; violent conflict was regularly provoked when 'many males pursue one female for sexual relations'. Males simply could not reach agreement on sharing females and the resulting altercations were always bitter: nothing led to 'more furious discord'. This was equally true of humans. Polyandry would lead to 'quarrels and contentions' among the husbands, as each would want to be 'first or preferred' in enjoying the wife. There were bound to be disputes in a polyandrous family for 'reason as well as revelation teaches us that jealousy is the rage of man and that there is nothing more insupportable to human nature'.[50]

Theorists also contended it would be impossible for a polyandrous wife to know which of her husbands she should obey if they gave conflicting orders. There would be 'discords, perturbations, and great inconveniences' when husbands 'should will things contrary and command their wives to do them'. Polyandry was not conducive to 'domestic peace'. Scripture taught that the husband was the wife's head; polyandry distorted natural order by giving her many heads. It was simply unnatural for a woman 'to contract marriage with two husbands when plenary duty and obedience is to be paid to each and therefore impossible to be performed to both'. A woman cannot serve two masters. Matters would only be worse if the natural order were inverted and the wife had authority over a number of husbands, for

> a man who has several wives can chastise a wayward and quarrelsome wife by word and blow, whereas a wife with several quarrelsome husbands cannot restore peace among them in this way.

A household could not properly be ruled by a wife, and if the husbands all ruled there would inevitably be dangerous conflicts of authority.[51]

In early-modern eyes, polygamy caused less problem to the smooth functioning of the household. The argument that polygamy was harmonious with female inferiority appealed to some medieval and early-modern theorists. Wives, like servants, were subjects and it was not 'contrary to natural right' that one lord have many servants. Wives were not demeaned by this: 'polygamy does not reduce wives to a state of slavery ... for the very weakness of their sex urges that they live under the tutelage, as it were, of their husbands'. But most moralists rejected this argument. Cardinal Robert Bellarmine admitted that a wife was subject, but rejoined that – as co-ruler of the household with her husband – her subjection differed greatly from that of a servant; polygamy *was* incompatible with preserving wives' proper status. Theodore Beza responded to Ochino's use of the argument that a body ruled by two heads was unnatural, with the scathing comment that it was no more natural for one head to have two noses than for one nose to have two heads. Beza insisted that polygamy interfered with the goal of companionship. There would be 'no tranquillity' but instead 'unending quarrels' stemming from 'rivalry' between the different wives and their offspring. The association of one male and one woman naturally led to love and affection because of common interests; the wives of a polygamist would have no such bond.[52]

Beza's belief that a polygamist's wives would be jealous and quarrelsome was widespread. Almost all European theorists asserted that monogamy was more 'conducive to domestic peace'. In a polygamous family, disputes would arise as the wives competed for a greater share in the husband's affections, while the children of different mothers would squabble incessantly. Theologians insisted that a husband would not be able to love numerous wives with the deep affection that God required from him. Theorists did not adduce much data on the actual practice of polygamy. Virtually all their examples of polygamous dissension came from a one-sided reading of the Bible. They supplemented this with a strong prejudice against Moslems in general, and the Turks in particular. In marital matters 'the law of Moses is a law for boys; the law of Christ, an impossible law; the Mahometan law, a law for pigs'. Despite limited anthropological evidence, all early-modern theorists dogmatically stated that 'the common experience of all ages teaches that there was always jealousy and contention between co-wives'. The household, they were sure, would function more peacefully and amicably with one wife than with many.[53]

Early-modern theorists contended that polyandry was naturally incompatible with all three ends of marriage – reproduction, remedy, and companionship. They therefore condemned it as clearly against the law of nature. Polygamy was more problematic. It did satisfy the primary, most important end of marriage – the reproduction and nurture of children, although it was widely doubted that it met the second end, and probable that it failed in the third. Nonetheless, many theorists were reluctant to admit that only Christian teaching rendered polygamy illegitimate. Thomas Aquinas had argued that polygamy was against the 'secondary' principles of the law of nature but not against the 'primary' ones. Some theologians followed Aquinas's lead and adopted this distinction, but many still struggled to decide whether the argument from reproduction was strong enough to render polygamy a reasonable form of marriage.[54]

In order to undermine the force of the argument for polygamy from reproduction, early-modern theorists drew attention to the fact that God originally created only one woman for one male (Eve for Adam), although the world was then in as dire need of population as it ever could be. God obviously had the power to make dozens of females for

Adam if He had wanted, but 'in His wisdom, goodness, and holiness, He would not make more'. God's original creation showed what was natural and it was monogamy. Others reasoned from modern populations that nature itself must favour monogamy, since equal numbers of males and women were born. Renaissance statistical knowledge rendered this point open to debate. Some thought that more women were born in Southern climes, while others believed that an excess of female births was compensated by higher female mortality.[55]

Those who wished to assert that polygamy was against the law of nature, despite many non-Christian societies' acceptance of the practice, emphasized that 'the most virtuous and civilized Gentiles' and 'the wiser amongst the pagans' knew that polygamy was against the law of nature. Polygamy was also characterized by early-modern theorists as naturally inequitable. Their assumption that both male and female desire the exclusive sexual attentions of their mate, made the natural and Scriptural maxim 'do unto others' seem incompatible with polygamy.* A husband would not want to share his wife with others, therefore he himself 'should be content with one'. Marriage was naturally monogamous, because 'in all things else we can allow a sharer, but in the interests of our love'. As usual, theorists' arguments embodied the belief that the husband and wife should be afforded equal rights in specifically sexual matters.[56] The marriage pact was 'of equal force' for both and so in 'commutative justice' must be as exclusive for both. Since 'it was never lawful for the wife to have more husbands at once; therefore it was never lawful for the man to have more wives at once'. Polygamy was unjust because it undermined the 'equal obligation of the husband and wife as far as it concerns the use of their bodies'. It would be 'iniquitous' to assert that the wife must give herself exclusively when the husband did not.[57]

The book of Genesis, demographic statistics, and natural equity were all martialled by early-modern social theorists in their attempt to prove that polygamy was unnatural – whatever the customs of non-Christian societies and their own stress on the reproductive function of marriage suggested. However, even if these arguments were true in general, they did not seem to rule out the possibility that in certain circumstances (in particular, when there was an especially pressing need for fertility) polygamy would be justifiable. Indeed, it seemed that once just such circumstances had arisen and the Jews had responded by adopting polygamy. Did Scriptural history prove that polygamy – although not the best – was a naturally acceptable form of marriage?

*'Therefore all things whatsoever ye would that men should do to you, do ye even so to them: for this is the law and the prophets' (Matthew 7:12).

Patriarchal polygamy

Early-modern theorists were inclined to the conclusion that polygamy
might sometimes be licit, not only by the logic of the arguments from the
reproductive purpose of marriage but also by the case of the Old
Testament patriarchs. The Bible told how Abraham and Jacob both took
more than one wife, while King David and King Solomon accumulated
wives by the dozen. These polygamists were the leaders of God's chosen
people, and God's judicial law seemed to be one which freely permitted
polygamous marriages. How could this be so if polygamy were against
the law of nature? The opponents of polygamy scoured the Old
Testament for some sign of its condemnation. It was common to argue
that Lamech's bigamy had been censured, but this censure was visible
only to Christian – not Jewish or Islamic – biblical exegetes (Genesis
4:18–24). Some pointed to Malachi 2, but the book's ecstatic tone rendered this text intractable, particularly as the prophet was patently more
concerned with idolatry than polygamy. Others cited Leviticus 18:18.*
The problem with this text was that it explicitly forbad only incestuous
polygamy, unless 'sister' meant any other woman (rather than the obvious sense of a blood-relation). Unfortunately for all these expositions, a
number of passages in the Old Testament mentioned polygamy without
moral condemnation: a clear example was Deuteronomy 21:15–17, where
the custom and law of polygamy were 'presupposed and not censured'.
Another biblical text appeared to evidence God's endorsement of
polygamy on a large scale. When Nathan called David to task for committing adultery with Bathsheba and smiting her husband, he stated that
David had no excuse for such conduct, after God Himself 'gave thee ...
thy master's wives into thy bosom' (2 Samuel 12:7–8). Since David had
collected wives and concubines like stamps, this hardly suggested God's
unswerving support for monogamy.[58]

Faced with the problem of finding an explicit condemnation of
polygamy in the Old Testament, where it seemed generally to pass 'without reproof', some theorists simply stated that it was obviously wrong,
whether Scripture condemned it or not. One averred, 'though it's true
that Jacob is nowhere condemned' (for bigamy and incest) 'yet it's plain
that he acted contrary to divine institution'. But for early-modern theorists, it went very much against the grain to describe the guides of God's
chosen people as sinners. Catholics insisted that only heretics would
dare state that Abraham, Jacob, and David sinned by marrying biga-
mously. Protestants confessed that 'it would be something of an
unsavoury speech to affirm that Abraham and the holy patriarchs were
adulterers'. Yet if polygamy were dismissed as unnatural and immoral,

---

*'Neither shalt thou take a wife to her sister, to vex her, to uncover her nakedness, beside
the other in her life time'.

this conclusion seemed to follow. How could Moses have permitted polygamy when nowadays it was thought 'filthy, abominable, and disgraceful'?[59]

Three different views on Old Testament polygamy were adopted by medieval and early-modern theologians. The first argued that polygamy was *not* against the law of nature and that therefore it had been legitimate to take many wives until Christ forbad the practice. A second view was that the patriarchs had received special divine dispensation to marry polygamously in order to increase the chosen people. Finally, the third position was that polygamy was unnatural; the patriarchs had indeed sinned, even if contemporary customs and their own good intentions made their actions partly excusable.

All three views had been outlined in earlier centuries. The Calvinist exegete, Andrew Willet, listed Ambrose, Durandus, and Tostatus as espousing the view that 'polygamy was lawful before it was forbidden by the positive law of the Gospel'. Others added that some held this to be 'the opinion of St Jerome and Augustine'. Augustine had argued that in the time of the Patriarch Jacob 'no law forbad a plurality of wives' and that 'Abraham is no way to be branded as guilty concerning this concubine [Hagar], for he used her for the begetting of progeny not for the gratification of lust'. Augustine's authority was such that any view he had held was considered at least plausible.[60] The German Protestants, Martin Bucer and Philip Melanchthon, were swayed by the example of the patriarchs into believing that bigamy was not against the law of nature. Melanchthon argued that Henry VIII could with a safer conscience take a second wife while Katherine of Aragon lived than divorce her, for 'Abraham, David, and other saintly men had many wives, and so it is clear that polygamy is not against divine law'. The Catholic biblical commentator, Juan Maldonado, was another who subscribed to the view that, before Christ, polygamy 'was allowed to everyone by common right'.[61]

These theorists were atypical; the great majority of early-modern moralists argued that polygamy was against the law of nature. They distinguished polygamy from polyandry, accepting that the former was far less heinous than the latter. They were eager to excuse the patriarchs, either fully (on the grounds of a divine dispensation[62]) or in part (on the grounds of ignorance, and good – that is, procreative – intentions). However, they did maintain that polygamy was sinful. The prevalent belief in the Middle Ages was that God had specially dispensed the patriarchs to have many wives because of the need to multiply the chosen people. Thomas Aquinas, Albertus Magnus, Bonaventure, Jean Gerson, Duns Scotus, and Vincent of Beauvais were just a few of the influential schoolmen who subscribed to this view and many early-modern theorists followed them.[63] The Church of England's official *Homilies* taught the ordinary people that:

> The plurality of wives was by a special prerogative suffered to the fathers of
> the Old Testament, not for satisfying their carnal and fleshly lusts, but to
> have many children.

These theorists accepted that although polygamy was against the law of
nature, God had dispensed the Jews from monogamy between the
Deluge and Christ's coming so that they would not be so outnumbered
by the surrounding pagans who were multiplying polygamously. God
had decided to fulfil His promise to Abraham to make him 'exceeding
fruitful' by permitting polygamy (Genesis 17:6). The covenant with
Abraham was a polygamist's charter.[64]

One problem with this interpretation was that the Bible gave no textual evidence of God issuing such a dispensation. For 'we do not read in
the whole of scripture' of such a dispensation. Christian commentators
therefore resorted to 'diverse conceits', and disputed whether each patriarch received his own personal 'peculiar inspiration from God for this
dispensation', or whether the concession was just passed down by word
of mouth from one patriarch to the next. A literal reading of the Bible did
not suggest a dispensation was needed or issued, but theologians persistently speculated about how God had granted it. The patriarchs, exegetes
argued (radically vague as to their textual support), 'by divine revelation
and other notice of the will of God ... were certified that they might
marry many or sundry ways at one time'. Although the Old Testament
simply took polygamy's legitimacy as read, Christian commentators
repeatedly asserted that the patriarchs had needed 'a dispensation especially by God given' because polygamy was 'against the law of nature'.[65]

Despite its respectable antecedents, most early-modern English moralists rejected the belief in divine dispensation and argued starkly that the
patriarchs had sinned in taking many wives. Many were probably simply following John Calvin, who in his commentary on Genesis argued
that the patriarchs' polygamy was wrong. Calvin did his best to excuse
Abraham and place all the blame on 'the foolish and presumptuous
counsel of his wife', Sarah, but concluded that Abraham himself was not
'free from fault'. Outraged Catholics condemned the 'Manichaean'
Calvin for flouting Augustine's authority, but English divines followed
his lead in droves. One English Calvinist, John Stockwood, translated
one of Johann Brenz's biblical commentaries, which – like many
Lutheran works – took the standard scholastic view of a patriarchal dispensation. Stockwood prefaced his translation with an 'Admonition' dis-
tancing himself from Brenz's view and asserting that he considered 'their
judgement most sound who think the fathers in this point to be condemned and to have sinned'.[66] Many English protestant divines thought
natural reason showed that polygamy was 'simply wicked, impious, and
unlawful'. Calvinist divines listed the traditional Catholic views, but
concluded the soundest opinion was that the patriarchs had acted
wrongly; polygamy was 'never neither simply lawful, nor for a time dis-

pensed with'. The puritan theologian, William Ames, insisted that 'polygamy – even that which prevailed with the ancient fathers – was always a violation of the laws of marriage'. Most commentators tried to excuse the patriarchs. Polygamy was incompatible with God's first institution of marriage, but in Abraham's time that 'law (which expressly forbad adultery, polygamy, and the like) was not yet written'. Biblical scholars pointed to the blindness induced by the prevalent customs of surrounding peoples. They asserted that the patriarchs were interested only in increasing the number of God's people, not in sexual pleasure. Only the unusually outspoken bluntly stated that the patriarchs 'knew well enough and could not be ignorant, but that therein they swerved clean from that pattern God had given them and from that rule that therein he left them'.[67]

Those convinced that monogamy was the only form of marriage in accordance with the law of nature argued that God had not *dispensed* the patriarchs, He had only *permitted* polygamy by imposing no penalty in Mosaic judicial law. To early-modern theorists, the distinction between permission and dispensation was not an empty one. A dispensation suggested that polygamy was not inherently evil; permission did not, and so they could state: 'Bigamy was permitted to the patriarchs, yet unlawful'. On this interpretation, polygamy was evil, but God in His wisdom allowed it nevertheless. God permits many wrongs to go (apparently) unpunished, but this does not make them right; polygamy was one of these. Indeed, God even took advantage of this particular sin. Polygamy 'was never lawful' and yet 'God permitted it for the time, that His church might increase'. Monogamy was the only legitimate form of marriage, for 'if anything be a law of nature that is one by the consent of all men'. Polygamy was not necessary even for the patriarchs; they simply 'had a mind to it' and God allowed it by 'a direct act of jurisdiction'. Natural reason, instilled in men's hearts by God, showed that polygamy was forbidden, 'yet there being no punishment annexed to the law itself, the Jews looked upon it rather as a matter of prudence than obligation'. For this large group of early-modern theorists, the light of nature *did* show the Jews that polygamy was wrong; they contravened natural law, knowing that the judicial law would not penalize them when they did so.[68]

In the later seventeenth century, secular-minded theorists of natural law returned to the earlier view that polygamy was not naturally evil. Just as in the case of the double standard on adultery, they parted company with traditional Christian teaching. Theorists trying to look at biblical texts historically and to interpret them without inordinate regard to Christian tradition did not regard polygamy as so grossly immoral that only a special exemption from God could justify its adoption. Striving to reason from first principles without reference to the teaching of the Gospel and the Church's interpretation of it, they could find no rational grounds for stating that polygamy was unnatural and iniquitous. Polygamous marriage fulfilled the primary end of the institution: it ensured the birth and nurture of children. Scholars trained in humanist techniques of exegesis regarded the Old Testament Jews as simply another example of the many races that found polygamy morally unobjectionable and socially efficient. John Selden and Samuel Pufendorf both tacitly adopted this view. Hugo Grotius accepted that God originally instituted monogamy as the best form of marriage, but argued that this did not make other forms illegal. They were legitimate until Christ forbad them. By the end of the seventeenth century, one legal theorist could state unequivocally that polygamy 'was not contrary to nature or the law of the Old Testament'.[69]

The majority of divines adhered to Christian tradition. Indeed, Calvin's authority rendered the condemnation of polygamy more wholehearted in late sixteenth- and seventeenth-century England than it had generally been in the Middle Ages. Whether harshly or charitably, most early-modern English theorists did condemn the patriarchs' polygamy as sinful. This condemnation was a signal indication of their conviction that polygamy was unreasonable and immoral. Despite the apparent sense of the Old Testament, most early-modern theorists proscribed polygamy. Even their perception of the natural facts of fertility, the example of Abraham, and the opinion of Augustine could not undermine the majority of thinkers' conviction that monogamy was the only natural and reasonable form of marriage. The privileged status of monogamy was riveted into Western Christian thought. Males and women alike must be held to taking only one spouse at a time. Once again, it was in the area of sexual relations that early-modern theory upheld sexual equality.

Notes

1. Towerson 1676, 206. For example, when John Knewstubs (1584, 110) complained that 'the adulterer depriveth a man of the comfort of his true and natural seed and posterity', he clearly did not mean an unfaithful husband, but the lover of a married woman. My text follows sixteenth- and seventeenth-century usage.
2. Herbert 1648a, sig. Y1r [scorned]; *A discourse of the married and single life* 1621, 69 [silly]; *Aristotle's masterpiece* 1694, 64 [forked]. See Ingram (1987, 249–75) on the legal prosecution of adultery. For the popular contempt for cuckolds, see Alleman 1942, 116–18; Underdown in Fletcher and Stevenson 1985, 127–9; Stone 1990, 210–12.
3. Polygamy and a double standard on adultery are almost necessarily connected. It is difficult to see how a man can be thought to commit a moral offence against his wife by simply copulating with another woman, when he does not do so by also marrying the second.
    I deal with the topic of polygamy on pp. 150–66.
4. For early-modern discussions of Jewish and Roman practice, see, for example,

Saravia 1593, 13; Thorndike 1844–56, IV, 286; Wood 1712, 300–1. See Corbett 1930, 133ff. and Brundage 1987, 10–55, for brief introductions to the legal double standard of the ancient world.

5. Some sanguine males did try to overturn Christian tradition and revise the definition of adultery in their own sex's favour. *A letter to a member of Parliament* (1675, 5) stated:

> Adultery is the lying of a single or married man with another man's wife, and not the lying of a married man with a single woman. Thus it was constantly apprehended among the Jews to whom God gave the law, *Thou shalt not commit adultery.*

Since the author was attempting to have the death penalty for adultery introduced and since it was doubtful that most Restoration MPs were chaster than their king, it was probably wise of him to attempt this change in the generally accepted definition if he wished to have any success. (He did not succeed.)

6. Augustine 1955, 55–132; Rubeis 1599, 18 ('communiter receptam'). In E.C. McLaughlin's phrasing (in Ruether 1974, 258): 'The medieval Church insisted in theory on a true mutuality and moral equivalence between the sexes in the area of sexual fidelity within marriage'. McLaughlin's essay provides an excellent introduction to medieval theological attitudes to women. Denis de Rougemont 1983 has argued that adulterous love was implicitly and mystically upheld in the medieval poems of courtly love, but it found no explicit theoretical defence.

7. Harrington 1528, sig. D3v [sharp]; Brenz 1566, 278; Huygens 1698, 394–5. For Catholic condemnations of protestant clergy for lack of sexual restraint, see Kellison 1605, 330–2; Frarinus 1566, sig. F2v; for Reformers' condemnation of husbands' infidelity, see Bucer 1972, 411–13; Latimer 1844–5, I (1844), 243–4; Bullinger 1849, 403; Beza 1591, 151 ff. See also Ozment 1983, 55–7.

8. Elton 1625, 262 [strong]; Gouge 1622, 219 [excel]; Raulin 1512, f. 29r ('magis teneantur', 'sapientiores', 'caput') [see Augustine 1955, 108–9, for the origin of Raulin's view]; Monner 1561, 118 ('Maior prudentia et animi magnitudo' 'infirmiori'). For the history of this view, see Aquinas 1872–82, III (1882), cols. 1182–3 (Suppl. 62.4); Gerson 1606, IV, 323. Willet (1614, 71–2) ever the good Thomist, repeated Aquinas's arguments unchanged. Compare Dod and Hinde 1614, 29.

9. Spalding 1992, 100; Durston 1989, 152–3; Goodman 1629, 259; Elton 1625, 262–3.

10. Baxter 1830, 161 [equal]; Allestree 1683, 303–04; Hopkins 1701, 204, 206 [burn]; Wood 1712, 300–1. For similar condemnations of adultery by either party, see for example, Willet 1608, 395–6; Rainolds 1609, 41; Perkins 1618, 690; Gouge 1622, 219; Mayer 1622, 57–9; Dod and Cleaver 1629, 206 [= 260]; Wemyss 1633, II, 140–3; Byfield 1637, 596 [= sig. Fff1v]; Lawson 1659, 205; Towerson 1676, 406; *Conjugium languens* 1700, 4.

11. Bellarmine 1602–3, III (1603), col. 1353 (I.XVII 'impares videantur'); Gouge 1622, 219 [tenor]; Towerson 1676, 406; Willet 1614, 71–2; Baxter 1830, 161 [abuse]. Others (consciously or unconsciously) distorted the Old Testament in order to make it appear that 'adulterous husbands' were subjected to equal penalties under the Judicial law – Gouge 1622, 219. See also Barker 1624, 263.

12. Willet 1612, 14–15 [Willet's statement was, of course, false; the Old Testament afforded ample warrant for punishing adulteresses capitally, whilst ignoring a husband's infidelity]; Swinburne 1686, 229–31 [unchaste].

13. Niccholes 1809, 274 [pommel]; Carter 1627, 6 [first]; Vaughan 1600, sig. N4r–v [breaketh]; *Fifteen real comforts* 1683, 31 [page]; Sordi 1602, f. 213r ('et periniquum'; 'virorum adulterium', 'impudicas uxores').

14. Ames 1639, 211; Dove 1601, 30. See also, for example, Bartolus 1570, f. 24v; Arelat 1536, f. 12v; Covarruvias 1597, 222, 233 (I.II.7–2.4., 7–6.5.); Rubeis 1599, 44; Sordi 1602, f. 231r; Gentili 1614, 519; Pascale 1618, 67.

15. Biderman 1621, 597. For civil law on criminal proceedings for adultery, see Beust 1597, 131; Mauser in ibid., 395; Schneidewein 1585, sig. O3r. For 'non ad imparia' and 'correlativa', see, for example, Arelat 1536, (Bertrandi) f. 119r, (Arelat) f. 6r; Benincasa 1562, f. 121r–v; Nevizanus 1570, 323; Pascale 1618, 70, 333; Sordi 1602, f. 26v, 45r, 212v. For English common law, see Stone 1990, 209; but see also ibid., 267, for the implementation of the law.

16. Whately 1624, 37; *Reasons for the passing* 1699, 9.

17. See Chapter 5, Sex and marriage.

18. Artificial insemination would not have been possible, given the absence of early-modern sperm banks, and the Blessed Virgin Mary was considered a unique case.

19. Hacket 1607, 16–17 [fires]; Brydall 1703, 26–7 [harlots]; *Fifteen real comforts* 1683, 37 [Christmas-box]; Nicholls 1701, 106–7; Trapp 1662, 274 [posterity].

20. Brydall 1703, 85–6 [cart-way]; Nevizanus 1570, 45; Chamberlayne 1673, 333 [forth]. Thomas Wood (1712, 59) pointed out that the law did not apply in impossible cases, for example, 'if the husband be not eight years old'. See also Smith 1982, 132; Mainwaring 1673, 43. Pufendorf (1934, 857, VI.I.10) volunteered his opinion that these laws were 'too favourable to women', while Lawrence (1680, 77) complained that 'entails on marriage destroy patriarchy and introduce into families hierarchy, gynarchy, and paedarchy'. On civil law, see Bartolus 1570, VI, f. 169r; Pascale 1618, 160. See Baker 1971, 146–61, for a more detailed description of entails and the law of inheritance; and Chapter 7, Divorce, for laws on divorce.

21. *Reflexions on marriage* 1673, 38 [channels]; Brydall 1703, 127 [That Brydall was here citing Ulpian, indicates how long this argument had been thought persuasive]. See also, Stone 1990, 242–3.

22. Aquinas 1872–82, III (1882), col. 1183 (Suppl. 62.4.); e.g. Gerson 1606, IV, 323; Settala 1626, 119 ('magis convenit ipsi mulieri'); Busenbaum 1663, 168 ('incommoda graviora, v.g. damnum veri haeredis, prolis incertitudinem'). See also Rubeis 1599, 29; Olizarovius 1651, 65.

23. Bullinger 1541, f. 39r–v; Poole 1696, I, sig. Dd2v. See also Beard 1612, 386; Gentili 1614, 21; Barker 1624, 259; Dod and Cleaver 1629, 262; Wemyss 1633, II, 142; Herbert 1648a, sig. Y1r; Lawson 1659, 206; Durham 1675, 353; Towerson 1676, 406–7; Beverland 1698, 81; Whaley 1698, 22–4; Huygens 1698, 397; Willet 1614, 72; Thorndike 1844–56, IV, 319; *Reasons for the passing* 1699, 9; *Conjugium languens* 1700, 16.

24. Howell 1809–28, XIII, 1358, 1361–2; *Fifteen real comforts* 1683, 35 [noseless].

25. Sordi 1602, f. 223r ('aequale est peccatum'); Fleetwood 1716, 253. Monner 1561, 121 used similar arguments. For the thesis that the double standard grew not fundamentally from concern about spurious issue, but from 'the view that men have property in women', see Thomas 1959, 209–13.

26. Pufendorf 1934, 871–4 (VI.I.17–18) [The same tendency is detectable in Selden (1991, 334–65, III.1–17)]. Kelly (1984, 42) has argued that 'the norm of female chastity' gained new importance in the Renaissance, and Wayne (in Hannay

1985, 24) that the 'one good [of chastity] had become the only good'. In contrast, Thomas (1959, 203) has pointed out that in the medieval church, 'the idea that unchastity was as much a sin for the one sex as for the other, steadily gained ground, and with the Reformation the attack on the double standard grew stronger'. Each view highlights different sides of the same coin. During the Renaissance, secular learning grew in importance and it rationalistically stressed the practical consequences of infidelity, which were perceived as more damaging when committed by the wife. In response, Christian theorists more forcefully reiterated the traditional arguments that adultery was primarily a moral offence against matrimonial vows, and in this respect, each spouse sinned equally.

27. See Ingram 1987, 249–75, for an introduction to the practice of the ecclesiastical courts in adultery cases.

28. This discussion will follow early-modern usage, despite its technical inaccuracy.

29. Lagus 1566, 405 ('semper detestata'); Settala 1626, 52 ('apud Turcas et Barbaras fere omnes nationes'). For Roman monogamous practice, see Corbett 1930, 143 (although it should be noted that Roman society saw little or no objection to a husband keeping concubines). Many other theorists noted that polygamy was common in other parts of the world; see, for example, Raulin 1512, f. 31v; Campanella 1968, 160–2; Poole 1696, II, sig. Bbbb4r.

30. Pocock 1870, 458–9. See Catharinus 1551–2, 168, for an attack on Cajetan's views.

31. Stockwood in Brenz 1584, sig. C3r; Ochino 1657; Lyser 1682. See Eells 1924, for Bucer's views. Luther was often accused by Catholics (e.g. Kellison 1605, 329; Bellarmine 1602–3, III (1603), col. 1323, I.X) of supporting polygamy on the basis of his comments in *The Babylonian captivity* (1959, 3–126). However, Luther never unambiguously defended polygamy in print. During the English Civil War, radical sectarians were accused of endorsing polygamy (Pagitt 1646, 12), but this may well have been slanderous vituperation. John Cairncross 1974, provides an entertaining account of these and other polygamist writings for the general reader.

32. This argument did not convince John Butler (1698, 21): 'it must not be in a natural sense, that a man can be [one] with his wife, but in a mystical; and if so, then why not with two or ten, as well as with one'?

33. Dugard 1673, 42–3. See Brenz 1566, 613; Bellarmine 1602–3, III (1603), col. 1325 (I.X); Campanella 1968, 164, for exactly the same argument. As Thorndike (1844–56, IV, 358) argued, 'polygamy and divorce being matters of so near kin that the one cannot be imagined to have been allowed when the other was not'.

34. Medieval theologians who pointed out that polygamy promoted procreation in a way that polyandry does not, include Aquinas 1872–82, III (1882), col. 1194 (Suppl. 65.1); Alexander 1948, IV, 361; Albertus Magnus 1890–9, 30, 294–7 (XXXII.3); Bonaventure 1889, IV, 748–50 (XXXIII.I.I).

35. Albertus Magnus 1596, sig. I2v, K5v–6v, M7v; Crooke 1615, 234; Chartier 1596, 10 [let]; Crooke 1615, 233; Primrose 1655, 273–4 ('gravidis uteri os ita clauditur'); Aristotle (1984, I, 1197, 'Generation of animals', 773b). For the infertility of whores, see, for instance, Raulin 1512, f. 7v, 32r; Mercuriale 1591, 13; Nevizanus 1570, 253; Campanella 1968, 158; Dubois 1596, 83; Settala 1626, 58. For this belief in the Middle Ages, see: Albertus Magnus 1890–9, 30, 258, 294–5 (XXXIII.3); Duns Scotus 1639, IX, 704, 710 (XXXIII.1); Gerson 1606, IV, 360. See

Mercuriale (1591, 36–7) and *Aristotle's masterpiece* (1694, 125) for the argument that sexual relations are not dangerous to the foetus during some stages of pregnancy; Settala (1626, 168), Rueff (1637, 667–8, 163–5, 191) and Blegny (1676, 12, 18) were less indulgent.

36. Cited in Hair 1972, 94 (Case 184); *Concubinage and poligamy disprov'd* 1698, sig. A6r [destructive]; Pettus 1674, 82 [vitiated]; *Conjugium languens* 1700, 9 [paths]. That frequent copulation induces sterility, see also Blegny 1676, 12; Lawrence 1680, 153; *The batchelor's directory* 1694, 101; Huygens 1696, 31–2, and 1698, 376–7; Wood 1712, 303–34.

37. Hay 1967, 153; Chartier 1596, 127 [menstrual]. For 'corruption' of milk, see also Mercuriale 1591, 94; Settala 1626, 167–8; *Aristotle's masterpiece* 1694, 155; Harris 1990a, 613; Mack 1992, 37.

38. Hay 1967, 91–5; Maldonado 1677, 431 ('impeditur', 'vir natura sua'); Hickey in Duns Scotus 1639, IX, 704 (XXXIII. I. 4 'a pluribus foecundari'); Bellarmine 1602–3, III (1603), col. 1327 (I.X in one year); Wemyss 1633, 208 [cross]. For medieval instances of this argument see, for example, Albertus Magnus 1890–9, 30, 294; Aquinas 1872–82, III (1882), cols. 1193–6 (Suppl. 65.1.); Gerson 1606, II, 37; Duns Scotus 1639, IX, 703–6 (XXXIII.I); Vincent 1624, 2236.

39. Schickard 1625, 70 ('nusquam expresse determinatus', 'Rabini'); Vitoria 1991, 171–2. Compare Albertus Magnus 1890–9, 30, 323 (XXXIII.28); Abelard 1855a, col. 1745.

40. *Conjugium languens* 1700, 6 [satyrs]; *Marriage promoted* 1690, 18 [infirm]; Campanella 1968, 168–70 [Mohammed]; Trapp 1662, 34 [Solomon]. A small sample of the assertions that too much sex is unhealthy and produces less or less healthy offspring includes: Albertus Magnus 1596, sig. B7v [sparrows]; Raulin 1512, f. 32r; Nevizanus 1570, 154–5, 214; Chartier 1596, 5–6 [greyness, baldness]; Vegio 1613, 15; *Christian parents* 1616, 28–9; Settala 1626, 65–7, 164 [more debilitating if passionate]; Sclater 1633, 51; Rueff 1637, 56; Drexel 1642, 235 [= K10r]; Rogers 1642, 177–8; *Reflexions on marriage* 1673, 132; *Aristotle's masterpiece* 1694, 10 [weakens male and semen]; *The batchelor's directory* 1694, 93, 100–1; Beverland 1698, 12–13; Huygens 1698, 376–7.

41. Wood 1712, 304. On the rarity of polyandrous societies, see Rubeis 1599, 39; Settala 1626, 56; Lawrence 1680, 11.

42. Harpsfield 1878, 84 [resembles]; Highmore 1651, 102 [bitch]; *Aristotle's masterpiece* 1694, 19–20 [unlawful]. For confusion of blood from second marriages, see Arelat 1536, f. 7v, (Garronibus) f. 99r; Gentili 1606, 124. On the medical science of 'stirring up' seed, compare Blegny 1676, 12. On the importance of a pregnant mother's imagination, see also Highmore 1651, 94.

43. Hay 1967, 95 [care]; Pufendorf 1934, 870, 866; Soto 1619, (IV.III.I 'natura ipsa'); Raulin 1512, f. 22r. The argument from certainty of issue had been used by Aquinas (1872–82, III (1882), col. 1196, Suppl. 65.I.) in his attack on polyandry and Bellarmine 1602–3, III (1603), col. 1327 (I.X); Settala 1626, 58–9, among others repeated it. Many theorists simply repeated Aristotle's arguments against Plato's ideas; see, for example, Freig 1578, 62.

44. Nevizanus 1570, 413 ('pluralitas ciborum'); Settala 1626, 54 ('non frena concupiscentiae' 'irritamenta maiora'); Beza 1587, 19 ('etiam castos alioqui', 'sicut copia cibi gulosus, et copia vini ebriosus facit'); Campanella 1968, 158–60 ('sicut in meretrice'); Baudius 1638, 376 ('non incendendae, sed restinguendae, sedendae, non irritandae'); Raulin 1512, f. 6r ('concupiscentiam in se diminui').

45. Wemyss 1633, II, 178 [diversity]; Poole 1696, II, sig. Fff3v [bridle]; Comber 1679, 30 [imperious]. See also Whately 1624, 26–7; Herbert 1648a, sig. X4r; *Concubinage and poligamy disprov'd* 1698, 76; Sclater 1650, 100.

46. Ochino 1657, 47–8; Pufendorf 1934, 866 (VI. I. 15). For example, also Delfino 1553, 15; Bellarmine 1602–3, III (1603), col. 1327 (I.X).

47. Antoninus 1474, sig. P8v ('propter multa pericula'); Settala 1626, 172 ('aut verbo, aut nutu' 'sponte, et alacriter') [see Campanella 1968, 196 for the same argument]; Herbert 1648a, sig. V7v.

48. Beza 1587, 18; Fonseca 1568, f. 64r [small]; Wemyss 1633, II, 22 [dwell]; Blount 1638, 106–7 [Turks]; *Concubinage and poligamy disprov'd* 1698, 82 [crimes]. Compare Vitoria 1991, 201; Hemmingsen 1566, sig. I7r; Bellarmine 1602–3, III (1603), col. 1328 (I.X); Campanella 1968, 164; *An essay towards* 1697, 24–5.

49. Ochino 1657, 53; Lyser 1682, 21 ('inordinatis et carnalibus affectibus' 'vix decem, vel viginti', 'mulieri, quae semper, si non est praegnans, appetit'); Pufendorf 1934, 870 (VI.I.17). For the increase of female sexual desire during pregnancy, see Albertus Magnus 1596, sig. C3r–v; Mercuriale 1591, 36–7; Crooke 1615, 221.

50. Settala 1626, 57 ('plures mares unam ad actum venereum insequi'); Nevizanus 1570, 56–7 ('nulla vehementior est discordarum causa'); Campanella 1968, 158–60 ('rixae et discordiae', 'primus aut potior'); *An Essay towards* 1697, 189–90 [jealousy].

51. Ochino 1657, 49–50; Huygens 1696, 301 ('contra domesticam pacem'); Vaughan 1706, 228 [plenary]; Hay 1967, 91 [blow].

52. For example, Albertus Magnus 1890–9, 30, 294; Maldonado 1677, 431 ('contra ius naturale'); Pufendorf 1934, 874; Bellarmine 1602–3, III (1603), col. 1328 (I.X); Beza 1587, 15, 19 ('infinitas rixas', 'aemulatione', 'nulla tranquillitas').

53. Pufendorf 1934, 874 (VI. I. 20); Turner 1685, 45–6; Raulin 1512, f. 21v–22r; Jansen 1641, 37; Nevizanus 1570, 75 ('quod lex Moysi, est lex puerorum: Lex Christi, lex impossibilium: Lex Mahumeti, lex porcorum'); *Concubinage and poligamy disprov'd* 1698, 9 [experience]. See also Hemmingsen 1566, sig. I7v; Allen 1600, 196; Whately 1647, 133; White 1656, II, 105; Comber 1679, 49–50; *An Essay towards* 1697, 23. This belief again had medieval antecedents: e.g. Gerson 1606, IV, 360; Albertus Magnus 1890–9, 30, 293; Baconthorp 1526, f. 164. For other condemnations of Turkish/Moslem sexual morality see, for example, Monner 1561, 14; Baynes 1866, 351; Wemyss 1633, II, 182; Archdekin 1679, 185; Comber 1679, 52.

54. For faithful repetitions of Aquinas's position, see Covarruvias 1597, 224 (I.II.7–3.2.); Campanella 1968, 168; Bellarmine 1602–3, III (1603), col. 1333 (I.XI).

55. Poole 1696, II, sig. Ff2v. For the argument from Adam and Eve, see Tapper 1582, 297–8; Mauser in Beust 1597, 328; Campanella 1968, 162; Bellarmine 1602–3, III (1603), col. 1328 (I.X); Maldonado 1677, 421; Niccholes 1809, 255; Willet 1605, 189; Duns Scotus 1639, IX, 703 (XXXIII.I.2); Carter 1627, 2; Wemyss 1633, II, 173; Richardson 1655, sig. E4v; White 1656, II, 105; *Marriage promoted* 1690, 3; Patrick 1695, 57; *An Essay towards* 1697, 104.

Scriptural support was also martialled by pointing to Noah's Ark, where God had commanded that one of each sex of animal (not more females than males) should be placed aboard. It was often asserted [although the text – Genesis 7:6 – is not in fact specific in the case of the sons] that Noah and his sons only took one wife each; – see, Agrippa 1540, sig. B4v; Beust 1597, 102; Barker 1624, 262;

Fleetwood 1716, 260. Discussions on whether there were equal or unequal numbers of males and women born can be found in Raulin 1512, f. 21v; Lotichius 1630, 57; Bodin 1962, 562; Ochino 1657, 62; Lawrence 1680, 5; *Concubinage and poligamy disprov'd* 1698, 77–9.

56. Comber 1679, 52 [virtuous]; Cock 1668, 160 ('sapientiores inter gentiles'); Hemmingsen 1566, sig. I7v ('debet una contentus est'); Needler 1655, 193; *Reflexions on marriage* 1673, 28–9. On polygamous societies, see also Bodin 1962, 558–62; Lawrence 1680, 11; Laymann 1625, 430 (X.III.VI.7); Wemyss 1633, I, 207. On natural equity, see Vincent of Beauvais 1624, 2236, (I.30.32); Gerson 1606, II, 37; Duns Scotus 1639, IX, 703–4; Bonaventure 1889, 750; Bellarmine 1602–3, III (1603), cols. 1326–7 (I.X); Kellison 1605, 329.

57. Delfino 1553, 14–15 ('valoris aequalitatis', 'commutativa iustitia'); Wemyss 1633, I, 271; Beza 1587, 21; Lancelottus 1564, f. 44v; Hemmingsen 1566, sig. I7r ('aequalem obligationem'); Bellarmine 1602–3, III (1603), col. 1328 (I.X 'iniquitas').

58. Campanella 1968, 162 ('praesupponit ... nec reprehendit'). For the conventional view on Lamech's bigamy, see Lorenzo da Brindisi 1935, 413; Carter 1627, 3; Bentley 1582, 11; Richardson 1655, sig. B3v; Hughes 1672, 58; and see Lyser 1682, 185–6 for a polygamist's response. For the acceptance of polygamy as legitimate, see also Genesis 4:19; 21:10; 22:24; 30:3,9; Deuteronomy 22:19; 1 Samuel 1:2,6; 25:43; 2 Chronicles 24:3. For the belief that Leviticus 18:18 forbad polygamy, see Perkins 1618, 677; Wemyss 1633, 271; Trapp 1662, 19, 329; Turner 1686b, 43–4; on the problems of using it in this way, see Kidder 1694, II, 106. For Malachi as a text against polygamy, see *Christian parents* 1616, 25. On the use of Nathan's words, see Wemyss 1633, 274; Abbot 1653, 27.

59. Kidder 1694, II, 106 [reproof]; *An essay towards* 1697, 21 [Jacob]; Campanella 1968, 162; Needler 1655, 189–90 [unsavoury]; Tapper 1582, 319 ('turpia et abominabilia').

60. Willet 1605, 189; Hall 1837, IX, 93 [Jerome]; Augustine 1977, 333, 325. Compare Bellarmine 1602–3, III (1603), cols. 1329–32 (I.XI); Durandus 1571 III, f. 376v–377r; Abelard 1855a, col. 1745; Hugh of Saint Victor 1951, 342.

61. Maldonado 1677, 435 ('licuisse omnibus iure communi'). Melanchthon argued that the patriarchs' example proved polygamy was not unnatural in Bretschneider 1835, II, 526 (no. 1000). However, he seems to have changed his mind on this matter, since later he condemned their conduct and argued that it was only 'tolerated' ('toleravit') by God; see 'De Coniugio' (Melanchthon 1834–60, XXI, 1059). For Bucer's views, see Eells 1924, 78 and 101.

62. A dispensation was 'a gracious releasing to some certain person or persons of the common written law' (Harpsfield 1878, 129).

63. Albertus Magnus 1890–9, 30, 297–9 (XXXIII.5–7); Aquinas 1872–82, III (1882), cols. 1196–98 (Suppl. 65.2.); Bonaventure 1889, IV, 751 (XXXII.I.III); Gerson 1606, II, 37; Duns Scotus 1639, IX, 705 (XXXIII. I.4); Vincent 1624, 2236. Protestant examples include Chemnitz 1978, 730–1 [although his views were ambiguous]; Latimer 1844–5, I (1844), 94, 113; Monner 1561, 12–13; Brenz 1566, 612 and 1584 sig. D6v–7r; Hunnius 1604, 59–60; Thorndike 1844–56, V, 206. Roman Catholics included Catharinus 1551–2, 266; Vitoria 1991, 201; Tapper 1582, 299; Covarruvias 1597, 224 (I.II.7–3.2); Campanella 1968, 168; Azor 1600, 634–5 (I.VI.xiii); Biderman 1621, 50; Becan 1625, 392; Laymann 1625, 430 (X.III.VI.7–8); Suarez 1872, 139–41 (II.XV.18–19).

64. *Certain sermons or homilies* 1756, 323; Thorndike 1844–56, IV, 283–8; ibid., V,

206, 566. For complaints against the doctrine taught by *Certain sermons or homilies* see Howson 1606, In controversiam, 64.

65. Maldonado 1677, 436, ('At non legimus in tota scriptura'); Hall 1837, IX, 93 [conceits]; Cleaver 1598, 242 [notice]; Harpsfield 1878, 162 [especially]. The Thomist, John Capreolus (1967, VI, 529), for example, stated that the patriarchs' dispensation was given first through 'internal inspiration', and secondarily deduced by other faithful men from 'the example of the holy fathers' ('per internam inspirationem', 'per exemplum sanctorum patrum'). Calvinists were more sceptical; William Ames (1639, 199), argued that 'because there is nothing in Scripture manifested unto us of any such dispensation, we cannot affirm anything for certain'.

66. Calvin 1948, 423–4; Jansen, 1641, 65 ('spiritu Manichaeo'), [the Manichaeans had accused Abraham of taking Hagar from uncontrolled sexual desire – see Bellarmine 1602–3, III (1603), col. 1331, I.XI]; Brenz 1584, sig. C3r.

67. Stock 1865, 181–2 [impious]; Willet 1605, 190 [sounder]; Ames 1968, 318; Gaule 1630, 54–5 [expressly]; Bunny 1610, sig. *3v [swerved]. Other examples of those holding that polygamy was simply sinful, even in the patriarchs, include Bullinger 1541, sig. A7a; Beza 1587, 7, 12–13; Hotman 1594, 229; Wilson 1615, 134; *Christian parents* 1616, 24; Perkins 1618, 671; Gouge 1622, 115; Elton 1625, 263–4; Carter 1627, 5; Bernard 1845, 109; Griffith 1633, 233: Whately 1647, 119; Abbot 1653, 27; Needler 1655, 189–90 [Needler's cryptic comment that it 'may be placed in the middle between adultery and holy wedlock' is apparently copied from Perkins 1617, 302]; Richardson 1655, sig. C4v; Trapp 1662 , 81, 329; Hughes 1672, 58; *The Batchelor's directory* 1694, 77 [second so numbered]; Kidder 1694, 2.

68. Dove 1601, 18 [permitted]; Wemyss 1633, I, 271–4 [increase]; Taylor 1828, XII, 264 [II.I.ix mind]; Turner 1686b, 43–4 [prudence]. The same view that polygamy was permitted, not dispensed, was adopted by (amongst many others) Beust 1597, 104; Rogers 1642, 165; White 1656, II, 105; Comber 1679, 51; Corbet 1685, 226; *An essay towards* 1697, 10–11; Hopkins 1701, 205. On permission as the allowance of a lesser evil, see Howson 1606, In controversiam, 18.

69. Selden 1991, 73–87 (I.8–9); Pufendorf 1934, 868–74 (VI.I.16–19); Grotius 1689, 239 (II.V.IX); Wood 1712, 52. Grotius's view was attacked by various authors, see, for example, Felden 1653, 130–2; Lawrence 1680, 149; Turner 1698, 20.

7

Subjection and consent

Marriage and consent

The previous chapters outlined early-modern beliefs on the reasons for woman's subjection – its nature, extent, and limits. Now we must discuss *how* women became subject to males. The simple answer is by marriage. Virtually nobody believed every woman was naturally subject to every male, but almost everyone believed wives were subject to their husbands. Even a queen regnant was subject to her husband in conjugal matters. A wife became subject to her husband when she consented to marry him. It is vital to understand the early-modern theory of marriage, because the distinctively *female* form of subjection was that of wives. Daughters were subject to their parents, but so were sons. Those theorists who believed that sons' obligation to obey ceased on reaching maturity applied this rule to daughters also. Patriarchalists who insisted that daughters were subject to parents their whole lives, also insisted that sons were. Children's duty to obey father above mother was deduced from her subjection to him as a wife. The obligations of servants were taught without regard to their sex. Male and female citizens must obey the male or female prince. The sexually differentiated form of power was the husband's authority over his wife.

In early-modern theory, when a woman married she accepted subjection to her husband. For a woman, marriage entailed obedience. If, and only if, she consented to marriage was she any more subject than the next man. Before she married, a woman 'is then solely of herself, and free from the command of any, and hath free liberty to do then for herself what she will'. Wives were subject to their husbands, but adult spinsters were independent. God 'puts this burden of subjection upon no woman, who takes not the yoke of subjection upon her self; which the Lord doth force upon none, but allows each woman to be her own refuser, and to choose for herself (if she can) such a man as she can yield subjection unto'. Only by consent to marriage did a woman become subject, and 'nobody' could be forced to marry.

Canon, civil, and common lawyers, protestant and Catholic theologians, Anglican and puritan divines – all insisted that the consent of both parties was necessary to a valid marriage. Indeed, consent to marry *was* marriage. The standard English textbook of marriage law insisted that 'nothing but a full, free, and mutual consent is the essence of matrimony', whatever form that consent took. Canon lawyers declared 'a present and perfect consent ... alone maketh matrimony'; and nothing was 'of the essence of matrimony, but consent only'. The belief that consent alone, and only consent, made marriage was inherited from Catholic theology and canon law. Early-modern Catholic casuists continued to maintain that 'wholly free and voluntary' consent was necessary to marriage. In Catholic eyes marriage was a sacrament, but this did not undermine the centrality of consent; the substance and form of matrimony was 'mutual consent', and the minister of the sacrament was 'not the priest but the contractors themselves', because they provided the substance and form. The whole essence of marriage lay in the 'expression of mutual consent'.[2]

For many centuries, mutual consent (however private) was sufficient for a marriage to be legally valid in canon law, even though one party could not enforce it against the other in court unless it was witnessed. Only after the promulgation of the decrees of the Council of Trent (1563) was the presence of a priest and witnesses requisite to a valid marriage. So established was the belief that marriage was made by consent alone that some theologians doubted the Church's authority to institute this 'reform'. English law was not reformed in this way until Hardwicke's Marriage Act of 1753.[3] Until then, marriage arose simply 'by express and free consent of both parties'. The protestant casuist, William Perkins, adopted the same formula, writing of 'the free and full consent of the parties, which is indeed the very soul and life of the contract'. A couple became husband and wife the moment they agreed to be so, for the 'efficient cause' of marriage was 'the consent of both parties'. The Catholic apostate to Anglicanism, Marc' Antonio de Dominis, stressed the civil magistrate's power over marriage, but he, too, accepted that marriage was nothing without consent. Men were naturally free and 'nobody should be subjected to another by marriage unless by free and voluntary consent'. John Locke followed in a long tradition when he described 'conjugal society' as 'made by a voluntary compact between man and woman'. Without regard to nationality or religious persuasion, early-modern theorists accepted that consent made marriage, and that no marriage could exist without consent.[4]

The need for free consent to marriage is now so widely accepted in the Western world that it seems almost redundant to insist upon it. This was not always the case. In ancient Rome, children could be married by the *paterfamilias* without their consent, although this changed later. A daughter's consent was not originally needed for marriage among the Germanic tribes; authority over her was simply transferred from father to husband. The Old Testament contained many examples of daughters being handed over by their fathers without the slightest evidence that the brides had any say in the matter. The Christian West's insistence on the need for children's (even daughters') consent was novel.[5]

There were both positive and negative reasons why Christian theorists insisted that consent was necessary to marriage. In positive terms, lawyers and theologians reasoned that compulsion 'frustrated the ends of marriage'. A couple compelled to marry would be unlikely to have children or to rear them well if they did; nor would marriage probably avert adultery where the spouses were antagonistic. Unwilling marriage led to 'evil results'. The necessity for free consent was also maintained in order to *exclude* certain possibilities. First, early-modern theorists wished to insist that it was consent – not sexual relations – that constituted marriage. A couple who copulated without intending to marry were not husband and wife; those who had consented were married, even if they remained chaste. Second, no one else could consent for either party; in particular, parents or guardians could not consent on their children's behalf. Indeed 'not even a pope' or a king could marry two people, if they themselves did not freely consent. Third, the consent must be free; violence, duress, or the fear of harm invalidated marriage.[6]

The belief that consent makes marriage had a long history in Western thought. Civil law had laid down the principle: *'Nuptias enim non concubitus sed consensus facit'* [For not copulation but consent makes marriage]. This principle was authoritatively established in Catholic canon law by Pope Alexander III (1159–1181). Church fathers and medieval theologians alike insisted on the necessity for consent and the immateriality of coitus.[7] Theologians were so insistent upon this point partly because the Blessed *Virgin* Mary and Saint Joseph were piously believed never to have had sexual relations. If coition were necessary to true marriage then Jesus Christ's earthly parents would have been unmarried – an embarrassing idea to early-modern minds. There was 'just and lawful matrimony' between Mary and Joseph 'to show that matrimony is the state which pleaseth God'. Canon lawyers also stressed consent over coition because it was easier (and more decorous) to prove in a court of law that mutual consent had taken place than that sexual relations had.[8]

In keeping with centuries of canon law, English writers upheld the belief that consent was essential to marriage but that sexual relations were not. Thomas Cranmer seriously disputed much of the traditional canon law of marriage in his attempt to prove that Henry VIII could rightfully divorce Katherine and marry Anne, but he, too, accepted that the substance of matrimony was brought about only by the conjugal compact 'and not by carnal copulation'. Mary Tudor's champion, Nicholas Harpsfield, disagreed entirely with Cranmer on Henry's divorce, but agreed that it was a rule

> not only of the Civil Law ... but the sure, constant, and certain rule of all the Fathers and Councils, of all the divines and canonists, received of the whole Church, that the consent and not the carnal copulation maketh the matrimony.

Although the question of whether Katherine had sexual relations with Arthur featured prominently in the debates over Henry's divorce, neither side believed them crucial to the existence of a valid marriage. 'For all reached accord' on the immateriality of coitus. Seventeenth-century Protestants, too, accepted that husband and wife were 'one flesh' without 'carnal copulation'. The radical protestant divine, John Rainolds, did not often stress his agreement with the papacy but he wrote that 'consent not carnal company maketh marriage, as the Civil lawyers, Fathers, and Popes do teach'. English matrimonial lawyers did not differ in their assessment: consent, not 'lying together' made marriage.[9]

Despite the fact that law and theology were unequivocal on this point, many early-modern laymen seem to have thought that copulation was necessary to a valid marriage. The plots of many Restoration dramas hinged on the belief that an unconsummated marriage was not binding. This belief was based on a misunderstanding of some of the finer distinctions of canon law. The simple doctrine that consent made marriage was complicated by the principle that sexual relations 'consummated' the marriage contract. The couple were married from the moment they took their vows, but the marriage was not perfect 'until there go with consent of mind and will, conjunction of body'. In Roman Catholic canon law, a consummated marriage was indissoluble under any circumstances. In contrast, in a few cases, the church might allow a couple who had consented to marry but had not consummated the union to go their separate ways.[10]

In English ecclesiastical law, consummation could be crucial in deciding whether a marriage was valid. This was the case in the marriages of children. Consent to spousals[11] or marriage could be given by a child of 7 years, but the consent was not binding unless confirmed by girls at 12 or by boys at 14 years of age. By legal convention these were the ages of puberty. If a child withdrew consent before the marriage was actually consummated, it could be adjudged null. But after 'consummation, every man knoweth that – albeit the matrimony had been before for lack of years not valuable – that yet thereby it commeth and is made perfect and of full force and valor'. Cases of under-age marriage were hardly a daily occurrence, but this was an instance where sexual relations were crucial to determining whether the bond of marriage existed.[12]

Sexual relations were also crucial in establishing a marriage in the case of spousals *de futuro*. If a betrothed couple had sexual relations, their engagement was converted into marriage. A betrothed couple's consent to coition was 'always presupposed' to amount to consent to marriage. Theologians and lawyers were aware that, in practice, this might not be true, but maintained that it was the charitable interpretation. A betrothed couple who copulated were either marrying, or agreeing to commit the terrible sin of fornication. Surely the generous view was that it was the former rather than the latter. Protestants and Catholics alike held that mere betrothal became marriage by exhibiting any external sign of marital affection, 'for example, copulation'. Lawyers declared 'spousals *de futuro* do become matrimony by carnal knowledge'. Consent, not coition, made marriage, but under certain specific circumstances coition was regarded as equivalent to consent.[13]

Parental consent

Early-modern theorists also argued that consent to marriage must be given by the couple themselves; neither their parents nor anyone else could consent on their behalf. Johannes Lyser complained how women often kept not only the suitor but all their relatives waiting for months because 'they want to be adored before they give consent'. Professor Stone has pointed out that, in practice, wealthy parents often had the power to persuade their children into distasteful matches, noting the '[a]uthoritarian control by parents over the marriages of their children'. In some cases, the combination of physical and economic pressure undoubtedly made it difficult for children to resist parents' wishes. At the theoretical level, however, moralists and lawyers were absolutely insistent that parents could neither consent on their children's behalf, nor use coercion to compel their sons or daughters to consent. This had been the position of medieval theologians and canon lawyers and was accepted by early-modern theorists, both protestant and Catholic.[14]

Catholics and Protestants did disagree on whether the marriage of a (minor) child without parental consent was valid, or void *ab initio*. This debate had rumbled on throughout the Christian Church's history. Catholic clergy had long defended sons' and daughters' freedom validly to contract (or not contract) marriage as a correlative of their assertion that children were free to enter the religious life. Most Catholic theologians maintained that marriage without parental approval was sinful and irregular (for children should obtain their parents' approval), but that it *was* valid.[15] In contrast, laymen in general, and Civil lawyers in particular stressed parental rights in these areas. At the Reformation, Protestants – initially reacting against clerical pretensions and eager for the assistance of lay magistrates – adopted the civil lawyers' stance. Most Reformed divines insisted that marriage without parental consent was a mere nullity, although some qualified this view if the marriage were consummated. In English post-Reformation law, marriage without parental consent (except in the cases of the forcible seizure of an heiress, or the marriage of a clergyman) was valid. Most English divines disapproved of this state of affairs and insisted on the need for parental consent. Theorists argued for the same conclusion regardless of the sex of the child or of the parent.[16]

Catholic theologians denied that the absence of either parent's consent rendered a marriage null. Roman law had invalidated the marriage of a child without a father's consent, but had given the mother no power. Protestant commentators on the civil law explicitly disagreed with Roman theory and argued that the mother's consent should be obtained. They were uncomfortable about arguing that a mother could absolutely forbid (particularly) a son's marriage, especially when she was a widow who had remarried, but in general carried their interpretation of the fifth commandment to its logical conclusion and insisted on maternal rights. Protestants granted, and Catholics denied, to father and mother the same rights in their children's marriages.[17] The great majority of early-modern theorists were also impartial in relation to the sex of the child. Roman Catholics stated that a daughter's marriage without her parents' consent was valid; Protestants maintained that a son's marriage without their consent was not. One protestant lawyer was typical in declaring that in relation to parental consent 'there is in divine law no distinction between sons and daughters'. Both Catholics and Protestants tended to be more troubled by the case of the disobedient daughter, but there was generally no legal–theological double standard.[18]

Roman Catholic casuists and canon lawyers insisted that a daughter's marriage without her parents' consent was valid and that neither she nor her husband owed the father compensation. They maintained that no daughter could be disinherited simply for marrying without parental consent, and any local secular statutes that allowed the father to disinherit her on such grounds 'do not have the force of law'. A father was obliged to endow his daughter – even when she married without his consent – unless she was both under 25 years of age and also had married a husband of disgraceful common reputation. They impartially applied this rule to a son. A father was bound to maintain his children even when they married against his will, but this was not so when, disregarding the father's consent, 'the son married a shameful wife or the daughter an unworthy husband'.[19]

Protestant theorists stressed the duty of both sons and daughters to obey parents who arranged a marriage on their behalf. Theodore Beza, for example, argued that children should not withhold their agreement 'without great cause'. English protestant divines acknowledged it was 'the parents' duty to provide matches for their children', and that parents should 'bear this always in mind, that the right and authority to place and bestow their children is given and attributed unto them of God'. The Old Testament demonstrated 'both the obligation that lies upon parents and the right that is vested in them to dispose of their children in marriage'. A child should accept the parents' choice of spouse unless there were compelling reasons against it, for 'it is the duty of children to submit themselves to their parents in their matches and marriages'.[20] The protestant casuist, Jeremy Taylor, followed the standard civil lawyers' description of a son's and daughter's duties when their parents arranged

a marriage. A child could only legitimately dissent when the parent proposed 'a dishonest or filthy person, unequal, or unfit; that is when it is
notoriously or scandalously so'. Either could refuse a shameful match,
for: 'Son and daughter in this have equal right'. He did qualify this by
adding that a son could refuse an ugly woman, whereas a daughter
could not reject an ugly male, as a son was 'more easily tempted and can
sooner be drawn aside to wander', and because 'the ugliness of a woman
will sooner pass into an incapacity of person than it can do in a man'.[21]

The sixteenth-century German jurist, Conrad Mauser, had used the
same categories and terms and reached much the same conclusions.
However, Mauser was somewhat more generous to the daughter than
Taylor, leaning to the opinion that she could refuse 'even a worthy man'.
Her opinion was more important than her father's, because she, not he,
had to endure 'all the burdens and dangers of marriage'. She should
never be 'married against her will'. Across the length and breadth of
Europe, early-modern theorists argued that parents' agreement could
not bind children to marry against their will, 'on account of the sanctity
of marriage'. This was 'clear by the law of nature'. A daughter should
follow her father's guidance in choosing a husband, but he could not
compel her to marry if she did not want to; the father had 'no power
over his daughter's body' only to 'raise her in the fear of God.' Parents
might make arrangements without their children's explicit consent, but
the child had a right to disagree if there were just grounds. Despite their
strong stress on children's obligation to obey, legal theorists also denied
that parents could compel children's consent. Of course, in practice,
some children were pressed into marriage and in no position afterwards
to enforce their legal right to an annulment. However, legal theory was
clear; parents could not cross the line from persuasion to coercion without rendering the marriage void.[22] Virtually all English divines and
moralists, even the strongest advocates of filial obedience, taught that
parents should not force marriages 'without the consent of the children
themselves'. The child just as much as the parent had a right of veto, a
'negative voice' in marriage. The child's consent was essential to any
valid marriage.[23]

Divines appealed both to Scripture and natural law to support this
limitation on the power of parents. A biblical text often employed was
Genesis 24, where Rebecca's consent was obtained to an arranged marriage with Isaac. The text was readily susceptible of this interpretation
for Rebecca's 'brother and mother ... call the damsel and enquire at her
mouth' (Genesis 24:55–8). Theorists' determination to uphold the need
for a daughter's consent to marriage was itself evidenced by selecting
this untypical text. Yet biblical exegetes also used far-less-promising
material; they argued that God had obtained Adam's and Eve's consent
to their arranged marriage. The text stated only that God 'brought her
unto the man'; Eve did not utter a word (Genesis 2:22). Yet one commen-

tator happily glossed Genesis to the effect that 'a *mutual* consent and con-
gratulation followeth likewise between the parties, lest any one should
tyrannically abuse his fatherly power and force a marriage without love
or liking'. Another maintained that Eve was 'not forced' but 'willingly
obeyeth, and in her coming freely consents to be in that place for which
she was created'. It was true that Adam expressed his whole-hearted
approval of God's design, but Eve's reaction was left unclear.
Nonetheless, commentators continued to insist that 'Eve was not
dragged, but *brought* by God to her husband. There must be a mutual
consent, or it is not of God'.[24]

Standardly also, Saint Paul was interpreted as requiring fathers to
obtain the consent of brides-to-be, even though Paul in the relevant
Scriptural passage (I Corinthians 7) considered 'much more the father's
inclinations towards her single life or marriage than the virgin's'. Indeed,
Paul never discussed anything but the father's will.* Catholic theologians
argued that this text showed how a father 'cannot forbid' his daughter
from marrying if she wants to, and that Saint Paul was asserting that by
divine law a daughter had absolute power to give herself in marriage.
Protestant expositions also stated that Paul was recommending the
father to consider if 'the mind of the virgin be inclinable to marriage',
and 'if his daughter have the gift of continence'. There was no indication
in the text itself that Paul cared in the slightest what the daughter wanted. Divines were so eager to insist on the need for children's consent in
marriage that their Scriptural readings became anything but unadorned
summaries. Just as this text was employed by divines to support the
need for maternal consent to a child's marriage (although Saint Paul
made absolutely no reference to mothers), so it was used to afford a
daughter the right to refuse an unwanted husband, despite its apparent
lack of potential for this purpose. European divines' moral prejudices
indicated that a daughter's inclinations should not be ignored in arrang-
ing marriage, and they appear simply to have read their own beliefs into
Paul's words.[25]

The belief that parents could not compel their children to marry clearly
found acceptance among some of the educated laity. Anne, Lady
Halkett, the very model of a pious and dutiful daughter, recorded in her
memoirs that 'though duty did oblige me not to marry any without my
mother's consent, yet it would not tie me to marry without my own'.
Even Protestants did not believe that parental powers extended to compelling children to marry against their wills. Almost all divines felt
extreme reluctance to let children disagree with their parents on so

---

*Saint Paul taught: 'Nevertheless he that standeth steadfast in his heart, having no necessi-
ty, but hath power over his own will, and hath so decreed in his heart that he will keep his
virgin [daughter], doeth well. So then he that giveth her in marriage doeth well; but he that
giveth her not in marriage doeth better' (1 Corinthians 7:37–8).

important a matter as marriage. They insisted that no child should marry against or without parental consent or knowledge. They often loaded with many qualifications and reservations their conclusion that the child could not be forced. They bemoaned the fact that 'the laws of our nation take no notice of the consent of parents'. Yet the final conclusion of virtually all theorists was that parents could never compel a child to marry against his or her will.[26]

A signal exception to this rule was Thomas Barlow, Bishop of Lincoln, in the later seventeenth century. He discussed a particular case where Roman Catholic ecclesiastical authorities had granted an annulment to a woman who asserted that her consent had been compelled by her father's threats. The woman's father had apparently not only insisted on the marriage, but also 'gave her two blows' and said that 'he would strangle her if she would not have' the suitor. The case was complicated by the fact that the woman had for some time cohabited (apparently willingly) with her 'husband'. Moreover, the woman was no child; at the time of the marriage 'nearer 30 than 12'. Thomas Barlow did not merely argue that the facts suggested that the woman had been under no serious duress. Rather, he maintained that a nullity should not have been granted because 'if the fear arise from the many and severe threats of a father, yet this cannot make the consent involuntary and so a nullity'. In Barlow's opinion, the father had 'a just authority by the law of God and nature ... to use threats and menaces, yea, and castigations and whippings too' to force his children to 'obey his just commands'. Barlow did accept the orthodox view that allowed a daughter to disobey if her father commanded her to marry 'an impious and unworthy person'. Nonetheless, in insisting that consent to marriage was valid when extracted by parental coercion, he was adopting a most unusual position.[27]

Many theorists argued that if the child were merely overawed by parental authority and obeyed from 'reverential fear', the marriage was nevertheless valid. This was the most pressure that the great majority of early-modern divines and lawyers would permit. Any physical coercion, or threats of disinheritance or of marrying the child to a still-less-attractive candidate, were generally deemed to invalidate a marriage. Early-modern theorists were almost unanimous in insisting that parental coercion was incompatible with the free consent required in marriage. Valid marriage required the voluntary consent of bride and groom; their parents could not consent on their behalf, nor compel their child to acquiesce in an arranged marriage.[28]

Coercion and consent

The invalidity of coerced consent to marriage was asserted in all cases, not only against parents. Early-modern theorists consciously followed medieval theologians and canonists in arguing that a marriage was invalid if either party had been induced to consent by force or fear. This belief formed part of a complex philosophy on the validity of coerced contracts. It is impossible to set out all the subtleties of those ideas here but a brief outline will demonstrate the context of the particular case of the marriage contract.[29] Compelled consent was standardly discussed in terms taken from Aristotle's *Ethics*. Aristotle distinguished two types of force or violence. 'Simple' violence was where somebody was entirely overwhelmed by physical force; if a strong person grasped a weak one's hand and manipulated it to sign a document, this would be simple violence. In such a case there would be no sense in which the weak person could be said to have consented to sign the contract. Aristotle contrasted simple with 'mixed' violence, where force or fear induced a person to choose a certain action. If somebody pointed a gun at another's head and posed the alternatives of signing or being shot, this would be 'mixed' violence, because in some limited sense the victim consented to sign. (S)he did so only when faced by an extremely unpleasant alternative, but nonetheless there was an element of voluntary action.[30]

The case of mixed violence posed moral problems to medieval and early-modern theorists because they believed that any voluntary element potentially rendered a promise binding. Promises made in order to avoid injury were 'not simply involuntary and forced, but mixed'; a voluntary element was intrinsic to choosing the lesser evil. The Dutch political philosopher, Hugo Grotius, in his influential *De jure belli ac pacis* argued that (naturally) a promise given under duress did bind, and cited Aristotle in his support. Early-modern theorists often illustrated this belief with the example of someone waylaid on the highway, who promised to pay his or her assailant money if released. Somebody making such a promise (simply in order to save her or his life) was obliged in conscience to keep it: *prima facie* the victim was morally required to pay.[31] At first sight this might seem a strange doctrine, but it was subject to a significant limitation. Namely, that although the contract was not void, it was rescindable by higher authority. In the highwayman case, the released victim could go to the magistrate, explain the circumstances, and be freed from the obligation. Grotius argued that the promise bound only without taking into consideration the power of the law to annul or diminish the obligation. This was the standard civil law interpretation. In civil law, a person coerced into a contract could 'retract it through the action that it was *caused by fear*'. The distinction between void and voidable was not an empty one, for it was important where there was no higher authority. A coerced promise was binding if both victim and assailant lived in an anarchic society lacking a settled government (i.e. in a 'state of nature'). Casuists also argued that a king was bound to keep a pact resulting from coercion by rebellious subjects; the obligation stemmed from the public good, for otherwise peace could never be attained. A similar case was that of sovereign states, or treaties between them would be ineffective. If Utopia were defeated in war by Ruritania and agreed to peace in exchange for tribute, Utopia would be bound to pay, even though clearly coerced into that agreement by military might. 'In treaties at the shutting-up of wars do not the weaker princes usually consent to things much against their will? Yet such concessions confessedly convey human rights.'[32]

The theory of compelled contracts was further refined by rules on just what constituted coercion. Coercion is not always unambiguous. A threat of death seems like a clear case of coercion. In contrast, if someone merely threatened to utter curses, we might doubt that a plea of coercion was justifiable. Any significant element of uncertainty as to whether the threat will actually be carried out compounds the problem: we are clearly more frightened by a Mafia boss's threat of murder than an enraged toddler's. Early-modern theorists used a canon-law formula which defined what constituted compulsion as, 'fear that would move a constant man'.[33] For fear to be such that it would 'move a constant man', it had to meet a number of criteria. First, a constant man would decide on rational grounds between the evil threatened and the damage resulting from the promise. He would not, for example, in order to avoid material loss, promise to commit murder, because the latter was recognizably the greater evil. Second, a constant man would be moved only by the threat of grave evils, not minor ones: the evils most generally listed as grave were death, serious mutilation, imprisonment, enslavement, and rape.[34] This list was not meant to be exhaustive but to suggest the level of threatened evil that would render a contract voidable. If the threat were only minor, the promise or contract remained binding. Third, the fear must be realistic and of an imminent evil.[35]

A serious problem arose in applying this general theory to marriage. Most Roman Catholic theorists held that the marriage bond was indissoluble. Many Protestants thought it was severed only by adultery. Those whom God had joined together, no man could put asunder; 'for once marriage is validly contracted, it is indissoluble'. If a marriage contract made under coercion were not void from the outset, no human authority could rescind it. Moralists and theologians therefore decided that marriage was a special case; if anyone contracted marriage under the influence of fear that would move a constant man, it was simply void. Consent to matrimony must be free consent; grave fear eliminated the possibility of such freedom.[36]

Medieval writers developed all the essentials of this theory and bequeathed them intact to early-modern theorists. In the early sixteenth century, William Hay outlined them in his lectures on matrimony. 'Just fear' he defined as 'grave, well-founded fear' that would move 'a strong-minded [*constantem*] man or woman'. A marriage contracted because of just fear was 'null and void and can be annulled'. Although consent produced by fear was true consent, it was 'not sufficient for marriage'. Even where the pressure resulted from someone's own fault, 'this fear both prevents and nullifies the marriage'. Anyone contracting because of apparently well-founded fear of a grave evil was not bound by the marriage vows. Protestant theologians adopted the same theory in its entirety. Theodore Beza's discussion of a marriage contract made 'by force or fear which would move a constant man' employed the established vocabulary. Such a contract, Beza argued, was of no weight, even if there were some element of consent involved and even if it were confirmed by oath. This was because a marriage contract 'required the freest consent'; it would be 'ineffectual' if produced by force or violence. If there were subsequent free consent, the marriage would become valid, but without such consent, the marriage was simply null. William Perkins put forward precisely the same ideas in English translation. Contracts to marry 'made through force or fear' that would influence a 'constant and resolute man' were 'mere nullities'. Such marriage contracts could not bind because they lacked 'the free and voluntary accord and assent of both parties'. Not just any degree of fear was sufficient to invalidate consent to marriage, but if it were a reasonably based dread of serious injury, the marriage was invalid. Other English protestant theorists also borrowed and translated the canonists' arguments to establish that 'compelled consent' or 'tyrannical coaction and compulsion' would 'nullify the contract'.[37]

English lawyers, like English divines, adopted the canonists' view and stated that 'marriage holdeth not when it is extorted by force, or by such a fear as may *cadere in constantem virum; quia matrimonia debent esse libera*' [move a constant man; because marriage should be free]. Someone who consented to marry under duress was not obliged to sue for divorce, for (s)he was never married at all; 'the contract of spousals or marriage made through fear is utterly void *ipso iure*' [by the law itself]. Eighteenth-century handbooks of law reiterated the standard view. The plea of compulsion was valid only where a person had been subjected to fear that would 'prevail upon persons of courage and prudence' – reverential fear of a father was not enough – but fear of death or imprisonment would be a sufficient justification for the marriage to be deemed null.[38]

In early-modern matrimonial law and ethical theory, consent was essential to a true marriage and indeed itself constituted it. A couple who agreed to be husband and wife were married by virtue of that consent. Sexual relations were irrelevant except insofar as they embodied consent to marriage. No one could consent on behalf of the couple, not even their parents. If violence or the reasonable fear of grave injury induced consent, the marriage contract was null and void. There was a high level of agreement amongst early-modern theorists that free consent was the only means by which a male became a husband, or a woman a wife. Naturally, the question arises of exactly what each spouse was thought to be consenting to. What was the state of matrimony?

Terms and conditions

A woman became subject to her husband by marriage. What (if anything) could she do if she found that subjection galling? Would not any rational woman ensure that the contract of marriage included terms and conditions that would allow her to terminate the union if she found the burden of obedience unbearable? If she objected to her state of subjection, could she not simply withdraw her consent and become once again a free woman? Would anyone rational enter a partnership that made no provision for the other partner's failure to fulfil its purposes, and allowed no escape if the other's conduct proved intolerable? Was it not only reasonable and equitable that a couple entering marriage should draw up terms to the contract between themselves? – terms that would make provision for how they should live together, and how they should part if either failed to abide by them.

These suggestions might seem perfectly reasonable and unobjectionable nowadays, but they did not during the sixteenth and seventeenth centuries. Most early-modern theorists believed that marriage was instituted by God and that He had established the regulations that should govern it. The fundamental terms and conditions of a valid marriage stemmed from divine and natural law, not from human law and not from the wills of the parties contracting. Nowadays, there might seem a certain contradiction in stating that marriage is founded in the couple's consent but denying that they decide its character. Now it seems natural to think that the parties to a contract define its terms. Early-modern theorists believed this of most contracts, but denied it in the case of marriage. For marriage was a contract instituted by God, and He had decided upon its necessary terms and conditions. Any person was entirely free to choose whether to marry or not, and equally free to choose whom (s)he married. Indeed the marriage would be invalid if (s)he were coerced. However, having chosen to marry, both spouses were bound by divine contractual terms.

The belief that marriage is not simply a private contract like any other survives into the modern world. Even today, in an age when sexual relationships are increasingly considered a matter of concern only to the participants, marriage is regulated by law. In most of the Western world, children are not allowed to marry; polygamy and polyandry are prohibited; a woman cannot legally marry her father, brother, or son even if she and he are both willing; and many states require a public ceremony to establish a marriage's validity. The marriage contract is treated as a proper area for public regulation, as indeed are other agreements by which sexual and reproductive services are bought or exchanged. Many states restrict sexual relations with minors and between those of the same sex. In general, people are not left free to decide at their own discretion the terms for contracts of (fe)male prostitution and surrogate motherhood. In all these spheres, government policy either invalidates certain freely made contracts or severely controls their provisions.

In the sixteenth and seventeenth centuries, sexual relations were not considered merely the private concern of consenting adults, so it is hardly surprising that marriage was not seen as such either. Moralists, lawyers, and divines believed that sexual relations were legitimate only within marriage. But this is not a significant constraint if any two people are entirely at liberty to decide between themselves just what constitutes marriage. Otherwise, any fornicator can simply legitimate his or her actions by privately marrying, for (say) two hours, someone who then agrees to a non-contested divorce. If no external law establishes the terms of marriage, the distinction (so vital to early-modern Christian theorists) between marriage and promiscuous fornication can easily disappear.[39] Early-modern definitions of marriage were couched in terms intended to eliminate the awful possibility of 'temporary marriages'. Marriage was defined as 'a coupling together of two persons into one flesh, not to be broken, according unto the ordinance of God: so to continue during the life of either of them'. Marriage must be monogamous, in accordance with God's law, and last for life. It was 'a conjunction of man and woman, containing an inseparable connection and union of life'. The permanently binding nature of the marriage bond was stressed in almost all definitions, and many added to monogamy and indissolubility the requirements of cohabitation and common property. Virtually all early-modern theorists argued that only marriages meeting such requirements were fully valid.[40]

The marriage service in the Church of England's *Book of Common Prayer* described 'holy matrimony' as 'an honourable estate instituted of God'. These words were meant to be taken quite literally. God (not man) laid down the rules for marriage, and those entering the institution had to abide by them. In their marriage vows, both bride and groom undertook to 'live together after God's ordinance in the holy estate of matrimony'. Medieval and early-modern divines, moralists, and lawyers were agreed that the contracting couple could not (even by mutual agreement) ignore any part of God's ordinance, let alone conduct themselves in ways contrary to it. 'For marriage is not a covenant of man, but a covenant of God.' Those ignorant of the Christian faith might not be aware of all God's laws about marriage but they were still bound by them.[41]

At their wedding, a couple took not a marriage *oath* or *promise* but marriage *vows*. The terminology was significant. A promise simply involved two people; either could release the other at will. Even in an oath, God was invoked 'only to be a witness'. In contrast, a vow was 'made unto God alone'. When two people married, each vowed to God (not simply to one another) their acceptance of marriage's rules and conditions. 'It belongeth therefore to God alone to release the obligation of a vow; and no man hath power to do so.' A husband could not release his wife from her marital obligations, nor could she release him from his. By consenting to marriage an obligation was incurred 'not only to a person but also to God, so that it is written "What God has joined together"'.* Marriage was called 'the covenant of God, because He is the mean between them, and by His authority and leading the two married parties come together'. In early-modern thought, marriages were indeed made in heaven; once the couple had consented, the contract between them was subject only to God's jurisdiction.[42]

Medieval and early-modern theorists acknowledged that it was possible for the couple themselves and/or human authorities – ecclesiastical and civil – to add additional terms to the marriage agreement, regulating how it affected secular matters such as property, 'upon account of civil contracts, portions, and dowries'. '[T]he law-making power of the kingdom' could intervene in matrimonial matters. Roman Catholic divines regarded marriage as a sacrament and therefore properly under ecclesiastical jurisdiction. Yet even they accepted that its sacramental nature was 'super-imposed' on a 'human contract'. Beyond doubt, marriage was also 'a civil contract between husband and wife'. Protestant theologians, too, accepted that marriage was a 'mixed contract', for 'as far as it pertains to human society it is clearly civil', whereas it was 'divine and ecclesiastical' where God's law interposed. Civil and ecclesiastical authorities could add to God's institution for good reasons.[43]

Theorists differed on whether the laws of marriage should be enforced by the civil magistrate or the Church. Most English Protestants accepted that the king in parliament held ultimate jurisdiction over matrimonial litigation. Although, in practice, the canon law governed marriage, matrimonial causes

> originally do belong to the civil magistrate, by whose munificence they have been granted to ecclesiastical persons, and unto them properly it doth belong to make matrimonial laws.

The civil lawyer, Alberico Gentili, argued that marriage was a secular matter and should be controlled by the civil magistrate not canon law. Others balked at such Erastianism. When Barebone's parliament attempted to secularize marriage entirely, many couples responded by celebrating a second wedding ceremony conducted by a clergyman after

---

*'What therefore God hath joined together, let not man put asunder' (Matthew 19:6).

they had first satisfied the law's requirements by marrying civilly. Theorists insisted that the Church should decide marital cases and denied there was any instance 'in the Christian world (before all things ran riot here in England since 1642) [where] the temporal power took cognisance of marriages'. It was widely believed that marriage was a divine institution and that God's law must regulate it.[44]

Most protestant divines gave the civil magistrate significant powers in matrimonial cases. However (with the exception of a few iconoclastic theorists such as Thomas Hobbes), everyone agreed that God had instituted several conditions for marriage which secular laws must not infringe. No state could institute a Platonic community of wives, for example, because 'the institution of matrimony' was God's ordinance, 'which all the power of man is as incapable of evacuating as of altering the seasons'. Orthodox theologians rejected, with horror, Hobbes' assertion that adultery was only defined by the secular laws of the state. They insisted that matrimony existed prior to the state; civil laws could regulate the institution, but they could not infringe the dictates of divine law or right reason.[45] Common lawyers naturally tended to stress the role of statute and common law. They depicted 'all the spiritual law' as 'but branches of the common law', but they, too, accepted that there were divine laws on marriage that should not be infringed. In a case in King's Bench in 1660, the judge stated of husband and wife that 'when they are once well married all the world cannot dissolve the marriage; for an Act of Parliament in this sense, against the law of God is void'. Lord Chief Justice Vaughan believed that God had established in Leviticus 18, certain degrees of consanguinity within which men could not legally marry and no human edict could alter this; 'no man can dispense with God's law, as the clergy in the convocation and most of the famous universities of Christendom have affirmed'. Marriage was not a human institution but a divine one and human law-makers could not legislate against divine law.[46]

If civil and ecclesiastical authorities could not legislate new terms for marriage, still less could private individuals. Towards the end of the seventeenth century, one theorist wrote with horror about the new practice of 'concubinage' arising in England. Couples were living together 'on promise of fidelity and constancy' but without proper marriage, in order to avoid its 'strict constrainment and restraint'. In his view, any couple that did 'not think themselves thus bound' for life was 'guilty of fornication'. If they did not marry according to 'the laws of church and state', they 'are not married as the laws of God require, but undoubtedly live in fornication'. Most theorists concluded that a couple simply were not free to define the terms under which they conducted a sexual relationship. God had not left the 'nature of marriage' to 'man's arbitrary determination'. The institution of marriage was divine, and any other arrangement was sinful fornication. Generations of moral and religious theorists had insisted that it was God alone, 'the author and institutor of marriage', who could order how people conducted their sexual affairs. Pagans mistakenly conceived marriage as only a 'civil arrangement' but the law of nature ('although very obscurely') showed marriage to be 'something greater than a human invention'. One puritan divine taught this charitable Christian view when he warned his audience:

> If we should transgress men's canons in our matrimony we might happily hear of it in the court; take heed of trespassing, especially of wilful breaking God's canons, Who can not only punish the purse and body, but fling into hell.

Attempts to innovate in marital arrangements were liable to meet with God's final sanction. Other protestant divines were equally convinced that 'men are to take women as wives to live together in God's ordinance ... and not to live as brute beasts to defile themselves'. Animals could couple as the mood took them; humans must obey God's laws and be bound indissolubly in marriage.[47]

In early-modern theory, one of the most important characteristics of marriage was permanence. God had made marriage indissoluble: no married couple could break the tie even by mutual agreement. Men could freely decide amongst themselves how to organize local and national government, corporations, guilds, universities, and virtually every other civil alliance, but God had laid down the regulations for conducting sexual relations. Marriage was 'not like other civil associations, which just as they are entered into by mutual consent are also dissolved by mutual consent'. God 'gave and prescribed certain laws about wedlock' because He 'hates roving passions'. God made marriage indissoluble to ensure that men did not make arrangements that allowed a variety of sexual partners.[48] Scholastic theologians declared the marriage bond was so strong that even the pope could not sever it; for Catholics, no greater strength existed. Protestant divines insisted that the parties to a marriage contract cannot 'be set at liberty by themselves or by any other power whatsoever'. A couple could not decide to marry temporarily and part if they found their temperaments incompatible: 'Marriages *pro tempore* [for a time], dissoluble by consent are not of God's institution but contrary to it'. The indissolubility of marriage was as reasonable as it was divine, for 'if all should have leave to marry others when they consent to part, it would bring utter confusion'. It was believed that society would soon collapse if any husband and wife could part and marry someone else simply because they got on badly together. The indissolubility of marriage was based on reason as well as revelation: 'since the laws of God and man founded upon reason and experience forbid a temporary contract and engage the pair for life'. To the medieval and early-modern mind, indissolubility was the rational way to draw a clear line between marriage and fornication.[49]

Some theorists debated whether natural reason would conclude that marriage must be indissoluble if God's law did not supplement its deductions[50]. In the thirteenth century, Duns Scotus had argued that the indissolubility of marriage stemmed only 'from the institution of God, the law-maker'. It was God who imposed the law that any agreement to have sexual relations must involve permanent union. Otherwise, it would be inherently acceptable for a couple 'to give themselves mutually for only one [sexual] act, or for only one hour, or for another short space of time'. Duns Scotus suggested that natural reason implied that the duration of a contract for sexual relations (like any other contract) could be determined by the consenting parties. The indissolubility of marriage was discovered by looking at God's law, not by unaided human reason.[51] Some later Roman Catholic writers also admitted that nobody 'ignorant of God's law' would realize that marriages must never be dissolved. The law of nature did not make indissolubility 'most evident', rather it had to be deduced by argument from first principles. Indeed, the indissolubility of marriage was so 'obscure by the law of nature' that it excited no surprise that pagans had failed to notice it. In case human logic failed, God had given a positive law for indissolubility to remove any doubts or excuses. John Locke inclined to the proposition that the permanence of marriage was not based on unaided reason. He insisted that any marriage should last long enough for 'the procreation and education' of the children to be 'secured'. After that, however, there was 'no necessity in the nature of the thing nor to the ends of it, that it should always be for life'. It was the restraint of 'positive law which ordains all such contracts to be perpetual'. The jurist, Samuel Pufendorf, likewise took the stance that naturally 'marriages may be dissolved by mutual consent', although he was eager to add that it would be 'both unbecoming and menacing' if this were done 'without any serious cause'. During the sixteenth and seventeenth centuries, the orthodox dogma of the permanence of marriage was increasingly beleaguered.[52]

In the early-modern period, the indissolubility of marriage was attacked from other quarters. Thomas More's *Utopia* is a difficult text to interpret but – if Utopia is construed as a state run in accordance with natural law – it implied that divorce for adultery or temperamental incompatibility was not against the law of nature. Martin Luther was disinclined to relax the marriage bond, stating that 'I so greatly detest divorce that I should prefer bigamy to it', but many Reformers were more radical. Martin Bucer portrayed marriage as a purely civil contract and allowed divorce for a wide assortment of reasons. Catholics condemned his views as indicative of the debauchery to which denying marriage the status of a sacrament led, but others found Bucer's arguments instantly attractive. Some unhappily married rebels, like John Milton and – the more obscure but no less eccentric – John Butler, hoped for the restoration of a unilateral right of repudiation. Church and state should not interfere for 'the absolute and final hindering of divorce cannot belong to any civil or earthly power against the will and consent of both parties, or of the husband alone'. Milton found Bucer's arguments appealing grist for his mill and cited him extensively.[53]

John Selden's *Uxor Hebraica* [The Hebrew wife] dealt with the history of marriage from the earliest times; with dauntingly detailed historical scholarship he showed that the indissolubility of marriage was a Christian invention, and one not even accepted by all Christians until a late date. Before Mosaic law, marriage arose 'by mutual consent' and a union that arose by consent could be terminated in the same way. Selden argued that reading of Jewish sources suggested that natural law, 'the law common to the entire human race from the very beginning', showed that 'spouses by mutual consent may couple and unite, and so by similar consent could disunite, and so a marriage could be broken off'. Natural law did not limit this right, but 'each party, either man or wife, in whatever way, could renounce their marital association at will, and thus in turn lawfully separate'. For the Jews before Mosaic law, for the Greeks, and the Romans, marriage was dissoluble at the will of the contracting parties.[54] John Selden was not asserting anything new when he pointed out that there was 'among the ancients ... a very liberal law and practice of dissolving matrimony'. Earlier Christian historians and theorists had been well aware of this and condemned it as a gross form of libertinism with disastrous consequences for social stability. What was new in Selden was the whole tenor of his work; one which suggested that indissolubility and divorce at will were merely alternative, equally acceptable forms of marriage law. In his view, only because divorce and remarriage were not permitted by human law in the West after the ninth century, did indissolubility become 'seemingly a divine law'. Easy divorce need not necessarily involve disturbance of the 'public peace'; it had been outlawed solely on account of the religious motives of 'avoiding the opportunities for lustfulness', and maintaining 'righteousness'. The tone of Selden's discussion implied that a system which allowed for divorce at the will of the parties involved was more obviously reasonable and natural. As in the case of the double standard on adultery and the permissibility of polygamy, secular-minded theorists like Selden moved away from the traditional teaching of the Church. It seemed to them that 'coition without wedlock, ... when by mutual consent, injures no property'. Just as Duns Scotus had predicted, reason – unassisted or unfettered by God's written law – led Selden and others to the belief that the duration of sexual associations should, like all other contracts, be determined by the agreement of the parties contracting.[55]

The more secular view of marriage increased in influence during the sixteenth and seventeenth centuries, but its devotees were in a small minority for most of the period. The great majority of early-modern theorists thought that the couple should not be free to decide when to part. As God had ordained that husband and wife were joined for life, the bond could not be severed simply by mutual agreement or by the unilateral action of one of the parties. Charles Blount, Earl of Devonshire, was an enthusiastic advocate of divorce and the husband of a divorcée, but he accepted that 'natural convenience' meant that husband and wife should not divorce 'without the cognizance of the magistrate'. Otherwise, they would 'be both judges and parties in their own case, which were unjust and unnatural'. Most early-modern theorists were horrified by the contention that since a marriage arose by consent, it could be terminated in the same fashion. If marriage were simply the civil contract Selden portrayed it as being, how could anyone 'find fault with the Irish marrying for a year and a day, or the Welsh divorcing for a stinking breath?' A married couple could only separate where God's law provided that such a division was legitimate, and it was God's ministers on earth who should interpret His law and decide when this was the case. Only the Church or the civil magistrate could act as God's executive and put husband and wife asunder.[56]

It is, therefore, to the early-modern theory of divorce that we must turn in order to discover whether a woman could escape her marital subjection. Most theorists believed that the marriage bond could not legitimately be broken just because one of the parties no longer liked it. A wife could not terminate her subjection at her pleasure, but did God's will offer her a route to emancipation? Under what circumstances could a wife legitimately divorce her husband? Were her rights in divorce theoretically and legally the same as his? Did moralists, lawyers, and divines institute a double standard on divorce and oblige the wife to tolerate an obnoxious husband while allowing him to discard an odious wife?

Divorce

For the much of the history of the Western Church, the marriage bond was pronounced indissoluble. Even in the case of adultery, neither husband nor wife, neither the innocent nor the guilty party, could obtain a divorce 'from the marriage bond' [a vinculo]. The most that the Catholic Church did (or does) permit was a separation 'from bed and board' [a mensa et thoro]; there was no right of remarriage after such a separation. A couple could separate 'for no cause except for fornication only', and even in this case 'neither of them may marry any other as long as both they live'. Medieval canonists and theologians were even-handed between the sexes; whatever the conduct of their spouse, neither husband nor wife could obtain a replacement.[57]

Even before the Reformation, some theorists – notably Desiderius Erasmus – had argued in favour of divorce a vinculo for adultery.[58] Continental Protestant Reformers rejected divorce at will, but the majority argued that God's own law provided that adultery or desertion released the innocent party, leaving him or her free to marry again. After the Reformation in most of protestant Europe, ecclesiastical law was amended to allow divorce *a vinculo* at least in the case of adultery and for the innocent party. In England, no such changes were made. The catalyst for the breach with the Church of Rome was Henry VIII's urgent desire to trade in Katherine of Aragon for Anne Boleyn. As 'some loose persons' have commented, 'the Reformation here in England ... came from Henry VIII's codpiece'. However, divorces *a vinculo* were no easier to obtain in post- than in pre-Reformation England. 'The married man is entangled like a fish in a net; he comes merrily in, but he is mightily perplexed when he cannot get out.' Canon law continued to dominate English law on marriage and there were no changes in divorce law comparable to those made on the Continent. Some tentative steps towards the reform of ecclesiastical law were made under Henry VIII, but these were abandoned, and 'the old canon law was confirmed by a statute ... and the same is still in use as in the dominions of other princes'.[59] Throughout the sixteenth, seventeenth, and eighteenth centuries, the church courts granted only separations from bed and board. Both ecclesiastical and common law denied either party the right to remarry. The only legal way to divorce one spouse and marry another was by Act of Parliament – a difficult, expensive, and uncommon procedure.[60]

It was not unknown for those divorced *a mensa et thoro* to remarry despite the legal niceties. In the reign of James I, the *cause celèbre* of Robert and Penelope Rich, both of whom remarried after such a divorce, almost destroyed the career of William Laud who officiated at the wedding ceremony for Penelope and her lover Charles Blount. Those divines who accepted that the marriage bond was dissolved by adultery, tacitly defended such conduct. Some couples went their separate ways without any legal formalities. The marriage of Charles Brandon, Duke of Suffolk, to Mary Tudor (sister of Henry VIII) was impugned on the grounds that it was 'notorious' that Brandon's first wife – from whom he separated 'of his own motion, without any form or manner of legal judgement' – was still alive when he wed Mary. The frequent complaints of preachers suggest that informal separations were common. Husbands 'upon dislike they take at their wives (or liking of others) make nothing to send them home to their friends and live separated from them'. The *Homilies* complained of 'the divorces, which nowadays be so commonly accustomed and used by men's private authority, to the great displeasure of God'. In the seventeenth century as in the twentieth, husbands who abhorred their wives could simply pack their bags and leave: 'A deserted lady or gentlewoman is become a common notion; ... now the dogs bark at the masters of families when they return, as if they were absolute strangers'. Separation at the will of the spouses was given theoretical defence by few, but practice in this case obviously diverged from official morality. The bolder married again following such a separation, but this was a dangerous course as bigamous marriages were a felony punishable by death.[61]

In sixteenth- and seventeenth-century England, complete, legal divorce was virtually unobtainable for any but a select few of the noble élite. Discussions by English divines of the relative rights of husband and wife in divorce were therefore no more than academic or (like those of their Catholic counterparts) really about rights in a separation. This did not discourage them from considering the theoretical problems at length. Early-modern theorists had few historical examples of egalitarian treatment of husband and wife in divorce. Roman law had initially allowed the husband, but not the wife, a right of divorce (although this did change to some extent in the later period of the Roman Empire when wives were also granted the right to divorce). Similarly, Old Testament judicial law allowed Jewish husbands to divorce their wives at the drop of a hat, but gave wives no equivalent rights. In one modern historian's words, biblical law showed 'no consideration for the woman's wishes as far as the future of the marriage was concerned, nor was there public supervision of divorce'. The Scottish Presbyterian, Samuel Rutherford, pointed out that Jewish law did impose one condition on the husband seeking divorce; for

> the law of divorce gave not power to all husbands to put away their wives, but only to the husband who could not command his affection to love his wife.

In Mosaic law, 'the right of repudiation belonged to the husband not the wife'. Before Christ, God permitted divorce to his chosen people for the 'hardness of [their] hearts' (Matthew 19:8) but the 'dispensation among the Jews was granted only to the men'.[62]

The New Testament offered no obvious grounds for allowing wives to divorce or separate from husbands. In the texts where Christ dealt with divorce, he merely ended or restricted the husband's right of repudiation; they provided no explicit grounds for believing that a wife had a right to divorce her husband (Matthew 5:31–2, 19:3–9; Luke 16:18; Mark 10:2–12). If anything, the implicit assumption was that only husbands could have such a right. In the longest discussion (Matthew 19), as John Milton pointed out, 'no word of this text binds women, but men only'. In Mark, where the wife was mentioned specifically, the apparent sense of Christ's words was to forbid divorce and remarriage altogether (without any exception for adultery) so it afforded wives no right to divorce. Saint Paul took his usual inflexible line on wives' obligations and only explicitly countenanced divorce for either party if an infidel spouse refused to live peacefully with his or her Christian partner.[63]

Not only did Scripture fail clearly to pronounce any right of a wife to divorce, it was readily susceptible of an interpretation that gave her no right but left some to the husband. Christ stated: 'Whosoever shall put

away his wife, except it be for fornication, and shall marry another, com-
mitteth adultery' (Matthew 19:9). Saint Paul taught that the wife was
bound 'as long as her husband liveth' (I Corinthians 7:39). An appealing-
ly simple reconciliation of these two texts is that husbands could put
away wives for fornication, but that a woman was stuck with her
unfaithful husband till death did them part. Saint Augustine vigorously
rejected this exegesis, but Saint Ambrose had apparently adopted it.[64]
Archbishop Catharinus tried to recycle the belief that after divorce for
adultery the husband could – but the wife could not – remarry. He cited
Ambrose and Cajetan, and argued that Scripture showed it was 'not licit
for the woman to leave and marry another even on account of fornication, but it was licit for the husband to do both'. However, Catharinus
had no success as the one holy Catholic Church continued to reject
divorce with remarriage in any circumstances. The German protestant
divine, David Pareus, was also inclined to believe that God gave wives
no right to put away unfaithful husbands; 'for God wanted only the
wife's fornication to be a cause of divorce,- not the husband's', but most
Protestants disagreed.[65]

Most early-modern theorists rejected the sexually differentiated view.
They followed patristic and medieval tradition and held that the wife
had as much right to separation from an unfaithful husband as a husband did from an adulterous wife. The canon law on adultery was clear
that what was 'illicit for women is equally illicit for males, and the same
submission is required in the same state'. Students of canon law were
taught that 'a wife can dismiss an adulterous husband or demand a
divorce in exactly the same way as a man his wife, because in divorce
causes husband and wife are judged to be equal'. Catholic casuists
advised that 'by the adultery of the husband the wife hath right of a
divorce'. When they felt called upon to discuss the sexually differentiated view, many authors simply trumped Ambrose's authority with
Augustine's.[66]

Scriptural commentators noted that the Gospel gave no explicit
grounds for wives to divorce their husbands. The failure of any
Evangelist but Mark to mention wives in discussions of divorce 'hath
made some doubt whether the woman in case of the husband's adultery
may sue a divorce from him'. But commentators argued from equitable
principles, not the letter of Scripture: 'the most judicious interpreters say
there is an equal right on both sides: I am sure the reason is equal on
both sides'. Unequal rights to divorce after infidelity were regarded as
unjust to women and incompatible with Saint Paul's teaching on the
mutuality of conjugal duty. Those who gave 'a prerogative in this case to
the man above the woman' were libertines, guilty of 'abusing the word
of God to their own damnation' and 'turning the grace of God into wan-
tonness'. Despite slender textual support, Scripture was glossed as
upholding the Church's tradition that husband and wife must be treated

equally in sexual matters. Protestants supporting divorce upheld the (unmentioned) wife's equal right to it from Matthew 19:9; Catholics opposing divorce denied the (unmentioned) husband's right to it by citing I Corinthians 7:39.[67]

Societies giving husbands greater rights were condemned. The moral platitudes of the Hellenistic philosopher Plutarch were generally acceptable to the Renaissance audience. But when he implied that wives should turn a blind eye to their husbands' concubines, the English translator Philemon Holland added an outraged marginal note.

> Plutarch herein smelleth of the corruption in his time: for a Christian dame and honest matron will not abide to put up such an injury, nor wink at her husband's folly in that case.

Historians protested at the 'great inequality' in the treatment of husband and wife in Roman divorce laws. This injustice was 'complained of' by the Church Fathers 'who thought the man and the woman upon the same foot and right by the law of God'. Both had an equal right to put away a fornicating partner.[68]

Most Continental Protestant Reformers allowed divorce with remarriage for adultery and wilful desertion, and gave wives just as much right as husbands. They contended that 'it clearly appeared from the word of God' that both husbands and wives equally could separate on account of adultery. Husbands should normally rule wives in conjugal matters, but spouses had equal rights over one another's bodies, 'and in the case of adultery it is this right that should properly be considered'. I Corinthians 7:4 was cited to clinch the point.[69] English divines were divided on whether divorce with remarriage should be allowed in the case of adultery.[70] Whether they allowed divorce a vinculo, or only separation, the great majority of English divines argued that the husband and wife had equal rights. This was true of the earliest English Reformers. John Hooper, Bishop of Gloucester and Worcester under Edward VI, considered 'this controversy between my contraries and me whether it be as lawful for a woman ... to put away her husband, an adulterer, as the man to put away his wife'. He answered that she had 'the same authority'. The husband equally broke the bond of marriage; his crime in adultery was as great. Hooper dismissed the objection that 'if it should be lawful for the woman to make a divorce with her husband, marriage could never be sure nor constant; for women would change still at their pleasures'. Hooper viewed capital punishment of the unfaithful husband as preferable, but in its absence the wife, too, could divorce. Either spouse could divorce an unfaithful partner because 'they are equally bound to each other and have also the same interest in one another's body'.[71]

Andrew Willet was typical of English Calvinist opinion in insisting on the right of either party to divorce in the case of adultery. Husbands could divorce adulterous wives, and wives had 'the same remedy' for persistently unfaithful husbands. Once the marriage was ended, the divorcée was no longer subject to her husband; it was 'lawful ... both for men and women to use this remedy against incontinency and to be married again'. Divorced husbands could remarry, Willet argued (citing Ambrose) and 'there is the same reason and liberty both of the man and wife in this case'. He also cited canon law that there should 'be one law altogether for the man and the wife'.[72] John Milton became an enthusiastic advocate of divorce when his own wife signally failed to make the grade. He took the Genesis view of women as made for males to its logical extreme, and maintained that if a wife failed in any significant respect to be a 'meet help' to her husband, he should be able to discard her. In essence, Milton wanted the Old Testament right of repudiation restored to husbands, without any interference from civil or religious authorities. It was clear throughout Milton's divorce tracts that he considered the husband's interests and rights far more important than those of the wife. However, even he conceded that if the husband were cruel, promiscuous, or 'the bond-slave of Satan', the wife could leave him 'according as the Gospel seems to make the wife more equal to her husband in these conjugal respects than the law of Moses does'.[73] Most later-seventeenth-century English protestant theorists adhered to the same egalitarian tradition. Throwing out or 'putting away' might be only the husband's right, because he was 'usually the owner of the habitation', but the wife was free to depart. 'And I know of no reason to blame those countries whose laws allow the wife to sue out a divorce as well as the husband.' The right to seek divorce was 'alike common to either party, though it appears not to have been so among the Jews'. The only difficulty was whether the guilty party, like the innocent, should be allowed to remarry. Husband and wife's equal rights followed from 'the mutual power which God hath given them', from Saint Paul's teaching (I. Corinthians 7:10), and from 'the practice of the primitive church'.[74]

One of the few parliamentary divorces was that in 1670 of John Manners, Lord Roos, from his flagrantly adulterous wife. Reasons of state as well as of conscience led some to oppose the passage of the bill, but during the debates no one denied that husband and wife should have equal rights in divorce. The 'vulgar error ... that men have a greater pre-eminence than women' was specifically rejected, for 'all Christians hold all privileges reciprocal between the man and the woman (though the Jews did not so)'. Even those who rejected divorce with remarriage employed arguments that implied the equality of the sexes in this matter. Scripture showed that 'the woman adulteress can never marry again' and therefore it was inequitable to let the husband, for 'it seems to want charity for the woman who, whilst living, may need marriage as much or more than the man'. They debated whether divorce with remarriage was ever permissible, regardless of the sex of the person divorcing.[75]

A few Anglicans argued differently. In his *Ductor dubitantium* the Anglican casuist, Jeremy Taylor, discussed at length the case, 'which nowadays happens too frequently', of 'women married to adulterous and morose, vile-natured husbands'. He was unequivocal that wives had just as much right to refuse to cohabit with unfaithful husbands, as husbands with adulteresses; the guilty party lost his or her marital rights. He added that the wife was under no obligation to leave the husband if it would lead her 'to be reproached with the noises of a divorce or to become an actual widow before death'. However, Taylor was not completely even-handed between the sexes. He maintained that the woman could not 'put away' her husband as he could her. She simply lacked the authority. Husband and wife

> have not equal powers; neither can the woman put away the man as the man can the woman . . . in government and divorces they are not equal.

For Taylor, equal rights were not equivalent to equal powers; either spouse could withdraw from the other, but only the husband could send his wife away.[76] Another Anglican theorist denied that husband and wife had equal rights in and to divorce. Henry Hammond, chaplain to Charles I, maintained that the husband could put away his wife for adultery but not vice versa. He founded his case on the standard argument that the 'family inconveniences' of a 'base brat' taking the legitimate heir's inheritance resulted only from wifely adultery. Hammond argued that the power to put away an unfaithful spouse did not stem from the mutual 'conjugal contract'. Husbands alone had the authority to repudiate, 'because the wife hath by promise of obedience made herself a subject and owned him as lord'. Most theorists rejected this argument from the wife's subjection to the husband and did 'not find this reason sufficient. For Saint Paul maketh the interest of the wife in the husband and that of the husband in the wife, both one and the same'.[77]

Hammond's and Taylor's stance was unusual, despite the fact that it was entirely compatible with the New Testament sources and was certainly closer to Old Testament law. The egalitarian view of Augustine, the canonists, and the schoolmen was deeply ingrained. General custom in the sixteenth and seventeenth centuries (and for long afterwards) continued to regard simple adultery by the wife as good grounds for divorce, while in the husband's case it had to be compounded by an additional offence such as cruelty or incest. The divergence between popular practice and official theory meant that the relative rights of husband and wife to divorce remained a matter of debate. Nevertheless, most moral and religious theorists rejected popular attitudes and historical precedent. They maintained that, in the event of adultery, husband and wife had an equal right to divorce or separation. Most early-modern theorists carried the equality of the sexes in sexual matters to its logical conclusion in the case of divorce.[78]

Notes

1. Carter 1627, 70 [liberty]; Rogers 1642, 260 [refuser]; Benincasa 1562, f. 147v ('nemo'); Agrippa 1540, sig. C3v.

2. *Baron and feme* 1738, 4; Swinburne 1686, 14 (and see 84–5); Laymann 1625, 359 (V.X.I.I.7 'magnam animi libertatem, atque spontaneum non illiberalem consensum'); Busenbaum 1663, 549–50 (VI.VI.II.I 'mutuus consensus', 'Minister, non est sacerdos sed ipsi contrahentes'); Bellarmine 1602–3, III (1603), cols. 1309–10 (I.VI 'in signis exprimentibus consensum mutuum'). See also Antoninus 1474, sig. M5v; Mauser in Beust 1597, 327; Monner 1561, 38; Benincasa 1562, f. 11r; Watson 1558, f. 172r; Nevizanus 1570, 245; Tapper 1582, 285, 305; Schneidewein 1585, sig. F2r; Gentili 1614, 100, 112; Olizarovius 1651, 17. A few Catholic theologians had argued that the priest's blessing (rather than the mutual consent of the parties) created the sacramental bond, but the view was never widely accepted. For a defence and history of this opinion, see Marca 1669, 65–83.

3. For the detailed provisions of Hardwicke's Marriage Act, see Stone, 1977 35–7. Even before the Council of Trent, secret or 'clandestine' marriages were officially disapproved and subject to ecclesiastical penalties. However, until Trent, they were valid and continued to be so where the Tridentine edicts had not been promulgated. As Harpsfield (1878, 34) stated, 'he that maketh a privy contract of matrimony transgresseth the law, yet doth the matrimony hold'. See also Covarruvias 1597, 139 (I.I.4.4); Vaux 1599, 130. For one instance of the argument against the Church's capacity to insist that marriage must be public, see Turriani 1563.

4. Harrington 1528, sig. A3r [express]; Perkins 1618, 681; Blount 1911, 322 [efficient]; Dominis 1617–20, II, 430 ('nec unus alteri pro coniugio subiicitur, nisi libero, et voluntario consensu'); Locke 1988, 319. Compare Willet 1634, 348.

5. As Goody (1983, 24–5) states, 'the consensual nature of marriage . . . was characteristic of the Christian church as a whole in principle from the earliest times'. On Roman law, see Corbett 1930, 54–7; on the Germanic tradition, see Bailey 1959, 117. For fathers' power to dispose of their daughters, see, for example, Genesis 29:23, 34:4–10, 16; Exodus 22:16–17; Deuteronomy 7:3; Joshua 15:16–17; Judges 1:12, 14:20, 15:1–6; I Samuel 17:25, 18:17–21, 25:44.

6. Covarruvias 1597, 139 (I.I.4.4 'impediretur finis ad quem tendit matrimonium'); Nevizanus 1570, 225 ('malum exitum'); Covarruvias 1597, 153 (I.II.2.1 'nec summus Pontifex'). The Council of Trent (Schroeder 1941, 189) placed under anathema any civil magistrates who 'directly or indirectly compel their subjects, or any others whomsoever in any way that will hinder them from contracting marriage freely'.

7. *Digest.* xxxv.I.15 (see also, D.xxiii.I.5; D.xxiv.I.66); *Decreti.*c.3.C.27.q.2. For a clear, brief history of the Christian Church's adoption of the doctrine that consent, not coitus, makes marriage, see Bailey 1959, 118–41; Alexander's role, Ibid., 128–9. Brundage (1987, 92): 'Patristic writers assumed, as Roman law did, that consent made marriage. They rejected the notion that consummation was an essential part of marriage'. In fact, the Christian view in the early-modern period was somewhat different from the Roman concept of marriage. Both believed that marriage was made by consent, but for the Romans consent could be implied by 'use', that is simply by lengthy cohabitation with signs of 'marital affect' on the part of the couple. For the canonist, consent meant a single act of contractual

commitment. Allowing marriage to be presumed from cohabitation would have opened the door to fornication. For condemnation of marriage by use, see Bartolus 1570, X, 108; Hotman 1585, 59–62; Covarruvias 1597, 150–1, (I.II.1.5); Hotman 1594, 170–1; Peleus 1602, 30; Bernard 1845, 108; Vacant and Mangenot 1903–46 (1927), IX col. 2135 ('Mariage').

8. Cardwell 1841, 524. Medieval and early-modern theologians expended a great deal of ink on maintaining the validity of marriage even without full sexual relations, despite the fact that 'it is hard to find any sort of example (albeit some few there be) of married folk that have abstained from all carnal meddling' (Harpsfield 1878, 155). On Mary and Joseph's marriage, see also Covarruvias 1597, 150, (I.II.1.1); Harpsfield 1878, 241; but compare Hutchinson 1842, 148. Protestants also used Mary and Joseph's example to show that consent without coition is sufficient for marriage: Pocock 1870, 361–2; Mauser in Beust 1597, 308; Stock 1865, 177. However, some later Protestants expressed doubts about whether they had refrained from sexual intercourse after Jesus' birth: see Trapp 1656, 10; Blackwood 1659, 23–4; Poole 1696, II, sig. A3v.

9. Pocock 1870, 353 ('et non carnali copula'); Harpsfield 1878, 241; Bellarmine 1602–3, III (1603), col. 1301 (I.V 'Omnes tamen in eo conveniunt'); Gouge 1622, 112; Rainolds 1609, 34; Ridley 1607, 49; Swinburne 1686, 14; Wood 1712, 49. See Kelly 1976, for a detailed discussion for the role of Katherine's (lack of) carnal relations with Arthur in Henry VIII's case for annulment. On common law, see also E.,T. *The lawes resolution* 1632, 52, 117; and on Continental law, Beust 1597, 169; Schneidewein 1585, sig. F2v.

10. E.,T. *The lawes resolution* 1632, 52; Lagus 1566, 397–8 ('Consummari vero dicuntur, cum coniunctione animorum sequitur coniunctio corporum'). On Restoration drama, see Alleman 1942, 127–8. On consummation, see Kornmann 1610, Linea, 97; Tapper 1582, 280; Fulbecke 1602, f. 24r–v. Some Protestants rejected the distinction between consummated and unconsummated marriage as an ungodly invention of Popish canon lawyers; for example, Beust 1597, 11. Roman Catholic theorists debated whether the pope could dissolve a marriage 'ratum, non consummatum', that is to say, one where the couple had legally contracted but not yet had coitus. Some canonists argued the pope could do so, most theologians held he could not. See the discussion in Azor 1600, II, 322ff. (II.V.iii). The canonist Covarruvias held that the safer opinion was that the pope could not (1597, 229, I.II.7–4.14). For modern accounts, see Smith 1964, 71; Brundage 1987, 375.

11. The term 'spousals' (Latin: 'sponsalia') really has no modern English equivalent. Spousals in the present tense ('I take you as my wife/husband') were the equivalent of marriage, even if made privately with no religious or civil ceremony. Spousals in the future tense [de futuro] ('I shall take you ...') resembled modern betrothal or engagement, but were considered far more binding in character both legally and morally. Since spousals were a normal precursor to the marriage ceremony, early-modern discussions treated them as equivalent for many purposes. See E.,T. *The lawes resolution* 1632, 52; Hotman 1594, 180; Pascale 1618, 321. On the changing attitudes to spousals during the early-modern period, see Ingram 1987, 189ff., and Stone 1990, 66.

12. 'Allegations in behalf of ... the Lady Mary,' in Atwood 1690, 12. This tract was written against Catherine Grey and her descendants' claim to the English throne, and maintained that Catherine's under-age marriage to Henry Herbert

(later Earl of Pembroke) was consummated, and therefore her subsequent marriage to Edward Seymour, Earl of Hertford, was bigamous and void. [This marriage was indeed declared invalid in 1562]. In the late seventeenth century, a lengthy legal case between John, Lord Decies and Katherine Fitzgerald revolved around the intricacies of the law on under-age, non-consummated marriage. For details, see Loftus 1677.

13. Perkins 1618, 673 [presupposed]; Archdekin 1679, 472 ('v.g. per copulam'); Swinburne 1686, 224. See Smith (1964, 74) for the history of this principle, and Quaife (1979, 45) for its popular acceptance. See also Duns Scotus 1639, IX, 793 (XXVIII.I.32); Major 1509, f. 153v; Cucchi 1565, f. 100v; Covarruvias 1597, 141, (I.I.4–1.1); Harpsfield 1878, 45; Beust 1597, 13; Schneidewein 1585, sig. F3v; Hotman objected to this presumption (1594, 177–8). Canon lawyers even argued that coition could amount to consent where the identity of the betrothed was uncertain. If, for example, a man agreed to marry one of three sisters without specifying which, and subsequently copulated with one of them, it would be presumed that he had married that particular one: Antoninus 1474, sig. 07r; Beust 1597, 93; Peleus 1602, 15.

14. Lyser 1682, 7 ('saepe per aliquot menses' 'velint adorari, antequam assensum praebeant'); Stone 1977, 180–91 (quotation at 184). See, for example, Aquinas 1872–82, III (1882), cols 1105–6 (Suppl. 47.6.); Guido [1500?], f. 51r; Lagus 1566, 202; Beust 1597, 4–5. In all these, the standard dictum that 'marriage ought to be free' ('Quia matrimonia debent esse libera') was employed. On the actual practice of marriage, see Houlbrooke 1984, 68–73; Houlbrooke 1985, especially 343, 350; Rushton 1986, 208–9; Carlson 1990, 448–50.

15. Catholics were themselves divided on the issue of the validity of a child's marriage without his or her parent's consent. In the broadest terms, theologians and canonists argued that it was valid, while civil lawyers argued that it was not. Humanists, strongly influenced by Roman law, followed the standard civil law arguments: see Jordan 1990, 54–5. Catholics with Gallican inclinations (for example, Hervet 1593 and de Coras 1605) defended French secular law which, after Henri II's edict of 1556, forbad marriage without parental consent of daughters aged under 25 years and of sons aged under 30 years. For the canonists' and theologians' position, see for example, Turriani 1563, f. 9v; Ludena 1563, f. 8v–9r; Schroeder 1941, 183; Azor 1600, II, 77 (II.II.ii); Bellarmine 1602–3, III (1603), cols 1368–9 (I.XIX); Campanella 1968, 118; Pascale 1618, 248; Olizarovius 1651, 17; Huygens 1696, 67. For the canon/civil law divide, see Bartolus 1570, X, f. 152v–153r; Lagus 1566, 398–9; Tapper 1582, 316.

16. Paul Baynes (1866, 350) described marriages made without parents' consent as 'but licensed fornication'. In contrast, Andrew Willet (1634, 801) thought that 'the marriage being once consummate, we say not then that it ought to be dissolved for want of the parents' consent, but the contract and espousals only'. Marriages without parental consent were held to be void by, amongst others, Chemnitz 1978, 764; Bullinger 1541, 7; Monner 1561, 40; Calfhill 1846, 241; Brenz 1566, 615; Batty 1581, f. 101r; Stockwood 1589, 15, 91; Freig 1591, 266–7, 271; Hotman 1594, 123–5, 198; Beust 1597, 178–93, 193; Hacket 1607, 4; Beard 1612, 380; Pareus 1612, 125; Perkins 1618, 682–4; Whately 1624, 32. This was the standard German protestant view, see Ozment 1983, 36–9. For the English law on this matter, see Carlson 1990, 446–51.

17. Monner 1561, sig. F2v–F3v, 'the consent of the mother is also necessary'

('etiam matris consensum esse necessarium'); Schneidewein 1585, sig. L4r; Hotman 1594, 194–5; Hunnius 1604, 418. See also Batty 1581, f. 99r; Herbert 1648a, sig. X4v–5r; and Chapter 3, Mothers and mistresses.

18. Schneidewein 1585, sig. L2v ('Nec distinguitur de iure divino inter filios et fil-ias'). Ambrosius Catharinus (1551–2, 256–7) and Gentian Hervet (1593, f. 14r–v) were clearly more upset with allowing a daughter to go her own way in mar-riage, than a son. Catharinus was unusual among Catholic theologians in wanti-ng the Catholic Church to render void, marriages made without parental consent (ibid., 255–60). Pierre Ayrault (1614, 71–3) was unusual in maintaining that par-ents had greater rights in relation to a son's marriage than a daughter's.

19. Bañez 1594, 190 (LXII.II); Covarruvias 1597, 173–5 (I.II.3–8), quotation ('nec vim legum habere') at 175; Sordi 1602, f. 4v ('filius indignam duceret uxorem, vel filia nuberet indigno'). See also Tiraqueau 1574, f. 65v; Pascale 1618, 247–8.

20. Beza 1591, 79 ('sine maxima causa'); Bernard 1845, 65 [matches]; Batty 1581, f. 100v [God]; Poole 1696, I, sig. E4v [vested]; Stock 1865, 41 [submit].

21. Taylor 1828, XIV 217–22 (III.V.viii).

22. Mauser in Beust 1597, 346–7, ('viro digno'; 'onera et pericula matrimonii', 'ne invita cogitur nubere, etiam digno'); Hotman 1594, 124 ('propter sanctitatem mat-rimonii'); Campanella 1968, 100 ('in iure naturali'); Tapper 1582, 316 ('sui cor-poris potestatem non habeat, 'instituendum in timore Domini'); Swinburne 1686, 4; Keck 1697, 22. See also, for example, Monner 1561, 40; Tapper 1582, 316; Schneidewein 1585, sig. L4v; Kornmann 1610, 78–80; Pascale 1618, 247–50; Gentili 1614, 345–62. See also Trumbach 1978, 103; Durston 1988, 53.

23. Dillingham 1609, 40; Allestree 1673, 162. See, for example, Agrippa 1540, sig. C4v; Chemnitz 1978, 764; Bullinger 1541, f. 10r, 21–2; Bale 1849, 198–9; Watson 1558, f. 180v; Stockwood 1589, 81–2; Gibbon 1591, 7–8; Topsell 1597, 14, 171–2, 221; Hacket 1607, 6; Pricke 1609, sig. H4v; *Christian parents* 1616, 141, 193–4; Cleaver 1598, 117, 115 Stock 1865, 41; Gouge 1622, 564; *A discourse of the married and single life* 1621, 45–6; Elton 1625, 133; Carter 1627, 141–2; Bernard 1845, 20; Fuller 1845, 18; Gaule 1630, 86, 111–12; Rogers 1642, 74–6, 100–1; Comber 1679, 65–6; *The batchelor's directory* 1694, 231–3; Hopkins 1701, 150; Bingham 1855, VIII, 48.

24. Swan 1635, 502 [*mutual* – my italics]; Hughes 1672, 23 [willingly]; Trapp 1662, 115 [dragged]. On the use of Genesis 24, see, for example, Tapper 1582, 312; Gibbon 1591, 8; *Christian parents* 1616, 195; Settala 1626, 297; Gaule 1630, 111–12; Willet 1634, 802; White 1656, II, 110. For the more standard Old Testament prac-tice of fathers handing over their daughters without any reported consultation, see endnote No. 5 above.

25. Woodhead 1687, 5 [inclinations]; Turriani 1563, f. 12r–v ('non potest pro-hiberi'); Dickson 1659, 51. Others who argued from this text that the daughter's opinion should be considered, include: *Christian parents* 1616, 196; Mayer 1631, II, 204; Ness 1695, II, 135; Poole, II, 1696, sig. Ggg1r; Lapide 1854–9, IX, 216–17. Even Woodhead (1687, 13–14) who noted the gist of Paul's remarks, accepted that a father 'might be necessitated to against his will by the virgin's incontinacibility [sic]'. Primrose (1617, 213) interpreted this passage in the more obvious sense that a father possessed 'the liberty of giving [his daughter] in marriage', but only in the context of his attack on the Roman Catholic tenet that a child could enter the religious life without parental permission. On maternal rights see Chapter 3, Mothers and mistresses.

26. Halkett 1979, 15; Fleetwood 1716, 32–45. Compare the case of Mary Rich, Countess of Warwick (Mendelson 1987, 66–77).

27. Barlow 1692, 122, 125, 24, 26.

28. On reverential respect, see Beust 1597, 176; Mauser in ibid., 348; Schneidewein 1585, sig. K3v; Saravia 1593, 22–3; Covarruvias 1597, 170–2 (I.II.3–6.5); Pascale 1618, 323–4.

29. The following discussions skirt over many of the disagreements amongst medieval and early-modern theorists about the binding nature of contracts made under coercion. I describe those principles that were broadly accepted, but this was a highly controverted matter and many exceptions could be found.

30. Aristotle 1984, II, Nicomachean Ethics, 1752–5 (1110a–11b). See, for example, Covarruvias's (1597, 143, I.I.4–1.11–12) lengthy discussion of the meaning of this section of the Ethics. Aristotle's own example of mixed violence was of a merchant at sea in a storm who threw his goods overboard to prevent the ship from sinking. In one sense, the merchant did not want to lose his property, but clearly preferred this to the alternative of losing his life. 'Such actions, then, are mixed, but are more like voluntary actions' (1984, II, 1752, 1110a).

31. Freig 1591, 263; ('non simpliciter involuntariae et coactae, sed mixtae'); Grotius 1689, 354 (II.xi.vii), 456–7 (II.xvii.xvii–xix), 838ff. (III.xix). For similar arguments, see Beust 1597, 16, 172; Monner 1561, 83–84. See Taylor 1828, XIV, 396; Bañez 1594, 71; Ames 1639, 227, for the highwayman example.

32. Benincasa 1562, f. 148r ('potest per actionem, *quod metus causa*, retrahere'); Grotius 1689, 355 (II.XI.vii); Beust 1597, 172; Huygens 1696, 66; Kettlewell 1691, 31 [princes]. For the case of sovereign states, see also Grotius 1689, 838–40 (III.xix.i–ii). Compare Hobbes 1991, 98 (I.14).

33. *Decretal. Greg.* I.XL.iv. This is a rather cumbersome translation of the Latin phrase 'metus qui in constantem virum cadat'. Virtually all serious discussions of this topic were written in Latin, and unfortunately there is no modern English word which combines the connotations of firmness, stability, and good sense conveyed by '*constans*'. The meaning of the phrase can best be deduced by reading the discussions in, for example, Aquinas 1872–82, III (1882), col. 1103 (Suppl. 47.2); Bonaventure 1889, IV, 699 (XXIX.I.I); Antoninus 1474 sig. n1v; Guido [1500?], f. 59; Major 1509, f. 154v; Baconthorp 1526, f. 163v; Delfino 1553, 27; Lagus 1566, 403; Nevizanus 1570, 228; Bellarmine 1602–3, III (1603), col. 1370 (I.XIX); Huygens 1696, 64.

34. 'Rape' is a translation of the Latin word 'stuprum', which again has no modern English equivalent. It was used of any disgraceful sexual act, for example, the abduction or defloration of a virgin. In fact (in the precise language of theology), strictly rape differs from *stuprum* [dishonour] in that it involves violence, the latter only harassment.

35. Juan Azor's (1600, 25, I.I.x) lengthy discussion was more complete than, but very similar to, those of Bartolus 1570, I, f. 125v; Antoninus 1474, sig. n1v–n2r; Guido [1500?], f. 59r; Delfino 1553, 27; Covarruvias 1597, 164 (I.II.3–4.1 and 2); Gentili 1614, 245–6. For the condition that the fear must be realistic, see Delfino 1553, 27, and Gentili 1614, 244–6.

36. Biderman 1621, 31 ('cum enim matrimonium semel valide contractum, sit insolubile'). Compare Covarruvias 1597, 170, (I.II.3–6.4); Bellarmine 1602–3, III (1603), col. 1370 (I.XIX). That grave fear is incompatible with free choice, see Lancelottus 1564, f. 45r; Cucchi 1565, f. 97r; Gentili 1614, 242; Pascale 1618, 318–19. For divorce, see Chapter 7, Divorce.

37. Hay 1967, 85–9; Beza 1591, 82 ('per vim vel metum qui in constantem virum cadat'); 83 ('liberrimum consensum requiri' 'inutiles'); Perkins 1618, 681; Ames 1639, 201; Willet 1634, 805. See Joyce 1948, 80, for medieval canon law. For the same account, see Covarruvias 1597, 164–75 (I.II.3); Campanella 1968, 98; Suarez 1872, II, 97 (V.XXII.9); Beust 1597, 172–5; Mauser in ibid., 327, 348; Monner 1561, 83; Schneidewein 1585, sig. K2r–v. Thomas Barlow was exceptional in applying Grotius, and Aristotle's theory of the binding nature of contracts made under 'mixed violence' to the marriage contract also; see Barlow 1692, 17–19; and Chapter 7, Parental consent.

38. E.,T. *The lawes resolution* 1632, 59; Wood 1712, 50–1 [prevail]. The case of forced marriage was one of the exceptional ones where a 'wife' could testify against her 'husband', since otherwise the successful execution of the crime would silence its chief victim and witness. See Grey 1730, 140; *Baron and feme* 1738, 447.

39. As Musculus (1548, sig. b6r–v) argued, fornicators and adulterers 'also couple their filthy, damned flesh together'. 'Therefore it must be perpended, what singular thing holy wedlock hath of itself, wherein it differeth from fornication and adultery.' See also Poole 1696, II, sig. Fff3r.

40. Cleaver 1598, 93; E.,T. *The lawes resolution* 1632, 51–2; Ridley 1607, 7. Almost all early-modern definitions were simply adaptations of the Roman civil law (*Digest.* XXIII.II.1) and canon law (*Decreti.*C.27.q.2.1) formulae. Compare, for example, Guido [1500?], f. 52r; Lagus 1566, 397; Lancelottus 1564, f. 39v; Campanella 1968, 62; Bellarmine 1602–3, III (1603), col. 1318 (I.VIII).

41. Dod and Cleaver 1629, 202. Catholic theorists afforded infidel marriages a limited degree of recognition, as lawful (*legitimus*), if they were in accordance with natural law, but not indissoluble (*ratus*), because they did not satisfy divine law. Protestants argued that Christian marriage did not essentially differ from pagan. See Joyce 1948, 492–9 for a more detailed discussion of these questions.

42. Sanderson 1685, 12–16 [vows]; Gentili 1614, 485 ('Et nec solo consensu hominum, sed & Dei contracta hic obligatio intelligitur, ut est scriptum, *Quod Deus coniunxit*'); Bastingius 1593, f. 220v–1r [together]. For the doctrine that a vow is a promise to God, see Delfino 1553, 74; Lagus 1566, 370; Settala 1626, 185.

43. Bartolus 1570, I, f. 6r; Craig 1703, 17 [portions]; Vaughan 1706, 207 [law-making]; Laymann 1625, 444 (V.X.IV.I.1 'superveniens', 'humanum contractum'); Maldonado 1677, II, 418 ('& contractus quidam civilis inter virum & uxorem'); Beza 1591, 246 ('quoniam quatenus ad societatem humanam spectat, plane civilis est'; 'plane divinus est et ecclesiasticus'). Catholic theologians thought marriage both a sacrament and a human, civil contract; see Azor 1600, 35, (I.I.xi); Bellarmine 1602–3, III, (1603) cols 1315, 1381 (I.VIII., XXI); Campanella 1968, 90; Kellison 1605, 215.

44. Willet 1634, 811 [munificence]; Gentili 1614, 1–89 (Book I); Beza 1591, 246–7; Coke 1662, 78 [riot]; see Durston 1988 (especially 52 and 57) for the unpopularity of radical puritan attempts to remove marriage from the religious sphere; and compare Comber 1679, 3. Lutherans tended to allow the civil magistrate extensive powers in all areas including marriage, but even they argued that 'Because marriage in the church has as its norm the word of God itself ... the church cannot simply refuse to deal with matrimonial cases and simply throw them to the civil magistrate' (Chemnitz 1978, 759). Compare Trapp 1656, 256. See Helmholz 1990, for a lucid and informative account of canon law in early-modern England.

45. *Treason, popery* 1680, 14–15 [seasons]; Cock 1668, 151–65. Erastian theorists, notably Hobbes, Marc' Antonio de Dominis and John Selden, began to suggest that marriage should be regarded as a *purely* human and civil contract, rather than a spiritual one. See Dominis 1617–20, 431, and Selden 1991, 268–77 (II.28–9). Their views were unusual for most of our period.

46. Keble 1685, 386, 188; Vaughan 1706, 214. Like many common lawyers, Vaughan extended the power of English law to declare what was God's law to a degree which would have horrified less Erastian Christians. (See, for example, ibid., 327–8). Nonetheless, where God's law was known, he accepted that no human law could stand against it.

47. Turner 1698, 57–60; Antoninus 1474, sig. l6r ('deus fuerit auctor et institutor coniugii'); Chemnitz 1978, 718 [author]; Baynes 1866, 351 [hell]; Bernard 1845, 9 [beasts]. See also Lawson 1659, 205; Swinnock 1868, 470.

48. Chemnitz 1978, 719–20 [associations]; Hughes 1672, 25; Barlow 1692, 4–5, 7–8. John Butler carried his unorthodox views about marriage into practice by divorcing and remarrying without legal approval, but he, too, accepted that 'the duties of marriage' are to be found in 'holy Writ' (Butler 1697, 7). For the same views, see Raulin 1512, sig. Ir; Turriani 1563, f. 7v; Brenz 1566, 279; Campanella 1968, 70; *Disputatio* 1641, 188; Huygens 1696, 87–8.

49. Nevizanus 1570, 86 [scholastics]; Cleaver 1598, 134 [liberty]; Baxter 1830, 159 [tempore]; Astell 1730, 17 [contract]. Compare Antoninus 1474, sig. m7v; Agrippa 1540, sig. B2r; Turriani 1563, f. 7r; Fisher 1597, 264–5; Schneidewein 1585, sig. G1r; Beust 1597, 109; Mauser in ibid., 319; Major 1509, f. 150r; Pocock 1870, 365–6; Cardwell 1841, 75–6; Cucchi 1565, f. 95r; Primrose 1617, 230; Gouge 1622, 116; Settala 1626, 46; Willet 1634, 307–8; Rogers 1642, 116; Trapp 1656, 81–2; Corbet 1685, 225; *The house-keepers guide* 1706, 103.

50. One Scriptural reason for doubting the natural indissolubility of marriage was Paul's teaching in I Corinthians 7:15, on which was based the so-called 'Pauline privilege'. Medieval and early-modern theologians interpreted this passage as permitting a Christian convert (male or female), whose spouse remained an infidel and refused to cohabit peacefully, to separate and remarry a Christian. Since it was generally believed that God did not dispense from the law of nature, this concession seemed to imply that only Christian (not infidel, natural) marriages were indissoluble. See Bailey 1959, 111–15 and 176, for a description of the development of this doctrine.

51. Duns Scotus 1639, XI.2, 805 (XXI.I.11 'ex institutione Legislatoris Dei', 'posset esse datio mutua ad unum actum tantum, et ad tempus unius horae, vel alterius horae satis brevis;'); and see also 816 (XXXIII.III.6). Scotus believed that God's original institution of matrimony was simultaneous with the creation of mankind, so fornication or temporary marriages had never been licit. He also believed that the good of the children meant that indissolubility was consonant with right reason. He did, however, believe that unassisted natural reason would not recognize that indissolubility. Duns Scotus 1639, IX, 671–2, (XXXI.I.5). Durandus (1571, 377–8, XXXIII.II–III) reasoned in much the same way.

52. Catharinus 1551–2, 239 ('si non instructi de lege Dei'); Delfino 1553, 5 ('evidentissima'); Tapper 1582, 280 [= 278] ('obscurum sit ius naturae'); Locke 1988, 321 (II.81); Pufendorf 1934, 875–7 (VI.I.20).

53. More 1989, 82–3; Luther 1959, 105; Bucer 1955, XV, 152–200; Milton 1959, 344. For Catholic condemnation of Bucer's views on divorce, see, for example, Tapper

1582, 286–7; Bellarmine 1602–3, III (1603), col. 1342 (I.XV); Kellison 1605, 30. For the orthodox view of Milton's proposals, see Pagitt 1646, 150. For Butler, see endnote No. 48 above.

54. Selden 1991, 200 (II.19); 400 (III.22); 433 (III.26).

55. Selden 1991, 402 (III.22), 507–8 (III.33); Vaughan 1706, 339 [wedlock]. For the more traditional view of easy divorce, see Maldonado 1677, 461–2; Comber 1679, 99. For one attack on Selden's views, see Thorndike 1844–56, IV, 293–305.

56. Blount 1911, 325; Thorndike 1844–56, IV, 328 [stinking]. *Concubinage and poligamy disprov'd* 1698, 30 'no man in his right wit did ever allow that a private man on his own authority should put a murderer to death, or divorce his wife and marry another'. Thomas Paget argued for divorce where one of the spouses was an idolater, but even he insisted that it must be done by judicial process (1650, 6).

57. Vaux 1599, 128. Joyce 1948, 380: 'From the days of Gratian until the opening of the sixteenth century the doctrine of the indissolubility of marriage was undisputed in the Western Church'.

58. Erasmus in his commentary on I Corinthians 7, spoke strongly in favour of divorce with remarriage in cases of adultery by the wife, and was an early advocate of the notion that separation *a mensa et thoro* was a purely human invention, unthought of by Christ. His arguments were translated and printed in England in about 1550 by Nicholas Less as *The censure and iudgement* of Erasmus on divorce. Other Catholics such as Cardinal Cajetan and Ambrosius Catharinus also argued in favour of divorce for adultery until the Council of Trent's edicts stilled all debate.

59. *An essay towards* 1697, 30 [cod-piece]; Grantham 1641, 6 [fish]; Duck 1724, xxxvi [statute]. See Ozment 1983, 83–92, for German protestant revisions of divorce law. On English marriage law, see Haw 1952, 59–67; Carlson 1990, especially 441.

60. Burn 1788, II, 460. For a thorough, scholarly, and readable account of the actual practice of divorce in England, see Stone, 1990 – parliamentary divorce at 301–22.

61. 'Allegations in behalf of ... the Lady Mary', in Atwood 1690, 5 [Brandon]; Stock 1865, 188 [dislike]; *Certain sermons or homilies* 1756, 121; Rogers 1642, 218 [dogs]. See Stone 1990, 307–8 on the cases of Blount and Rich; for tacit defence of such conduct, see, for example, Willet 1634, 785. The act punishing bigamy capitally was 2 James I. c.11. On voluntary separations, see also Agrippa 1540, sig. D2–3; Abbot 1608, 49–50; Webb 1653, 89. For informal divorce and remarriage, see Gillis 1985, 99; Hair 1972, 106 (case 224); Ingram 1987, 178ff. Such informal separations were never legally valid: see Chamberlayne 1673, 336.

62. Falk 1966, 113; Rutherford 1843, 196; Jansen 1641, 591 ('ius enim repudii competebat viro, non uxori'); Laud 1847–60, VI (1860), 617 [only]. See Corbett 1930, 242ff., for the Roman law of divorce.

63. Milton 1959, 327.

64. Augustine's views are laid out in 'To Pollentius on adulterous marriages': Augustine 1955, 55–132. (Pseudo-)Ambrose's solution can be found in Ambrosiaster 1845, XVII, cols 45–508, especially 218 and 225. Ambrose is now thought to be "Ambrosiaster", see Chapter 3, Restrictions on inferiority, endnote No. 16 of Chapter 3.

65. Catharinus 1551–2, 275–92 (280, 'mulieri quidem iuxta Scripturas non licere

discedere et alii nubere, etiam ex causa fornicationis: viro autem contra licere utrunque'); Pareus 1631, 481 ('Deus tamen non viri, sed uxoris tantum fornicationem causa divortii esse voluit'). Bernardino Ochino (1657, A dialogue of divorce) also argued that a husband could divorce an adulterous wife but not vice versa, using arguments consistent with his support for polygamy, but his views were generally rejected on both points.

66. *Decreti* C.32.q.5.c.20; Hay 1967, 65 [dismiss]; Lessius 1621, 211 [right]. Ambrose's opinion was dismissed as inferior to Augustine's by, for example, Delfino 1553, 19; Campanella 1968, 180–4; Chemnitz 1978, 753–4; Fulke 1601, 504; Maldonado 1677, 468–9. On medieval thought, see, for example, Hincmar of Rheims 1852, cols 656, 708; Bartolus 1570, X, f. 154r; Antoninus 1474, sig. Q1–2; Durandus 1571, II, f. 380r. The tag that adultery was equally illicit for male and female because 'eadem servitus pari conditione censetur', was often repeated in early-modern discussions of the topic: see, for example, Tapper 1582, 290.

67. Poole 1696, II, sig. Q1v [interpreters]; Dove 1601, 51 [wantonness]. For Catholic argument from equality, see Tapper 1582, 290; Kellison 1605, 330; Bellarmine 1602–3, III (1603), col. 1337 (I.XIV); Maldonado 1677, 466; for the protestant view, see also Mayer 1631, II, 200.

68. Plutarch 1603, 318; Bingham 1855, VIII, 86–7. Byfield (1637, 583) likewise insisted that a wife's 'subjection doth not bind her to consent to or conceal his whoredoms, wherein he breaks the covenant betwixt them and defiles the marriage bed'. See also Tilney 1568, sig. B7v.

69. Beza 1591, 180–1 ('manifeste fatetur ex Domini verbo'); 170–1 ('agitur autem proprie de hoc iure in causa adulterii'). See also Monner 1561, 141; Gualther 1570, 412r; Schneidewein 1585, sig. P1r–P3r; Freig 1591, 285; Bucer 1955, XV, 194–6; Brenz 1566, 275–9.

70. For example, see Cosin 1845–51, 489–502; Willet 1634, 777ff.; Blount 1911, 337–8 and Stock 1865, 164 for defences of divorce with remarriage. John Howson's tract *Uxore dismissa propter fornicationem aliam non licet superinducere* [It is not licit to marry another wife after dismissing one on account of adultery] 1606, offered a particularly vigorous (and lengthy) defence of the conservative, high-Church Anglican view. Others included Prideaux 1664, 299–301 (6.VI); Dove 1601; Laud 1847–60, VII (1860), 618.

71. Hooper 1843, 382–4, 378–9; Perkins 1618, 690. See Thomas 1959, especially 203, and Bailey 1959, 211–31 on early-modern English protestant views on divorce.

72. Willet 1634, 780–1, 783, 785 [The canon law cited was *Decreti*, c.1.C.29.q.2]

73. Milton 1959, 591, 700.

74. Baxter 1830, 161 [habitation]; Towerson 1676, 402–5 [common]. See also Andrewes 1854, 107; Gouge 1622, 219; Mayer 1631, I, 105, II, 194; Blackwood 1659, 263; Corbet 1685, 238–9; Poole 1696, II, sig. H4r; Durston 1989, 98.

75. Harris 1912, 322. See Stone 1990, 309–12 for a more detailed account of the Roos case.

76. Taylor 1828, XII, 143–51 (I.V.viii), 482 (II.III.iii). In arguing that a wife could continue to live with an unfaithful husband, Taylor was following scholastic tradition. Canonists and casuists had attributed to the spouses equal rights, but not equal duties: a wife was not *obliged* to put away an openly unfaithful husband, whereas – to avoid public scandal – a husband must dismiss an openly adulterous wife. Wives were excused because they generally lacked the requisite author-

ity, and because husbands' infidelity did not place in doubt the legitimacy of the family's heirs. In civil law, a husband who failed to dismiss an openly adulterous wife could be adjudged guilty of *lenocinium* [pimping or pandering]. See, for example, Antoninus 1474, sig. Q2v; Durandus 1571, II, f. 380r; Nevizanus 1570, 69; Sordi 1602, f. 200v; Pascale 1618, 304.

77. Hammond 1847, 139–44; Thorndike 1844–56, IV, 319.

78. See Stone 1990, 360–2; Ingram 1987, 182–3; Amussen 1988, 57. English ecclesiastical law on divorce *a mensa et thoro* was theoretically equitable between husband and wife in its treatment of sexual infidelity as a ground for divorce. See Ridley 1607, 8, 56; Burn 1788, II, 457–63.

8

Contract and subjection

Subjection in marriage

Most early-modern theorists believed that marriage was indissoluble except when God's law gave a right of divorce. Apart from a few unusually irreligious theorists such as John Selden and Thomas Hobbes, no one believed that marriage was a contract like any other whose terms and conditions were at the discretion of the parties consenting to it. Neither wife nor husband could contract a marriage that would last only as long as one were satisfied with the conduct of the other. A woman could not legitimately remain a wife for as long as, and no longer than, she found her husband satisfactory. She was bound and subject to him for as long as he refrained from infidelity or desertion.

God was believed to have built rules into the institution of marriage that no couple could ignore. Contractual conditions contrary to the divine ends of marriage would not bind any couple even if they agreed to them. For example, they could not agree to marry and then prevent the conception or birth of children. By God's law, procreation was the primary end of marriage. Similarly, no couple could agree that one or both should be free to commit adultery, since averting fornication was another divinely ordained object of marriage. Such conditions were 'contrary to the essence of matrimony'.[1] Those marrying obliged themselves to observe God's laws 'concerning the immediate and principal end of marriage'; they must abide by the house rules or sin mortally. Early-modern theorists believed that anyone consenting to marry chose not only a particular mate but also a particular state – the state of matrimony. It involved obligations pre-ordained by God that were quite independent of either spouse's will, or both of them combined.[2]

The majority of early-modern theologians thought that God had commanded wives to be subject to their husbands. They believed that Scripture was as clear on the wife's obligation of obedience as it was on procreation and fidelity. To them, God's words to Eve – 'thy desire shall be to thy husband and he shall rule over thee' (Genesis 3:16) – did not seem to admit of debate. Every time they mentioned marriage, Saint Peter and Saint Paul ordered wives to obey their husbands. 'A wife then, say those Apostles, is one that is subject and obedient to her husband as her head.' The Fathers waxed lyrical on the need for a wife's subjection, and medieval theologians reiterated their exhortations. Pagan philosophers adopted exactly the same position. Faced by such a cloud of witnesses, the subjection of the wife was built into the concept of marriage. The Augustinian definition of marriage as a relationship of subjection, 'a certain friendly relationship of the one ruling and of the other obeying', was still thought persuasive.[3] Early-modern theorists found it enormously difficult to read the Bible without reaching this conclusion.

> The commands are so many and so express, that there is scarce any other duty which the Scripture doth urge with so much insistence and earnestness, with such pressing reasons and enforcing motives as this of the wife's obedience.

No wife could deny her duty to obey without rejecting God's word, for it was 'out of the Apostle Paul that the woman ought to be in subjection to her husband'. Most early-modern exegetes did not believe that the wife's obedience was an optional extra; it was an essential part of the marriage.[4]

In Renaissance theory, when a woman consented to marry and live after God's ordinance she accepted subjection to her husband. It is important to clarify two aspects of this doctrine; both stemmed from the couple entering an institution whose rules were established by God, not by themselves. First, the woman's obligation to obey her husband existed simply by virtue of the nature of marriage, *not* because of any promise to obey that she might or might not have made. In the Anglican marriage service, a woman did vow to 'obey' and 'serve' her husband, but if these words were omitted (say, in a clandestine service) she was not one jot less obliged to obey. In just the same way, spouses agreed in the Church of England's prescribed service to marriage 'as long as ye both shall live', but marriage was no less indissoluble if they did not explicitly agree to this. Indissolubility and wifely subjection were part of God's institution and therefore immutable and inescapable. Second, because no woman could be married without her own consent, her subjection resulted from her consent. However, this did *not* mean that the husband derived his power from the wife's consent. Virtually no Renaissance theorist thought that a wife granted her husband the power he possessed over her. God gave power to the husband by making marriage an institution in which husbands ruled. This distinction was crucial since so much followed from it. A woman who gave her husband power, might give it only on certain conditions, judge whether he had fulfilled them, and withdraw the power if he had not. Since *God* gave husbands power, no such limitations were possible.[5]

In early-modern eyes, marriage was an institution like the armed services in modern society. In a society without press-gangs or conscription, no man can be forced to serve. The soldier's obligation to obey his officers therefore results from consent. However, the officers' power of com-

mand is not given to them by the soldier, but by the state. Once he has
entered the military, the soldier cannot withdraw because he dislikes
officers' use of their power. When enlisting, the soldier cannot determine
his own conditions of service and cease to obey if they are not met. The
officers' power of command and the soldier's duty to obey exist quite
independently of any private contractual arrangements made between
them. The officers' power is derived from law; the soldier's consent
merely renders him subject to it. This analogy was used by early-modern
theorists arguing that husbands should always rule, not agree to obey
their wives.

> It is not humility but baseness to be ruled over by her whom he should rule.
> No general would thank the captain for surrendering his place to some com-
> mon soldier: nor will God [thank] the husband for suffering the wife to bear
> the sway.

God was the general, the husband was the captain, and the wife the 'common soldier'. The private should not command the captain even if she
were more intelligent and capable, nor should the captain privately consent to accept her orders. The regulations of the institution laid down that
captains (husbands) command, and common soldiers (wives) obey.[6]

Wives were under husbands' 'power and domination by the ordinance
of God'. God wanted wives to obey their husbands; that was why He told
Eve to obey Adam and why the Apostles reiterated the point. 'Wives submit yourselves unto your own husbands', pronounced Saint Paul,
because 'the husband is the head of the wife' (Ephesians 5:22–3). Early-
modern theorists accepted that Paul was speaking on God's behalf. God
made 'the man to be the head' and the wife should be 'glad to acknowledge the pre-eminence and superiority without disdain for the ordinances
of God'. All theorists agreed that marriage arose by consent; none of the
endless tracts that detailed wives' moral duties suggested that a wife need
only obey because she had consented to do so, or need obey only as long
as she found her husband's performance satisfactory.[7] On the European
Continent, Roman Catholic theorists put forward just the same theory of
husbandly power. The 'sacrament of matrimony' made a husband his
wife's head. He ruled by God's decree 'not by the will of the wife'.
Marriage was based on voluntary contract of the spouses but 'if they do
contract marriage they cannot impede this superiority'. The subjection of
the wife to the husband stemmed from God's will and existed independently of (and even despite) the couple's own inclinations or agreement.[8]

Wives should obey their husbands because 'God commandeth it'.
Even queens, if they married, were (as wives, though not as magistrates)
'subject by the law of God'; no mere social edge like being born to rule a
whole realm could interfere with this basic divine duty. The same lesson
applied to each and every wife, however much more noble, wealthy, or
talented than her husband:

Let thy birth, thy education, estate, endowments exceed his never so much: yet the ordinance of God hath subjected thee to thine husband with all thy perfections.

More-intelligent or better-educated wives could no more disobey their husbands than citizens now can disregard the orders of stupid policemen. The same arguments were used in the case of children. Even children with 'more learned education' must honour and obey their parents, 'for still that does not lessen the natural superiority'. Sons must obey their mothers, although 'the infirmity of their sex and the imperfections of their age are combined together'. Many children

exceed their fathers in wit and wisdom ... yet is this no good reason that they should take upon them their fathers' authority. The wife may not therefore be a master because she has more knowledge sometimes than her husband, but she must obey and the husband is to rule, because that God hath willed that it should be so.

Generals are deemed normally to be more competent than private soldiers, but privates cannot mutiny and take command in the generals' place even when they are lions and the generals are donkeys; 'God hath ordained subjection to an husband as an husband, be he what he may'. The properly ordered family (or army) was a meritocracy, but the obligation of subjection did not diminish when the commander was stupid or ignorant.[9]

The power of the husband was part of the divine institution of marriage; this was what Saint Paul's teaching on a husband's headship meant. The 'husband's power does not arise from the wife's submission or subjecting herself to her husband'. Submission was simply an act of the will; if submission were the only grounds of a wife's obligation to obey, then by another 'act of her will she may when she list set herself free'. Women were free 'before they marry', but afterwards the husband possessed power, 'from the law of God and nature', which made husband and wife 'one mystical person of which the husband is the head; and ... therefore the husband is the director and ruler of his wife'. The woman's consent was necessary to her subjection, but the power to rule came from God not from her consent.[10] The husband 'hath not his authority ... from the wife who chose him for her husband', but 'from God, who hath made the husband the head of his wife'. Women were completely free in their choice of husbands, 'but after it they are not free as to their subjection, nor can they disobey them without disobeying God'. Political patriarchalists often bolstered their arguments for the divine right of kings by pointing to the divine right in husbands that even contractarian theorists accepted. Wives gave 'voluntary subjection and obedience' but God appointed husbands 'heads and governors'. God's law of nature gave power to husbands so that 'marital and husbandly power is natural, though it be not natural but from free election that Peter is Ana's husband'. A woman could chose which male she would marry (just as soldiers could choose in which regiment to enlist), but once the choice was made, the nature of her subjection was determined by laws independent of her will.[11]

The wife's subjection was no more a function of her husband's will than of her own. His duty 'to rule his wife' was as inescapable as his duty to maintain, love, and be faithful to her. A husband who allowed his wife to take on the 'honour of headship' was sinning. He could let her act as his lieutenant, but must never 'resign or give over his sovereignty unto his wife'. God 'made man head of the woman' and God's will was flouted when a husband made himself 'equal with his wife in power'. Husbands' authority was ordained by God; neither jointly nor severally could spouses legitimately make contrary arrangements. Works of domestic instruction coupled insistence on wives' duty to obey with condemnation of husbands who let wives wear the trousers. A husband 'must not be his wife's underling contrary to the order of nature and ordinance of God'; 'the man that suffereth his wife to take his place hath already transgressed the order of God'. A husband neglecting to rule his wife was like a parent not teaching a child, or an owner failing to train a pet; such actions were irresponsible, not generous. God had placed wives under strong, sensible males' guidance because women were weak, silly, and emotional: His trust was 'not basely to be betrayed'. God had not left husbands free to chose whether to rule their wives; He had commissioned them to do so and would not take kindly to any shirking of responsibilities.[12]

Princes and parents could not absolve subjects or children from their natural duties, nor 'the husband release his wife from the obligation to be subject and pay the [marital] debt only to him'. These duties stemmed from divine and natural law, and no human agreement could alter them – 'any pretended derogation from them is void'. Jean Bodin argued that any matrimonial contract purporting to release a wife from subjection was of no effect, because it was 'contrary both to the laws of God and men' and 'to public honesty'. If made, it was 'not to be observed and kept' and indeed, 'no man can thereunto be bound by oath'. In a case of 1663, English lawyers showed that they held exactly the same opinion. When a wife tried to enforce her contract with her husband that 'she shall not be subject to him', her case was thrown out of court because the contract was 'void, and contrary to the law of God and nature and public honesty'. The husband's power over the wife was ordained by God and she could not 'deliver' herself from it.[13]

The belief that husband and wife could not bindingly contract that she should not be subject was long the standard one. Most sixteenth- and early-seventeenth-century political theorists accepted that certain duties were naturally and divinely entailed in the relationships between parent and child, husband and wife, and ruler and subject. But during the seventeenth century, some theorists of natural law began to move away from this belief. Just as John Selden and Hugo Grotius tentatively insinuated that Christianity did not offer the only possible legitimate form of marriage, so some political philosophers began to rethink other social relationships. Thomas Hobbes was merely the foremost representative of this tendency to reappraise what modes of association reason and agreement might lead men to establish. Abandoning traditional assumptions about the 'naturalness' of familial affections and of human sociability, Hobbes attempted to derive the rules of all co-operative and authoritative organizations from man's instinct for survival and security. Very few theorists could stomach such strong meat as the unabashedly absolutist conclusions that Hobbes drew from his dissection of social conventions. But many questioned more searchingly what obligations 'nature' taught, and reconsidered critically what duties abstract reason would lead man to accept in social life. In the course of this re-examination, a few theorists questioned the belief that a wife's obedience to her husband was ordained by the law of God and nature and that no private contract could impede it.

Thomas Hobbes used contractarian arguments to reach conservative conclusions by insisting that a contract made under duress was naturally binding. Where government was absent, conflict would be endemic, and those weak males and women who lost battles would agree to subjection to save their lives. In a Hobbesian state of nature, supremacy between husband and wife would be decided by struggle. Individuals would fight to gain power and establish control of families; the losers would agree to accept subjection. Males generally won wars with females, became the heads of families, and established commonwealths in which husbands held power over wives, but in exceptional cases – such as that of the Amazons – contracts might establish female dominion.[14]

Liberal contractarians reached very different conclusions from Hobbes about government, because they argued that a contract made under duress did not bind. However, although Whig theorists were motivated to combat the very absolutism Hobbes supported, some of them also began to maintain that a wife *could* contract marriage without subjection. Such a contract would be abnormal, but not necessarily invalid. John Locke argued that God's punishment of Eve applied to all women and entailed 'that subjection they should ordinarily be in to their husbands'. However, he argued that this penalty was not legally binding on women, any more than the painful childbirth to which Eve was also condemned (Genesis 3:16). God was merely describing what would normally be the case, not commanding what ought to be. Just as a woman could avoid painful labour if an anaesthetic were available, she could avoid subjection 'if the circumstances either of her condition or of her contract with her husband should exempt her from it'. Locke took this stance in the

Two treatises, although elsewhere he insisted on the 'natural superiority of the man' and women's 'natural subjection to the men'. His eagerness to gainsay absolutist notions of subjection led him to an extreme position on possible contractual relations between the sexes that little accorded with his ideas of their natural relationship.[15]

James Tyrrell was another liberal contract theorist. He toyed with the idea that marriage might 'subsist by a bare compact or the power of friendship alone', without any power of punishment in the husband. He rejected it because he believed that the family must have one head possessed (ultimately) of punitive power. Even so, he contended, power was only normally in the husband, 'being commonly stronger both in body and mind'. Where that was not so, 'the subjection will likewise of course cease'. In unusual circumstances, the wife 'may and does' govern her husband, 'his family and estate'. Scripture generally mandated husbandly rule, but husbands might forfeit 'this prerogative' by incapacity or negligence. Another late-seventeenth-century political philosopher, Richard Cumberland, argued similarly. Because husbands were generally more able and because Scripture endorsed their superiority, any woman who did not specifically opt out of wifely subjection 'does tacitly contract to submit herself'. But this general regulation only applied where there was 'no contrary covenant'. If a wealthy or gifted woman would not marry unless she were the ruler, and if the male accepted this, 'she would have right by the law of nature to the same dominion which now is in the husband'. Such a contract would not be invalid according to the Gospel or the law of nature for '[g]reater strength of either body or mind is not universal in men'. 'The Gospel has done no more' than suggest a 'prudent', 'general regulation'. Scripture was only indicating what distribution of power between the spouses would normally be beneficial; it did not immutably mandate the wife's subjection.[16]

These theorists cited and interpreted Scripture in a novel and heterodox fashion to serve their anti-absolutist purpose. From time immemorial, biblical scholars had taught that wives' subjection was 'the sentence of God given on Eve and all Eve's daughters, which may not be revoked'. Countless commentaries had insisted that God's directive for Eve to obey Adam 'must be an ordinance to all her daughters forever'. Whig theorists broke with the ancient and virtually unquestioned orthodoxy that Genesis 3:16 and the Petrine and Pauline Epistles unambiguously prescribed wives' obedience. Eager to destroy every plank of Tory political philosophy, they denied that God dictated the duties of subjection inherent in any social relationship. Whigs insisted that the extent of any obligation to obey arose only from consent and contract; Scripture merely described the normal pattern of subjection, it did not immutably prescribe it. Liberal contractarians consistently extended this assertion to conjugal power. Wives were not merely (as had always been accepted) placed under husbands' divinely decreed power by consent. They held that wives were subject because of their consent, and that the nature and extent of that power was determined by contract.[17]

This shift produced a problem in political theory – that is, why should free women agree to such subjection? This had not been a problem for conventional theorists, since any woman who wanted to marry was morally obliged to accept subjection to her husband as part of the divine institution of marriage. The change also highlighted a problem implicit in earlier writings about the political rights of women. If conjugal power were not political, why should a woman's conjugal subjection deprive her of political power? Furthermore, why did not unmarried (and therefore free) women possess the same political rights as males?

Conjugal subjection and political independence

Some contract theorists allowed that a woman might contract marriage without subjection to her husband. Why, then, should any woman consent to a marriage rendering her subject? This was obviously not a practical problem for early-modern Englishwomen since the law afforded them no choice. An Englishwoman could remain legally *sui iuris* by not marrying; but if she married, her obligation to obey and her status as a *feme covert* were unavoidable. No private marriage contract would stand against the law. The social and economic facts of life in early-modern England in any case meant that only a tiny minority of women were in a position to bargain with males on equal terms. Similarly, in practice, no Englishwoman – married, single, or widowed – was politically free in the sense of being able to wield political power. With the sole exception of queens regnant (a small and select group), no woman could hold a public office of political power. By law, a woman could not stand for election to parliament, nor exercise the franchise.[18]

For patriarchalist and, indeed, absolutist theory in general, the exclusion of women from political power produced no contradictions in their theory. English political patriarchal theory held no sexual bias except that implicit in male primogeniture. The Anglican cleric, Hadrian Saravia, wrote full-blown patriarchal theory under Elizabeth I and it was published by the royal printer as a mark of official approval. If a woman inherited the throne she ruled. The powers of a queen regnant were as great as those of a king regnant.[19] In patriarchal theory, God did not give political power to all free men; both male and female were born subject. God gave power to the prince and if (s)he chose to rule autocratically or to delegate power only to some of the male sex, (s)he was not acting unjustly towards women or the excluded males. The prince was God's vicegerent upon earth and against Him only could (s)he sin.

For contractarian theorists who vested power in the people's consent, the case was altered. If all males were born free, so also were all women.

If no male could justly be subjected without his own consent, so, too, could no female. Patriarchalist writers gleefully pointed to the contradictions in contractarian theory that appeared to result. If all were 'naturally free' and 'naturally equal' then 'women and children ... have as plausible a pretence to the government as the greatest peers in the realm'. Sir Robert Filmer scoffed at the notion of an original contract. Even if everyone had once consented to government, new people were being born every moment and if contractarians were right, they could not be subject to government without their own consent. To deprive even children of their right to consent would be unjust;

> not to speak of women, especially virgins, who by birth have as much natural freedom as any other, and therefore ought not to lose their liberty without their own consent.

An unmarried woman had consented to obey no one, so why (on contractarian reasoning) should she be bound by law? Another patriarchalist, George Hickes, pointed out that contractarians maintained that all rational people share in political power. Why then should not women's consent to the political process be necessary? Women, too, were 'useful members of the commonwealth' and 'necessary for human societies'. He could see no 'Salic law of nature' that excluded women from power. Rhetorically, Hickes became an enthusiastic advocate of women's equality.

> Who gave the men authority to deprive them of their birthright and set them aside as unfit to meddle with government; when histories teach us that they have wielded sceptres as well as men, and experience shows that there is no natural difference between their understandings and ours, nor any defects in their knowledge of things but what education makes?

Hickes was no proto-feminist. He aimed at discomfiting liberal contract theorists by extending their argument *ad absurdum*. In his view, if contractarian arguments involved asserting that, naturally, political power should be shared with women, they were patently preposterous.[20]

How did (or could) contractarians cope with these accusations of inconsistency? Few responded at any length to patriarchalist charges about the place of women in their theory so the contractarian reply must be constructed from principles they accepted and expounded on related issues. Such an exercise runs a significant danger of anachronism even if the more facile pitfalls (such as assuming that all the theorists were hypocritical misogynists) are avoided. Nonetheless, the enterprise is certainly more justifiable than reiterating charges of inconsistency without serious examination. In the case of *married* women, certain principles are clear. Liberal contractarian theorists did not equate conjugal and political power. A husband's power over his wife was domestic, not political, both in its nature and its extent; he possessed only the power necessary to run the family and household. He was not his wife's political ruler.

Early-modern theorists also conceded (when discussing the case of queens regnant) that a conjugally subject woman could be her husband's political superior. Given these tenets, what barred a married woman from exercising political power? In order to answer this question we must look at early-modern attitudes to political independence in general.[21]

It was generally accepted in early-modern England that a certain irreducible minimum of autonomy and self-sufficiency was a prerequisite for free political action. Anyone who was wholly dependent on another person for their position and advancement could not be expected to make political decisions freely. Even the Levellers (the most politically radical group in serious politics, who wished to extend the franchise extremely broadly in seventeenth-century terms) agreed that servants and semi-feudal retainers had interests so closely related to their masters' as to be incapable of voting disinterestedly. Apprentices, servants, and those dependent on charity should be excluded from the franchise, 'because they depend upon the will of other men and should be afraid to displease them'. This belief underlay the consensus that only property-owning heads of households should be eligible for political office and allowed to vote. However slight was the agreement between patriarchalists and liberal contract theorists, both accepted that government was a matter for the heads of households, that 'the fathers of families, or freemen at their own dispose were really and indeed all the people that needed to have votes'. Servants' interests were too closely tied to those of their patron for them to hold independent opinions on political issues.[22]

In early-modern terms, what was true on this issue for servants was doubly true for wives. If servants were to be excluded from the franchise because they did not possess enough property, were closely tied to their masters by economic interests, and were strongly motivated to obedience, how much more so was this the case for a wife? Under English law, at marriage all a wife's property except her jointure became her husband's, she ceased for many purposes to have an independent existence at common law, and she solemnly vowed to obey her husband. English law virtually ensured that – at least in economic terms – no woman's interests would ever significantly differ from those of her husband; 'the interests of married people' were 'exactly the same in all things whatsoever'. After marriage they were partners in everything, and law and theory required and reflected this; 'there is so strict union between a man and his wife that the law counts them one person and consequently they can have no divided interest'.[23] Since one of parliament's major functions was to vote taxation, independent control of property was seen as an essential qualification for the franchise, but (in the words of one modern commentator) a husband's powers 'made it virtually impossible for a wife to be able to make independent decisions about her own property'.[24]

Political assumptions firmly established in Western thought since

Aristotle, grouped wives, children, and servants together.[25] Not only political patriarchalists held these attitudes; 'Commonwealths-men' also presupposed 'that the wife is included with her husband, the child with the parent, and the servant with the master'. All three were similar cases. 'All inferiors in families – wives, children, and servants' were consciously joined by theorists defending their exclusion from the franchise. Since neither children nor servants had 'any property in goods and land', there was 'no reason' for them 'to have votes in the institution of government'. Wives were excluded 'as being concluded by their husbands and being commonly unfit for civil business'. Married women did possess submerged rights in marital property and the rights of trustees in their jointures. But these rights were not deemed sufficient to exclude them from the large category of those too dependent on heads of households to be capable of autonomous political action.[26]

This reason for the exclusion of wives from political power explains why married queens' regnant were excepted. They did own land in their own right (a whole realm, in fact), and therefore had an interest in the kingdom that was 'permanent and fixed' and 'without dependence'. For the good of her subjects, a queen's realm (unlike an ordinary woman's possessions) did not pass into her husband's control at marriage; the realm was like an office 'fixed in her own person'. A queen marrying a commoner did not make him king, nor could she 'give him the realm as dowry'. As one theorist asserted, no monarch could 'part with, resign, or make away to another' 'the sovereignty and regal dignity', because 'it comes to them as a trust of which they are indefeasibly tenants for life'. When Mary Tudor married, parliament carefully emphasized that she continued to rule 'solely and as a sole queen', and that Philip did not gain 'any right ... by force of the said marriage'. Later, seventeenth-century theorists stressed that this Act was aimed at ensuring Philip 'was precluded and shut out from the having or exercising the regal power'. By retaining sole control of their property and realm, queens regnant retained their independence, and therefore the normal arguments about servants, wives, and children did not apply. Dismissing the argument from dependence and inseparable interests in the case of married queens was consistent with the rest of liberal contractarian thought.[27]

Early-modern contract theorists applied to wives the same arguments they used about dependent males. The sole and independent possession of freehold land in county elections, or comparable property in borough elections, was generally regarded as a necessary qualification for the franchise.[28] No married woman satisfied this qualification, just as no child, servant, or landless labourer satisfied it. Therefore, wives – like children and servants – should not vote. In strictly logical terms, the argument was not inherently sexually biased. Of course, the laws of England relating to property-holding were sexually biased; the laws enforcing male primogeniture and the absorption of a wife's property into her husband's estate at marriage discriminated solely on the basis of sex. However, contractarian theorists were not in the business of social engineering by the massive reform of property law. Most people in England were not freeholders.

> He that considers the constitution of the House of Commons in England must know that the members thereof do not represent *all* the people of this land that are under government. For ... the knights are to be chosen only by the freeholders: And how many that are free born subjects are not *liberi tenentes* [freeholders]?

Contractarian theorists made no sustained attack on the established franchise restricting political power to independent property-owners, and married women did not enter this category.[29] The House of Commons imposed laws on 'mean peasants, servants, women, and children, who are undoubtedly the major part of the nation; and yet were never allowed to have any vote or suffrage in the election'. Patriarchalists questioned how it could be consistent with contractarian principles of natural freedom and equality to give 'freemen or freeholders any power or jurisdiction over women and children or the meaner sort of people'. They rightly observed that the political inequalities resulting from the established distribution of property went unchallenged by contractarians. Faced by such criticism, Members of Parliament simply reiterated that 'we represent the valuable part [of the nation] and those that deserve a share in government'. Politics should be a matter not for

> copyholders, whose lord represents them, nor of the meanest freeholders, much less of servants or women or children, but ... of all that have a share in the government or are fit to have a share in it as to the legislative part by choosing representatives.

Poverty and lack of independence were thought sufficient grounds for excluding men from the franchise.[30]

Liberal contractarians did not propose that the laws on property or taxation be altered in order to ensure that those (males) excluded from the electorate by poverty should be placed in a position of sufficient independence to vote. Perfectly consistently, they did not suggest the revision of property laws relating to married women either. However, charges of inconsistency cannot so easily be avoided in the case of adult, landowning widows and spinsters (rather than that of wives). Since these women were legally *sui iuris* and economically independent, it followed logically that – if they satisfied the necessary property qualifications – they too should be enfranchised. This was not the case and no contractarian theorist seems to have attempted to justify their exclusion (as opposed to that of married women) or to suggest that the law should be altered. There were very few wealthy widows and spinsters in early-modern England (and, as the main target of male pursuit, rich spinsters rarely remained single for long). Perhaps this was why their anomalous position in contractarian theory attracted little attention. Certainly, social prejudice against women's involvement in public affairs was not so internalized that theorists simply could not recognize there was a problem, for – as we have seen – patriarchalist polemicists clearly referred to the anomalous position of 'women, especially virgins'.[31]

Liberal contractarians – on a consistent application of their own principles – should have conceded political participation to wealthy spinster and widow freeholders. No contractarian suggested that weak, foolish, and emotional males (who met the property qualifications) should be excluded from the electorate; they insisted upon juridical equality. The exclusion of landowning widows and spinsters was inconsistent in a way that the exclusion of wives was not. But this was not the only problem with their theory. The essence of the attack on patriarchalism, made by liberal contractarians, lay in an insistence that political obligation stemmed from consent. Since everyone was born free, political subjection legitimately arose only when one person consented to obey another. If one person were not naturally under another's power, (s)he could only become subject by agreement. A woman consented to obey by marrying, but what rational grounds could a woman have for doing this. Why should women marry? The following sections answer this question in early-modern contractarian terms.

Force and contractualism

Liberal contract theory was not based on (implausible) history. Early-modern contractarians did not maintain that all members of English society had, in fact, instituted the present form of government. Nor did they maintain that every adult in the country had explicitly consented to obey the existing government.[32] Rather, they insisted that governments were legitimate only if their subjects would rationally consent to obey them. If England's government protected the lives, liberty, and property of its subjects, they should obey. Government was established because reasonable men agreed that limits on personal freedom, which were necessarily involved in subjection, were preferable to the dangers and inconveniences of anarchy. Contractarian theorists believed that rational people would not agree to obey an absolute regime, irresistible even when acting directly against subjects' vital interests, since this would be worse than a state of nature without any government. Rational men, on first establishing government, would insist that it possess no right to make laws depriving them of their lives, their freedom, or their property without their own consent. A government failing to meet such conditions lost its authority, need not be obeyed, and could legitimately be resisted.

Liberal contract theory had deep roots in Western political and moral beliefs and was widely adopted in early-modern Europe. It was not universally accepted: political patriarchalists and conservative contractarians opposed its conclusions. But no early-modern writer rejected liberal contractarian theory because of women's position within it. Yet some modern critics have argued forcefully that the position of women is problematic even on the liberal contractarians' own terms. Why should women consent to the form of government advocated by liberal contractarians? Liberal contract theory suggests good reasons why male heads of households should consent to government. In the anarchy of the state of nature their lives, liberty, and property are not safe; government secures these. Yet women appear not to have correspondingly good reasons for consent. Apparently women's lives, liberty, and property are *not* secured by government in the same way, since they may be deprived of them by laws from which they cannot dissent. Women

> have nothing to do in constituting laws or consenting to them, in interpreting of laws, or in hearing them interpreted at lectures, leets, or charges; and yet they stand strictly tied to men's establishments . . .

But if there were no rational grounds for consent, no adult Englishwoman was morally obliged to obey the government.[33]

In response to this point, it is worth noting that contractarian theorists did not require women's explicit consent either to a particular existing government or to the (hypothetical) original contract, since, by marriage, women consented to obey their husbands, to grant them control of common property, and to accept their judgement in political matters. Husbands would exercise political power in the interests of their wives as well as of themselves. Obviously, this rejoinder merely forces the question one step back, making the problem one cf why a free woman would ever agree to marry. Why sacrifice political and legal autonomy when it might be retained simply by remaining single? Women who did not consent to marriage would stay in the state of nature when males made the original contract. They would not be morally bound to obey the government or its edicts.[34]

It is important to establish first that liberal contractarians could not coherently, and did not actually, argue that women in the state of nature would consent to subjection in marriage (and therefore in the polity) because they would be coerced into doing so. One brutally obvious account of the origin of exclusively male government is that physically (mentally and emotionally) stronger males subjugated weaker women by force; in the language of early-modern theorists, 'conquered' them. Thomas Hobbes' theory was virtually this. 'For there is not always that difference of strength or prudence between the man and the woman as that the right can be determined without war.' Wars were fought and males (being usually, though 'not always', stronger) generally won them. As a consequence, 'the most part of commonwealths have been erected by the fathers not by the mothers of families'. No liberal contractarian

could have argued in such Hobbesian terms without sacrificing a major
plank of the defence against absolutist theory.[35] Other political absolutists had argued that victorious princes gained legitimate power; that
'a king may be lawful king by conquest merely – without the consent of
the people – is so evident in Scripture it cannot be denied'. Some theorists asserted that victory in battle was a providential expression of
God's will. Others portrayed anarchy as the only alternative to subjection to every *de facto* government. They insisted that in a state of nature,
contracts made because of force or fear were binding. Civil authorities
could not rescind such contracts because – by definition – there were no
magistrates in the state of nature. If a strong man forced weak ones to
promise him political subjection, their promise was binding no matter
how distasteful the conditions (s)he imposed. Even absolute government
gave more security than anarchy; men would rationally consent to severe
limitations on their freedom rather than live without any government at
all.[36]

Liberal contractarians rejected these arguments. Force gave no legitimate claim to power. John Locke compared obtaining political power by
force to piracy and robbery.

> Should a robber break into my house and with a dagger at my throat make
> me seal deeds to convey my estate to him would this give him any title? Just
> such a title by his sword has an unjust conqueror who forces me into sub-
> mission.

Conservative contract theorists like Hugo Grotius and Thomas Hobbes
would have argued that – in a state of nature – the robber did indeed
gain good title. John Locke took the opposite view. An unjust invader
can 'never come to have a right over the conquered'. If (s)he conquered
in a *just* war, the conqueror obtained limited power over those who had
forfeited their rights by waging war unjustly against her or him. But this
power could not extend beyond that the conqueror already possessed in
virtue of a pre-existing (contractual or hereditary) right. To have admitted that males compelling females into subjection obtained legitimate
power would have been incompatible with this critique of absolutist theory. Liberal contract theory logically involved the proposition that if
males in a state of nature used force to compel women to accept their
rule, they would obtain no legitimate power. The women could rightfully resist the unjust government and rebel when a favourable opportunity
arose. Samuel Pufendorf repeated the standard early-modern view that
males usually were physically stronger and more intelligent than
women, but argued that this in itself did not give them authority over
women. Rights over a woman (like rights over a male) could be gained
only 'by her consent or by a just war'. Males could not establish political
rule over free women except by obtaining their voluntary consent.[37]

This was also held to be true in the case of marriage. Indeed, more true, since consent to marriage under duress was almost universally agreed not to be binding, whereas this was debatable in other contracts.[38] Furthermore, it was generally agreed that – precisely because women were weaker – a woman could legitimately claim that she had been coerced on the grounds of less force than could a male.[39] Valid marriage required the full and *free* consent of the woman. A male could never coerce a woman to marry him – the marriage would *ipso facto* be void. Free women could not be pressed into marriage and then dragooned into a commonwealth under pretence of their conjugal duty to obey. If males in the state of nature applied force or fear that would move constant women, their 'marriages' would be null and the 'wives' under no obligation to accept subjection.

Natural inferiors

A liberal contractarian could not explain female subjection by arguing that males had forced them into the state or into marriage. To accept the role of force was to allow the absolutism of conservative contractarians to enter by the back door. Moreover, reasons for consent (other than force) had to be found, not only for women but for all those – male or female – excluded from participation in the political process; that is, for all but property-owning, independent heads of household. For liberal contract theory to remain viable, a rational justification of why the disenfranchised should agree to political society had to be provided. All were born free and each must consent to be legitimately subject, yet many were excluded from full political citizenship. Some contract theorists attempted to explain and justify why servants and slaves should obey government.

The description of the state of nature as one of political equality was in no way thought to entail equality of abilities or talents.[40] Some modern commentators have construed the early-modern contractarian assertion that all men were free and equal in the state of nature as implying that all men have equal strength and intelligence.[41] Quite the contrary was true. As Professor Scott has perceptively argued

> demands for equality have rested on an implicit and usually unrecognized argument from difference; if individuals were identical or the same, there would be no need to ask for equality.

Early-modern contractarians insisted on political and juridical equality because they believed that all men did *not* have equal talents. Their argument (spelling out the implicit beliefs) was: despite the fact that some men are weaker, less intelligent, and more emotional than others, they have equal political rights in the state of nature before the institution of government.[42]

Early-modern theorists defined the state of nature as one where people had not yet established government or incurred any political obligations. People in the state of nature were juridically equal and free from political subjection. Political freedom and equality was a philosophical postulate rather than a description of reality, because contractarian theorists thought that most men had incurred political obligations. By swearing allegiance, voting in elections, inheriting property under existing laws, and accepting the protection of magistrates, they had tacitly acknowledged the legitimacy of the government that ruled them. Apart from the absence of government, contractarian theorists wished to characterize the state of nature realistically; this was why authors such as John Locke supported their arguments by references to the Americas – a land where (in their view) there were people living under no settled form of civil government. People in the state of nature possessed just the same biological and psychological characteristics as those in civil society. Since they believed that people were naturally unequal, they constructed the state of nature as one where the stronger and cleverer would thrive, while the stupid and lazy would not. Inferior men would rapidly become the economic (and when the state was founded, therefore, the political) dependents of the wise heads of households. Natural inequalities would be expressed in social and political hierarchy.[43]

In early-modern discussions, household servants were perceived as the key group of the disenfranchised. Contractarians had to fit servants into their description of the consensual origins of government. Samuel Pufendorf argued that servitude should be seen – initially at least – as voluntary. In the simplest state of nature there was approximate social and economic equality, but as life grew more complex, differences arose between 'the more sagacious and more wealthy' and 'the more sluggish and the poorer sort'. Both groups saw the advantages in the latter working for the former, until gradually the less intelligent and less able became permanently attached as servants to the families of the wise. In exchange, their masters provided 'sustenance and all other necessities of life'. Servitude was therefore originally based on 'the willing consent of men of poorer condition'.[44]

When early-modern political theorists discussed servants, their first thought was of 'such as have little use of reason, and are only fit to be governed and not to govern'. To the stupid they added the poor; those 'who through want or penury, or a competent estate or family of their own become mercenary hired servants who otherwise were free'. People sometimes became servants involuntarily – for example, those sold by their parents or captured in war – but it was natural inferiority that most commonly led to servitude. There were plenty of stupid people around to fill the role of natural servants because the 'numerous rabble that seem to have the signatures of man in their faces are but brutes in their understanding'. A few men obtained 'dominion and government' because 'many being compelled by poverty or by the dullness of their intellects did proffer their labour and service to great men'. The natural élite of the strong and intelligent became heads of households, the rest agreed to this 'upon condition of being maintained by them and supplied with necessities'. Poor and foolish men preferred economic security to individual autonomy. Aristotelian notions of a natural élite came prominently into play, and theorists reasoned that the 'debased nature' of some men made them 'born for drudgery' and 'content to increase their pains that they may lessen their cares; and upon such terms become servants to some of their brothers'. Original political equality was brief and transitory or never existed, except hypothetically. Natural differences in intelligence, industry, and ability necessarily materialized in social inequality. Only the propertied should be entitled to vote because only they were sufficiently independent, but 'there will always be more people in a nation than the proprietors of lands'. Political subjection followed easily from economic inequality as the 'poorer sort' became 'domestics'.[45]

The analysis of why – naturally and consensually – the stupid and indolent would become servants held as true for women in the early-modern mind. Both in the state of nature and in established states, weak, foolish, intemperate women needed strong, wise, prudent males to protect and govern them. Weak males became servants, women became wives. Both would consent to obey heads of households – even though this entailed exclusion from political power – since their overall position would be better than if they attempted to survive autonomously. The early-modern belief in female inferiority led them to expect that women, like 'the sluggish', would not prosper in a state of nature. Early-modern theorists wanted to describe the state of nature realistically and so they portrayed it as just like the daily reality that surrounded them. Accounts of the state of nature portray it as a copy, socially and economically, of the contemporary world; it simply lacked law and government.[46]

Early-modern society was based on human and animal muscle power, not on the extravagant consumption of fossil fuel. The vast majority of people worked in agriculture and there were virtually no mechanical, labour-saving devices. Nor was there any large-scale industrial production. Except in a few crafts and the professions, physical strength was important. Women were believed to be physically weaker than males and to suffer from many ailments 'to which males are not liable'. Moralists thought that women's health hung by a fragile thread, 'for infinite are the diseases, and those strange and wonderful beyond the common course of nature, which the womb of a woman doth make her subject unto'. Woman's susceptibility to disease was not considered accidental but essential to her physiology, for the 'womb is the root, seed-plot, and foundation of very near all women's diseases'. The woman's 'soft body' was less suited to labour. Since women were believed to be less strong they earned less money; in early-modern France and England, daily wage rates reflected the belief that in the same time a woman would be less productive than a male.[47]

Bureaucracy and the professions were not open to women and so there was no evidence to suggest that women could prosper through them. 'Most writers evidently thought it unnecessary to prohibit occupations so obviously unsuitable to women.' With the exception of a small number of governesses and midwives, there were few employment opportunities for women in areas where skill or education were more significant than physical strength. 'Women at all levels of the society were' in any case 'an educationally deprived group compared with men'. Although a spinster had the same legal rights to own property as a male, the structure of employment made it unlikely that she could attain satisfactory economic independence. Women working successfully in crafts were usually married: 'Rarely did single women or the poor support themselves through craftwork'. To early-modern theorists, daily observation suggested that women needed males to support and maintain them in comfort.[48]

There was no police force, the streets were not lighted, and seventeenth-century males were just as capable of rape as their twentieth-century counterparts: 'if the rampart of laws were not betwixt women and their harms, I verily think none of them being above twelve years of age and under an hundred, being either fair or rich should be able to escape ravishing'. Early-modern society was subject to a high level of casual violence and since fists, knives, and swords were the only generally available weapons, physical size and strength tended to give victory. 'Why is it a thing so shameful to kill a woman? *Answer*. Because she is weak and not able to resist'. It was 'a dangerous thing' for women 'to travel alone or work alone without any company, for the weakest are soonest oppressed, and women are easily conquered'. Physical prowess was significant because it 'promises protection to those who want either heart or strength to defend themselves: This makes the authority of men among women, and that of a master-buck in a numerous herd'. Then, as now, women were less likely to commit serious crimes, and they 'seemed likely to depend on men to settle their quarrels when they could. Compared to men, women were unlikely to resort to violence'. Once again, everyday life evidenced to early-modern theorists that women needed males to protect them.[49]

Daily experience showed that single women rapidly slipped to the bottom of the economic heap. 'A single woman makes a sad figure in the world.' Deserted wives were frequently numbered amongst the destitute. Widows were the objects of pity. Widowhood was 'a state of misery', 'by God himself reckoned as a condition the most desolate and deplorable'. Widows were particularly indigent, because they were 'deprived of support by the death of their husbands'; 'for many women widowhood brought a severe drop in their standard of living, sometimes reducing them to acute poverty'. However, widows were in a happier position than spinsters, for connections with children and their deceased husband's family meant that a 'widow had allies at various levels which the spinster did not'. Widows were more likely to have inherited wealth, since in most husbands' wills 'the provisions for widows were generous'. A spinster was 'looked on as the most calamitous creature in nature'. Early-modern theorists thought that common experience showed that women living alone – spinsters even more so than widows – were physically, economically, and emotionally less prosperous and secure. A woman wishing to be 'masterless' must be foolish, for God 'reckons them most miserable when they are most at liberty'. Early-modern theorists thought that women were weaker and therefore probably would not prosper alone; this would be particularly true in anarchic conditions. Women – like weak, stupid, and lazy males – would accept dependence on strong, intelligent male heads of households in order to improve their physical and economic state. Those attractively endowed would marry heads of household, the less fortunate would become servants or the wives of servants. Both weak males and females, motivated by self-interest, would willingly accept dependence.[50]

Even within early-modern theorists' own terms, the analysis that weakness would lead women to accept subjection in marriage seems open to one major objection. If women were so economically and physically insecure that they would regard subjection in marriage as better than attempting to go it alone, in what sense could their consent to marriage be deemed free? Surely fear and deprivation amount to duress.[51] And if consent were not free, would not the marriages be void and the 'wives' not subject?[52]

Considering this objection properly involves looking once more at the theory of coerced consent. Early-modern theorists did not simply state that any form of coercion rendered a marriage contract null and void. Rather, they held that three conditions must obtain in order for duress to invalidate a marriage contract or to render any other contract rescindable. First, the force or fear must be applied by a party to the contract. Second, it must be employed with the intention of forcing the other party to contract. Third, there must be a threat of some evil, *not* a promise of some benefit: bribery did not invalidate a contract. Further explanation of each of these qualifications can best be offered in the form of examples.

1. If I purchase a gun because I do not believe my life is safe on the streets of Washington without one, I cannot claim that I have no obligation to pay the gunsmith because I was induced by fear of death to make the contract. A (reasonable?) fear of death may indeed have motivated me, but the gunsmith did not cause my fear. He was (in some sense) taking advantage of that fear to make a profit, but this differs essentially from him threatening to shoot me if I do not purchase his wares. The

contract is a valid, non-rescindable one. Seventeenth-century theorists used the example of a man captured by enemies in war and 'condemned to death'. If a woman agreed to pay a ransom for his release on condition that he would promise to marry her, he could not later renege on the contract because fear of execution induced him to make the promise. The eager bride does not herself coerce the prisoner, and his enemies (who do coerce him) are not parties to the contract.[53]

2. If a doctor at Immigration Control refuses to issue me with a visa unless I submit to and pay for an X-ray examination for tuberculosis, and I agree to this, I cannot then refuse to pay his fee even if my very life depended upon entering the country. The intention of the threat to deny admission was not to force me to pay his fee but to exclude all those with an infectious disease from the country. Once again, early-modern theory deemed such a contract valid. Casuists employed the example of a judge agreeing to revoke the death sentence on a rapist if he promised to marry his victim; the rapist could not rightly argue that the promise did not bind because he was impelled to make it by fear of death. The judge's intention in sentencing him to death was to punish rape, not to force marriage.[54]

3. If a publisher offers an impoverished author like myself an advance of a million dollars to (say) write a historical monograph, I may well feel in no position to refuse. Having accepted the money, I cannot then decline to write the book on the grounds that the offer was so tempting in my straitened circumstances as to leave no real choice. The publisher does not compel by a threat of evil, but offers an (almost irresistible) benefit to induce commitment. Like the gunsmith above, she in some sense takes advantage of my unfortunate position (poverty, the absence of two-million-dollar offers) but she offers gain and does not threaten harm. Such a contract would be valid and non-rescindable in early-modern theory. A seventeenth-century example was that of a woman travelling through bandit-infested country, who agreed to marry a male if he would protect her from assault. The male offered her only a benefit (protection), and did not himself threaten her with any injury; the marriage would be valid.[55]

Conditions of this kind were regarded as necessary to almost any viable theory of contracts, but particularly to contracts made in a state of nature. By definition, the state of nature was one without government. No civil association could arise if the presence of *any* fear rendered all contracts invalid or rescindable. In contractarian theory, government was necessary because rational people doubted the security of their lives, liberty, and property in the absence of a society with the power to enforce laws. If they could deny their obligations once rule was established, simply because of the very fear that impelled them to desire government, anarchy would be permanent. Conservative contractarians took a still tougher line and insisted that a contract, made in the state of nature, bound even if it was entered under coercion, and even if the person using coercion – for instance, a conqueror forcing defeated people to accept her rule – did so with the intention of compelling them to contract. However, both liberal and conservative theorists accepted that at least the three conditions outlined above must obtain before the plea that a contract was void or voidable was legitimate. Unless all three conditions applied, forced consent 'not wholly free, but on consideration to prevent a worse thing is yet so much freedom as transfers things among men and makes an human right'. The acceptance of these conditions dated back to medieval theory on the validity of contracts in general, and marriage contracts in particular. In the case of women contracting marriage (as, indeed, in that of weak males agreeing to servitude) these three conditions were crucial.[56]

In liberal contract theory, a woman who was assaulted and threatened with death, enslavement, or rape if she did not marry her oppressor was in no way bound by any marriage contract she made. If, however, conditions were such that women felt threatened by other people, and so agreed to marry suitors who would protect them (at the price of obedience), they would be deemed morally bound. Similarly, if women suffered economic deprivation under conditions of uncontrolled competition, and married in order to obtain a higher standard of living, their marriage contracts would bind. All three qualifying conditions were thought to apply. The woman *is* under a form of duress, but it is not her suitor who coerces her. He was not responsible for her weakness and deprivation, and certainly had not created these conditions in order to force her into marriage. He 'bribes' her with the goods of protection and support to persuade her to marry him; he does not menace or threaten her. The husband (like the gunsmith or the publisher) would be taking advantage of existing circumstances to obtain some benefit – in this case an obedient wife – but this, in sixteenth- and seventeenth-century ethical theory, would not invalidate the contract.

The liberal contract theorist's argument for excluding women from political power was based on these axioms: most women are weaker, less intelligent, and less well-balanced than males. Women suffer deprivation unless, by agreeing to marriage (or service), they obtain male providers and protectors. In exchange for protection, husbands (or masters) require obedience. A dependent who has pledged obedience is not sufficiently autonomous to hold political power, for entering a master's family involves accepting 'the ordinary discipline thereof'. Therefore, government is established and controlled by male heads of households. Implicit in this account of the consensual origin of female subjection is (obviously) the premise that women are weaker. However, the account does not involve denying the rationality of women in the state of nature; they are just as rational as the weak males who agree to servitude. Nor does this account assume the legitimacy of using force against women where it would not be legitimate against a male. Women's and weak males' natural inferiority and their consequent deprivation induce them voluntarily to accept subjection. No one needs to compel them. The natural élite become heads of households and form commonwealths; everyone else willingly consents to subjection to them.[57]

Amazonis rediviva

Twentieth-century commentators have suggested that early-modern contract theory ignored women's abilities to organize themselves. It is argued that social contract theorists merely assumed, without argument or support, that individual women would be reduced to dependence on individual males, whereas males would find security in (male-dominated) civil associations for mutual protection. This, it might seem, is to apply an inconsistent, intellectual double standard. Women – like rational males – value autonomy. Rather than accept obedience to males, women in the state of nature might join together in their own societies and jointly provide the defence they could not provide singly. There is no reason, some modern feminists argue, why this solution should have been inconceivable to the early-modern theorist. After all, Amazonian legends raised the possibility of such gynocracies even in sixteenth- and seventeenth-century literature. Ignoring this possibility, modern critics proceed, shows the inconsistent male bias implicit in the early-modern account. There are two responses to this point: one crudely practical, the other fundamental to early-modern beliefs about human psychology and sexuality.[58]

No early-modern political philosopher dealt in any depth with the feasibility of independent female-ruled communities, probably because it did not occur to them that a community of women could defend itself against male enemies. Early-modern warfare did not consist in launching smart-bombs by the push of a button at 60 000 feet. It required physical strength and endurance. Muskets weighed so much (up to twenty pounds) that they could only be fired when leaning on a 'rest'; ignited by matches, they were unreliable in wet or windy weather and the musketeers had to be protected by pikemen. To be effective, pikes needed to be about 16 feet long and were so heavy and unwieldy that officers had to watch the pikemen continually to ensure they did not cut a foot or two off their length to ease the burden on long marches. Musketeers and pikemen were subject to attack by cavalrymen who generally carried a small pole-axe, pistols, a carbine, and a sword – weapons which also required physical strength to be wielded effectively. It was difficult for early-modern armies to recruit males equal to the strains of such warfare. Believing as they did that women were significantly weaker than males physically, it is doubtful that early-modern theorists would have been able to entertain seriously the belief that women could ever win a war against males.[59] When early-modern commentators did consider women in the context of warfare, it was only to dismiss them as not worth killing, for 'women and children, being unable for war, are not feared to fight the conquerors and therefore are usually spared'. When one military theorist, James Turner, discussed women's role in armies, he did conclude that they could be useful but only because 'they provide, buy, and dress their husbands' meat while their husbands are on duty', and 'bring in fuel for fire, and wash their linens'. It was clear that he bore no prejudice against women's close proximity to military operations:

> At the long siege of Breda made by Spinola it was observed that the married soldiers fared better, looked more vigorously, and were able to do more duty than bachelors. And all the spite that was done the poor women was to be called their husbands' mules by those who would have been glad to have had such mules themselves.

It was equally clear that it never occurred to Turner that women could have been of the least use in the military operations themselves. Early-modern theorists believed that nature shaped women to be housewives, not soldiers.[60]

A community of women might, of course, hire mercenary male soldiers, but the Italian wars had shown the gross unreliability of mercenaries. They were often more dangerous to their paymasters than to the enemy. Alternatively, a community ruled by women might arm their own male offspring and servants, and hope these would serve loyally without using their military might to overthrow the gynocracy. But the compliance even of citizen armies could not lightly be taken for granted – particularly not in England after 1647 when the New Model Army's disobedience to its parliamentary masters was too recent and too painful a memory. For a female community to hire stronger males to defend them was to mount a tiger. No early-modern theorist would have doubted that had they done so, the natural order would have reasserted itself in short order and the erstwhile mistresses become once more 'meet helps'. The Amazons had triumphed briefly, only because the local males were wimps; such unnatural reversal of roles did not often occur.[61]

The other early-modern response can be extrapolated from their ideas of natural psychology. Wholly female communities would not arise, because women lust after males. Men were thought naturally inclined to form heterosexual couples. Renaissance and Restoration theorists believed that males and females naturally desired sexual relations with one another; 'the Lord made man a creature prone and ready to associate [with] another sex'. God had instilled sexual desire in mankind in order to ensure that the species was propagated. In man, like 'in every kind of animals', nature itself instilled 'a desire of congress for procreation sake'. 'Is it not this mutual love of males for females and females for males that

multiplies their species and preserves the world?' The continuing existence of mankind was proof to early-modern theorists of the naturalness
of heterosexuality. Sexual desire was portrayed as not only a 'strong and
natural propensity ... in each sex', but also 'absolutely the most violent
and predominant passion to which humanity is subject'. They believed
that pleasure in 'this mixture of the two sexes' was so obvious that there
was no need for argument 'to convince man that one was really made for
the other'. Sixteenth- and seventeenth-century theorists, like the population at large, 'took it for granted that sex was uncomplicated and enjoyable for both sexes'. In early-modern thought, heterosexual desire was
deemed normal, strong, and universal.[62]

Seventeenth-century writers believed that females, like males, desired
sexual relations. Although the patristic notion of women as sexually
voracious was being superseded by a view of them as naturally modest,
there was virtually no trace of the belief that women have little or no sexual libido. Young women's 'eager gazing and desiring to associate themselves with men' were seen as signs 'that nature prompts them to desire
what is ordained their due'. Renaissance medicine taught that female
sexual desire was specifically directed towards attracting semen. Women
'wanted to receive semen' so they could conceive. Physicians maintained
that women deprived of sexual relations might suffer physical illness; for
'if a woman do not use copulation to eject her seed, she oftentimes falls
into strange diseases, as appears by young widows and virgins'.[63]

Women were not made for 'a single, monastic life'; 'it is so natural a
thing for a woman to love a man' that she did not need 'to be prompted
or instructed in this point'. Even pagan Aristotle had realized that there
was 'in most men and women naturally an inclination and propensity to
the nuptial conjunction'; and 'nature teacheth that maids have been
made for marriage'. It was 'as natural to desire marriage and use it' as to
eat when hungry. The capacity for celibacy was rare, and to those 'who
cannot abstain, marriage is as necessary as meat, drink, and sleep'.
Having dissolved the convents, Protestants were particularly eager to
deny that perpetual virginity was either natural or desirable, but even
Roman Catholics portrayed voluntary virginity as a divine gift; 'the
chiefest bar and let in women to a monastical life is desire of marriage
and hope of children'. Women, like men, were afflicted with carnal
desires and 'the woman's sensuality is as hardly satisfied as the man's'.
Indeed, some thought it beyond doubt that males were better able than
women 'to resist lust'.[64]

Sixteenth- and seventeenth-century theorists were aware that some
women could find sexual satisfaction alone or with members of the same
sex. Saint Paul had fulminated against the 'vile affections' of women who
'did change the natural use into that which is against nature' (Romans
1:26). The sexual conjunction 'of woman with woman' was abhorrent not
only to Paul but 'to nature itself'. Lesbianism and masturbation (like

male homosexuality) were regarded as uncommon, unnatural deviations; nature 'confined our inclination to the other sex'. Most women
desired males not other women.

> What female comfort can one woman find
> Within the bed with other woman-kind?

Homosexuality in either sex was deemed aberrant, and 'sodomites and
such like monsters' rare deviations from the normal pattern of nature.
The notion that 'all sex is rape' does not seem to have occurred to any
early-modern theorist; Christian divines' distaste for sexual relations was
based on them being enjoyed too much, not too little. Even the Amazons
had needed to banish all males in order 'that none of them might seem
more happy than the rest by enjoyment of their husbands'.[65]

Early-modern theorists deemed homosexuality 'unnatural' in the
moral sense also; that is, contrary to God's law of nature. All women
(like all males) would know by natural reason alone that to satisfy their
sexual inclinations in this way would be to commit mortal sin and court
eternal damnation. Medieval and early-modern divines categorized both
lesbian practices and male homosexuality as 'sodomy' and insisted that
they were mortally sinful. Other sins infringed natural reason, but only
sodomy was 'against the order common to animals and men'. Theorists
listed as 'not only "against natural law" but "against the natural order"
... pederasty, buggery with animals, or lesbianism'. Even animals
eschewed homosexuality. Lesbian communes would not arise in the state
of nature. Early-modern moralists believed that unaided natural reason
would lead anyone desiring sexual relations to recognize that marriage
was their only proper forum. Roman Catholic casuists taught that a
starving woman could legitimately steal to preserve her life but she
could *never* prostitute herself. Reasonable women would marry not for-
nicate. Women's sexual instincts would incline them to want male partners; natural reason would tell them (just as it did males) that
promiscuous fornication, polyandry, and adultery were immoral.
Women would enter the permanent, procreative, monogamous institution of marriage.[66]

Protection and subjection

This account of early-modern contractarian thought moved without
comment from women's natural need for help and protection to marital
subjection and obedience. The critical reader might argue that this
involves a concealed – and fallacious – inference. Even if a rational
woman would look for a male protector and supporter, why should she
agree to obedience? Not only women profit from association in marriage.
Even in seventeenth-century terms, the male also benefited – gaining a

housekeeper, a sexual partner, a vessel in whom to beget children, and a helper in raising them. Since the male could not (legitimately) obtain these goods without the woman's consent, she was in a position to bargain. Why should the negotiations result in wifely obedience? Would not a rational woman hold out for (at least) an equal partnership? Why should not women in the state of nature accumulate wealth and power independently and (like queens regnant) retain autonomous control of these after marriage?

The answer to these questions again rests on the early-modern belief that women were physically, intellectually, and emotionally inferior to males. Early-modern theorists thought it rational that in any partnership the wiser and stronger should rule, and thought this was usually the male. Although 'such a thing may happen as that the woman, not the man, may be in the right ... ordinarily it is otherwise'. Saint Peter called woman 'the weaker vessel' (I Peter 3:7) to remind wives that they were 'weaker than the husbands, and that both in body and mind as women usually are'. Indeed, women naturally inclined to subservience; 'ye that are wives, you have this conscience of your own infirmity ... ye crave a head, ye crave to be under a superior'. God gave males as heads and guides to women, as a 'benefit granted unto their weakness'. Rejecting male guidance was like a cripple throwing away his crutch, like a disabled driver refusing to use his privileged car-parking space.

> Man's experience is woman's best eyesight and she that rejecteth it is like a
> sealed dove – soars high for a while, but at length comes tumbling down
> and lights in a puddle.

Because of the pride inherent in every member of the fallen human race, any woman naturally preferred rule to obedience. However, rational women would consider their weakness relative to males, and discern that they would be better off in a position of subjection. 'God and nature do attest the particular expediency' of obedience, 'by having placed that sex in a degree of inferiority to the other'.[67]

Even more than her mental handicaps, woman's physical weakness was thought crucial in determining where government would lodge. Early-modern Western thought, steeped in feudal concepts, held that protection and obedience were intimately linked. It was axiomatic that 'protection draws with it obedience'. Men were called upon to obey the government that protected them (not the most just or the most representative). Protection and subjection were coupled in the family as in the state. Traditionally, a husband's duties were said to include defending his wife. Whatever services wives could offer, protection was not commonly thought to be amongst them, 'for wives should be defended by husbands', not husbands by wives. The stronger protected the weaker, not vice versa.[68]

All early-modern theorists accepted that the 'weaker sex' needed defence. Early-modern accounts of a husband's responsibilities pointed to his obligation 'to rescue his wife if in jeopardy'. Women could not 'make shift for themselves' and so were dependent on males 'for provision and protection'. This was true not merely for a few exceptionally feeble specimens of womanhood; weakness was 'stamped upon the whole sex'. A husband must be 'careful to protect' his wife and 'provide for her maintenance'. Central to their view of woman's natural condition was 'the weakness and feebleness of that sex, being more helpless in danger than ours, and less able to relieve themselves'. This weakness cried out for 'ready aid and succour' from the male. Like Eve, all women had a 'natural yearning' for a husband to provide the 'sustenance, defence, protection, society, security, and pleasure' that a single woman lacked. Inferior woman knew that alone she was destitute and defenceless, and longed for a husband to provide the protection she needed.[69]

Habitually – almost instinctively – early-modern theorists associated protection and subjection. The husband was 'to protect, to defend and govern his wife'. Scripture called the husband 'the veil of his wife'* and this 'implies subjection on her part' and 'imports protection on his'. By making males stronger, God had given them 'the power to rule and govern', so naturally the wife should 'submit herself' 'for her own good'. All early-modern theorists believed that sensible women would see submission was to their own advantage. Because 'the husband is the head of the wife for her good, whom he ought in all things to defend, cherish, and comfort', it was clearly 'expedient' for her to obey him. God gave the husband as 'a saviour' to his wife by 'maintaining, protecting, and defending her'; all that God and her husband asked in return was that she should obey, 'and therefore the wife – if she regard her own good – should not grudge to be subject to him'. Common sense would tell women to obey if they wanted protection.[70]

No early-modern theorist doubted that husbands would expect obedience in exchange for protecting and supporting their wives. Daily observation showed that 'the love of the husband depends for the most part upon the due subjection of the wife'. Wives should 'by a loving obedience procure protection'. 'Women cannot rationally expect that their husbands should affect them, unless they obey', argued the puritan divine, George Swinnock. He helpfully composed a prayer for wives that spelt out the logic of subjection.

> The law of nature teacheth me this lesson: the body is ruled by the head. The law of nations also: those that receive protection from others yield subjection to them.

Women's inferiority placed them in need of protection and of direction.

*Genesis 20:16 [the Authorized Version renders the Geneva Bible's 'veil' as 'a covering of the eyes'].

In the early-modern mind, the two concepts were so closely linked as to be almost inseparable. Liberal contract theorists linked protection and obedience as readily as all their contemporaries. They wished to base the husband's power over the wife on a contract between them, but accepted that it would be an 'unequal league'. The husband, 'the principal', would provide 'maintenance and protection'; in exchange the wife owed 'all due respect and obedience'. In the marriage pact, like any other contract, there was an interchange of benefit and obligation, so it 'partakes of the nature of an unequal league, in which the husband and wife each owe the other something; the former protection, the latter obedience'.[71]

Weak males would willingly accept servitude involving disenfranchisement. Women likewise would recognize that they were too weak and silly to achieve a high standard of material comfort or to protect their property and persons without turning to strong males for help. The natural form of association between male and female was marriage. Women could obtain a far higher position within the household than servants on account of the couple's common property, and sexual and reproductive co-operation. The husband trusted his wife as junior partner because she shared his bed and raised their children. The servant had no such advantages.

> In all countries whatsoever – even where by their laws the husbands rule over their wives by the most rigid authority – there is always a difference to be observed in the treatment of those whom they have made consorts of their bed, and of those whom they have admitted only to the services of their family.

Sexual union placed the wife higher in the family's hierarchy than servants; she became to her husband 'a companion of his bed and government'. Women as a sex were inferior, but sex gave some of them advantages that weak males lacked.[72]

Woman's mental inferiority and emotional instability placed her in need of a husband to guide and govern her; a married couple should be ruled by the wise, temperate husband, not the silly, passionate wife. Physical weakness placed her in need of a husband to maintain and protect her; accepting protection involved owing obedience. Women, knowing they could not thrive without male protection and direction, would willingly embrace subjection in marriage. In a state of nature unprotected by law and government (just as in contemporary early-modern society), women would welcome dependence rather than suffer the insecurity, poverty, and deprivation that were the natural lot of the single woman.

Divine justice

In early-modern contractarian theory, women consent to subjection in marriage to improve their physical, economic, and social security. Marriage is an escape from poverty and danger. Dependence is the only refuge of the weak. This might sound redolent of injustice to modern ears. Simply mentioning inequality and poverty suggests, to certain twentieth-century analysts, the existence of an unfair social system. Only exploitative, male-made institutions could make women so desperate: any inequality is *prima facie* unjust, and the subordination of a whole class, race, or group is almost certain to be. No early-modern theorist thought that oppression and subordination were the root causes of inferior attainment; they were confident that female inferiority stemmed from nature, not environmental conditioning. There was no presumption that poverty and inequality arose from social injustice. To the early-modern mind, inequality was natural, and natural inequality could not be unjust since God had willed it.

Most modern political thinkers regard nature as politically and morally uninteresting. Natural differences are considered descriptively not prescriptively. Contemporary attitudes hold that if naturally some people are stronger or more intelligent, this should not lead to them having greater social status or political power. Even if there are natural inequalities, they are to be corrected; society should aim to ensure that the lame, the halt, and the blind are guaranteed as much equality as their disabilities permit. Natural differences are regarded as fortuitous or random. To the early-modern Christian thinker, nothing was less true. God made nature and natural differences. Divines instructed those members of their flocks who thought it rather unfair and unfriendly of God to give some men massive moral and natural advantages that He was 'not obliged' to make His creatures 'perfect'. God was doing people a favour by creating them at all, and they had no right to complain if there were 'some of greater health, some of less; some of greater strength, some of less'. There was simply no 'obligation on the blessed God to prevent' natural 'defects' in his creatures. Far from believing that gross natural inequalities were unfair, they insisted it would be 'audaciousness ... in the creatures to set rules to the Creator, and to bind Him that He shall not show Himself more bountiful to another than to oneself'. God made man, and had a perfect right to make him any way that He wanted to; 'no creature can challenge anything, but must enjoy only that which God is pleased to bestow upon him'. God expected and intended that political and social differences would follow from natural inequalities. Penury and relative deprivation were part of God's plan; 'poverty was introduced by the highest Creator of things, the great and good God himself'. Early-modern divines taught poor people (not that their condition was unjust, but) to thank God for their hardship: 'Thou, Lord, hast been

pleased to deny me the good things, the conveniences of this life; blessed
be Thy name for it'.[73]

For early-modern thinkers it was God, 'the universal monarch of all
the world' who was 'the true fountain' of 'honour', 'titles', and 'pre-emi-
nences'. This was because 'divine providence' gave superiors their gifts
and talents. Greater strength and intelligence were a gift from God, not
from random fortune; 'the shape of our bodies, the abilities of our minds'
were 'all the effects' of God's spirit 'in the works of Nature'. Men were
created 'unequal in degrees of excellency', so that the 'abilities God hath
furnished them with' should be 'a direction to us in the choosing of men
to any place of employment – public or private, civil or ecclesiastical'.
Stronger muscles, a higher IQ, greater diligence – all were divine indica-
tions that the brawny and brainy should stand higher on the social heap.
Government was 'founded in nature' on the basis of 'superiority and
inferiority'. Natural and political inequality came 'from God; – the law of
nature being no other than the law or appointment of God'. Christian
catechisms taught that God 'established a distinction of degrees of honour, both in nature and also for godly policy and communion of life'.
Divine law commanded that 'all degrees of superiors' should 'be honoured of their inferiors', 'in respect of the order which God hath set in
nature or in policy'. Men differed naturally: 'if he be a superior he ruleth,
if he be an inferior he is ruled and learneth to comply with his fellow
subjects'. Far from holding it self-evident that God created all men equal,
early-modern theorists insisted that God intentionally made them
*un*equal so that proper hierarchy would be established.

> God did so order it that ... there should be an imparity, not only of excellen-
> cy and dignity, but of power amongst them: for without imparity there can
> be no order.

Natural inequality should not be corrected, but confirmed and reinforced
so that society would operate in a properly hierarchical and orderly way.
Theorists employed the abstract concept of initial political equality in
order to deduce philosophical conclusions; empirically, they believed
people were anything but equal. These natural inequalities were pre-
scriptive not meaningless. They were 'arbitrary' in the strict sense of
stemming from wilful choice and the wilful choice in question was
God's.[74]

God had designed nature with its own internal hierarchy: plants were
superior to minerals, animals to plants, men to animals. In each case,
nature established superiority so that 'the less perfect serves the more
perfect'.

> Furthermore, the male by his nature is more excellent than the female: and
> therefore he should rule and she should obey.

Sex was another gradation in the Great Chain of Being; women stood above brute animals but below males. It was 'not violence or human arts, but God' who introduced different ranks among men so that government could be instituted to conserve the race. An inferior male should not protest about his subjection, 'just as no woman can impute a fault to God because she is inferior to the male by nature'. God intentionally made woman second-rate and since God (by definition) acts justly, she could not complain about it. God introduced natural inequality so that social gradation would result. Women – like the inferior males who became servants – must not 'murmur that men being all of one and the same nature are so unequally placed, one to rule and another to obey; for this also is of God, even His good ordinance which thou mayest not resist nor contemn'. From the beginning of the world, God had instituted the 'three principal estates and ranks' of husband–wife, parent–child, and master–servant. In each case, He made the former superior and the latter inferior: 'For if all were equal, no policy could stand, nor order on the earth but a confusion'. Moreover, God in His goodness was generous to women. Inferior males had little choice but to become the servile dependents of indifferent masters, but God had bound wives to be subject 'not unto your enemies or unto strangers, but unto a kind and loving husband'. God made women weaker so that males could and would rule over them. That was the way God planned it, that was the way God wanted it to be. God the Creator 'might have done otherwise' but did not. He made women weaker, and thereby endorsed and approved the resulting social and political inequalities which early-modern thinkers saw writ large everywhere.[75]

The assumption that woman's natural inferiority was divinely intended was implicit in the work of virtually all early-modern moralists. God made woman 'the weaker vessel' and so established her subordination in 'the order of nature'. Natural differences between the sexes were part of 'the wise administration of Almighty God'. Early-modern writers linked protection and obedience; equally instinctively they coupled nature and God: 'Nature and the God of nature in every kind hath given pre-eminence unto the male'. Woman should 'not be displeased with her condition, though she be inferior in sex'. Her inferiority did not result from male oppression, but from God's intentionally discriminatory modelling of the two sexes; He 'appointed that distinction in their being'. In early-modern theorists' discussions, sexual distinction – and all its social and political implications – stemmed from divine appointment. God made women naturally weaker than males, because He wanted males to rule over them.[76]

Woman's subjection was 'naturally grounded upon inequality'. Had God made the sexes equal, wives might have been tempted to insubordination, 'because things equal in every respect are never willingly directed one by another'. Having spotted this, God made woman by nature 'not only after in time, but inferior in excellency also unto man'. The 'natural imperfection of the woman not only in her corporal parts but in those likewise which are intellectual', rendering woman 'weak and less able to govern and defend herself than man', was not morally meaningless but part of God's purpose. The Church Fathers taught that 'law and equity' do 'require that they which excel in reason should exceed in rule'. Scripture taught that women were the 'weaker vessels', and Aristotle discerned that their 'virtue and ability' only equipped 'them to be subjects and not sovereigns'. Genesis showed that woman was made for subjection; the Apostles commanded women to accept it; by the light of nature, Aristotle recognized that women's physique entailed it. Every authority – biblical, evangelical, apostolic, and classical – concurred that woman was made to be a subject wife. Even women couldn't fail to perceive the logic of their situation.[77]

Early-modern theorists believed that rational women would voluntarily consent to subjection in marriage. Reason and observation alike showed women they could not survive and prosper without male protection. Woman's natural inferiority meant she would always suffer economic deprivation and physical insecurity when single. Her weakness forced her into a dependent and subordinate role, and this must be fair, equitable, and just, as God had ordered nature knowing it would produce this very result. He made women second-class humans because He wanted them to be second-class citizens. Women's self-interest would demonstrate to her the need for a male protector and provider. Protection would draw with it obedience. In early-modern political thought, rational women voluntarily consent to subjection in marriage because they are driven to do so by abstract reason and empirical observation, by instinct and self-interest, by God and nature. Once married, the woman would be a meet help to her husband in fulfilling her God-given role of reproduction; she would be saved through her child-bearing. The head of household's wife would discharge her Aristotelian role of managing domestic affairs, and governing her children and the inferiors who had settled for domestic service. Her husband would lead the family they formed – the fundamentally natural association of wife, children, and servants – into the political community.

Notes

1. Busenbaum 1663, 550–1 (VI.VI.II.I 'contrarium essentiae matrimonii'). For the same views, see Antoninus 1474, sig. O8r; Beust 1597, 18–19; Covarruvias 1597, 156 (I.II.3.1); Fulbecke 1602, f. 23r–v; Swinburne 1686, 134.
2. Huygens 1696, 62 ('legum circa proximum et principalem finem matrimonii').
3. Gataker 1623, 'A wife indeed', 14 [head]; Augustine 1955, 9; Covarruvias 1597, 152, (I.II.1–1.6 'amicabilem quandam coniunctionem alterius quidem regentis alterius vero obsequentis'); see McLaughlin in Ruether 1974, 213–66, and

Brundage 1987, for brilliant introductions to the place of women in patristic and medieval Christian thought. Some modern commentators have argued that early-modern theorists' stress on the duty of wives to obey their husbands was provoked by incipient feminism amongst women (see, for example, Kelso 1956, 10; Turner 1987, 119). It is difficult conclusively to prove or disprove this thesis but if female revolt can be deduced from divines preaching that wives should be subject, wives have been revolting throughout Christendom since the year dot.

4. Hopkins 1701, 167 [commands]; Snawsel 1631, 33 [Paul].

5. Allestree 1673, 183, dismissed as absurd the idea that a woman's duty of obedience could be 'precarious and liable to be subtracted upon every pretence of demerit'.

6. Whately 1617, 19. For the same analogy see Swinnock 1868, 511.

7. Harrington 1528, sig. D2v [domination]; Hooper 1843, 382 [glad]. Findley and Hobby (in Baker 1981, 12) rightly point out that 'the recurrent metaphor describing the relationship between husband and wife in this period was that of the head and body'. However, it is difficult to see why this should be regarded as distinctively part of 'the move from a society of blood to a society of sexuality, where the mechanisms of power came to be addressed to the body', when it had been repeated endlessly in all Christian writing since Saint Paul first employed it; see, for example, Albertus Magnus 1890–9, 30, 335 (XXXIV.7); 375 (XXXVI.3.5).

8. Delfino 1553, 45 ('sacramentum matrimonii'); Suarez 1872, 166–7 (III.III.2 'non ex voluntate uxoris', 'si matrimonium contrahant hanc superioritatem impedire non possunt'). For the divine right of husbands, see also Baynes 1866, 337; Gataker 1623, A wife indeed, 17; Carter 1627, 49; Goodman 1629, 253; Elton 1637, 554; Dickson 1659, 121; Keble 1685, 482.

9. Tilney 1568, sig. D8r [commandeth]; Stubbs 1968, 11; Rogers 1642, 281 [perfections], 261; Nicholls 1701, 32 [superiority]; Stock 1865, 43 [infirmity]; Stockwood 1589, 82 [exceed]. For the argument that even socially inferior, foolish, or immoral husbands, parents, or magistrates must be obeyed, see Bastingius 1593, f. 72v; Ridley 1548, sig. M6v–M7r; Ayrault 1614, 27–8; Whately 1617, 37; Cottesford 1622, 11–13; Gouge 1622, 273; Davenant 1627, 428; Dod and Cleaver 1629, 168–70; Tuvil 1635, 9; Rogers 1642, 254; Leighton 1844, 123; Comber 1679, 94; Bernard 1684, 192; Craig 1703, 207; The house-keepers guide 1706, 70–1.

10. Coke 1662, 76–7.

11. Hickes 1682, 2–3; Baxter 1830, 145 [appointed]; Rutherford 1843, 69 [Peter-Ana]. The two arguments were often tied together by patriarchalists; that is, just as a wife names her husband but his (irresistible, non-negotiable) power is from God, so a people name their king, but his power, too, is immediately from God and cannot be withstood or abated by the people. See, for example, Saravia 1593, 12; Maxwell 1644, 87, 99, 122; Rider 1680, 14–15; Kettlewell 1691, 13–14; Lloyd 1691, 18; Nicholls 1701, 44.

12. Olizarovius 1651, 53 ('officium est, uxorem imperio regere'); B.,Ste. 1608, 51, 48–9, 42; Wall 1690, 6; Byfield 1637, 641 [underling]; Bunyan 1860, 438; Hopkins 1701, 166 [betrayed]. The house-keepers guide 1706, 3: 'If masters then or parents do not govern, but let servants and children do as they list, they do not only disobey God and disadvantage themselves, but also hurt those whom they should rule'. See also Bernard 1845, 15; Stock 1865, 173–4; Trapp 1647, 230; Swinnock 1868, 508–9, 511.

13. Pareus 1612, 12 ('maritus uxorem, ne subsit, et soli sibi debitum praestet';

'Frustra igitur derogatio praetenditur'); Bodin 1962, 20; Keble 1685, 482 [the case was *Manby and Richardson v. Scot*]; Poole 1696, I, sig. B4r.

14. Hobbes 1991, 139–40 (II.20). For a complete account of Hobbes' political theory, see Sommerville 1992.

15. Locke 1988, 173 (I.47); Locke 1987, 221–2.

16. Tyrrell 1681, 111, 109; Cumberland 1727, 350. Similar arguments can be found in Pufendorf 1934, 853–5, 859–63.

17. Hacket 1607, 8 [sentence]; Tuvil 1635, 33 [forever]. See also Chapter 2, Fallen woman. In Whig theory, even obedience to parents was deemed to obtain only as long as children were unable to fend for themselves; thereafter any superiority could arise only by contract. See, for example, Tyrrell 1681, 16–18, 32; Locke 1988, 310–13 (II. 65–7); Sidney 1990, 88–9.

18. Hirst (1975, 18–19) shows that English practice was not always as consistent as English theory; the discreet participation of aristocratic women and propertied widows was not unknown in English parliamentary elections.

19. Saravia 1593; Stephenson and Marcham 1937, 328–9, An Act concerning the regal power (1554).

20. Rider 1680, 3; Filmer 1991, 142; Hickes 1682, 22.

21. See Chapter 4, Capital and corporal powers, for political versus conjugal power, and Chapter 3, Subjection to women, for a queen regnant's political power over her husband.

22. Woodhouse 1938, 83; Tyrrell 1681, 83 [see also 74–7]. For the limitation of the franchise to heads of household, see Dawson 1694, 29; Mackenzie 1684, 30; Cumberland 1727, 33. Even the Levellers proposed the extension of the franchise only to such 'housekeepers' 'as are assessed ordinarily towards the relief of the poor' – i.e. property-owners (Woodhouse 1938, 357). Apprentices in seventeenth-century England were usually young and almost always resided in the household, as part of the family of their master. As Levine (1977, 148) has stated, early-modern England was a society 'which equated social maturity with economic independence, which regarded laborers as unfree, and which subjected laborers cum servants to patriarchal household discipline'.

How far Levellers wished to extend the franchise was extensively debated by historians after the publication of Macpherson 1964. His thesis that Levellers supported a restricted ballot, excluding all wage-earners, aroused considerable controversy. For a summary of this debate's literature, see Koerner 1985. From the point of view of the contractarians' attitude to women, it is fortunately unnecessary to decide on the correctness of Macpherson's interpretation. Both Macpherson and his critics agreed that the Levellers *did* wish to exclude from the franchise 'those servants included in the persons of their masters by virtue of patriarchalism' (Koerner 1985, 53). In seventeenth-century England, this category clearly included the great majority of women.

23. *Marriage asserted* 1674, 33 [interests]; Allestree 1673, 177 [divided].

24. Okin 1983–4, 135.

25. For such grouping, see, for example, Aristotle 1984, Politics, 1988, (1253b); Tuvil 1635, 1; Maine 1875, 312; Camerarius 1581, 67; Burgersdijck 1631, 289; Settala 1626, 11, 30 ; Wemyss 1633, III, 83; Niccholes 1809, 281; Saravia 1593, 48; Sedgwick 1624, 77; Vauts 1650, 47; Lawson 1659, 187–9; Swinnock 1868, 395. Pierre Ayrault (1614, 22) added cattle to the three traditional categories.

26. Hickes 1682, 23 [Commonwealths–men]; Hughes 1672, 898 [inferiors]; Tyrrell

1681, 83–4 [unfit]. As Butler (1978, 139) has stated, Tyrrell, Gee, and Sidney gave power only to male heads of household; 'In accepting this sort of "democracy" these theorists were not very far from Filmer'.

27. Woodhouse 1938, 82 [dependence]; Suarez 1872, 187 (III.IX.6 'officium quod incumbit propriae personae'); Tiraqueau 1574, f. 45r ('neque in dotem dari'); Ferguson 1695, 44 [indefeasibly], 31 [precluded]; Adams and Stephens 1908, 287–8 [Act for the marriage of Queen Mary to Philip of Spain, 1554 I. Mary 3. c.2].

28. During the seventeenth century, political pressure led to extension or contraction of the franchise at different times in both county and urban elections. For an account of the property qualifications and these disputes, see Hirst 1975, especially 29–105. Hirst's fine study (ibid., 102–3) has shown that even when political interest motivated some gentlemen to allow extending the franchise to 'the scum of the people', they were eager to insist that those obviously dependent on others for their support should be excluded.

29. Lathom 1683, 20–1.

30. Rider 1680, 4 [mean]; Jones 1988, 109, 255 [valuable].

31. Filmer 1991, 142 [Filmer used 'virgin' as a synonym for spinster]. After all, as Houlbrooke states in his scholarly and judicious study of *The English family 1450–1700* – 'The childless widowed heiress was uniquely independent' (1984, 209); and as Erickson has pointed out: 'wealthy widows were the least likely of all widows to remarry' (1993, 196).

32. As Foreness (1683, 13) stated of the contractarians, 'if you ask them when, where, and by whom, or how this pact and covenant was made or consented to, you will find them as silent (or at least speak as little to the purpose) as the man in the moon'.

33. E.,T. *The lawes resolution* 1632, 2 [a 'leet' is a manorial court of record]. For powerful expositions of the argument that the position of women within liberal contract theory was anomalous, see, for example, Okin 1979; Nicholson 1986; Pateman 1988.

34. Brennan and Pateman (1979, 183): 'liberal theorists still have to confront and answer a very embarrassing question, namely, why it is a free and equal female should always be assumed to place herself under the authority of a free and equal male individual'.

35. Hobbes 1991, 139–40 (I.20). Pateman (1988, 44) has interpreted Hobbes' assertion that males are 'not always' stronger and more prudent than women to mean that 'natural individual attributes and capacities are distributed irrespective of sex'. In contrast, my interpretation is that Hobbes shared the belief of virtually all his contemporaries that women are generally weaker than males – there were merely occasional exceptions. See Chapter 3, Restrictions on inferiority.

36. Maxwell 1644, 158. On conquest theory in general, see Sommerville 1986a, 66–9, and Sommerville 1986b *passim*. For a discussion of whether William III and Queen Mary 'conquered' James II, see Blount 1693.

37. Locke 1988, 385 (II. 176–7); Pufendorf 1934, 853, (VI.I.9). Locke did argue that a conqueror in a just war gained the right to kill those who resisted, by force, his assertion of his legitimate authority, and a right to reparations for the costs and damages of waging the war, but the conqueror did not gain any right to rule by victory itself (Locke 1988, 386–92, II. 177–85).

38. See Chapter 7, Coercion and consent.

39. For the view that women can plead duress on lesser grounds, see, for exam-

ple, Covarruvias 1597, 165 (I.II.3–4.9.); Bartolus 1570, I, f. 125v; Nevizanus 1570, 229–30; Gentili 1614, 244; Huygens 1696, 64–5.

40. See Chapter 2, Natural subjection.

41. Shanley (1979, 86) has argued that Tyrrell's assertion that males are naturally superior to females 'seriously undercut his contractarian impulses'. Okin, (1979, 199) has stated:

> Given his initial premise of human equality and egoism, there was no way that Hobbes could logically arrive at the institution of the patriarchal family, on which his political structure is based, for this institution depends on the assumption of the radical inequality of women.

Pateman (1988, 94) has argued that in Locke and Pufendorf: 'The contradiction between the assumptions of contract theory and appeals to natural strength was immediately obvious'.

In fact, there is no inherent logical contradiction in arguing, as early-modern contractarians did, both (i) that most women are naturally weaker than most males, and therefore tend to be reduced to a subordinate position, and (ii) that in terms of abstract principles of justice, males and females have equal political rights.

42. Scott 1988, 173.

43. Brennan and Pateman (1979, 185) have suggested that social contract theory was ambiguous, because it failed to say whether the state of nature was a 'logical fiction' or a 'sociologically and anthropologically realistic condition'. In fact, early-modern contract theorists intentionally constructed the state of nature to be both. Political equality was a 'logical fiction'; natural inequality was (in their view) 'sociologically and anthropologically realistic'. It is difficult to see how the state of nature could have served any useful purpose in their arguments without being both a condition removed from actual existing political obligations, and also one where people were recognizably human.

Locke (1988, 269, II.2) described the state of nature as: 'A state also of equality, wherein all the power and jurisdiction is reciprocal, no one having more than another'; 'equal one amongst another without subordination or subjection'. He also accepted (1988, 337, II.105.) that in such a state some were afflicted with 'defect of mind or body', 'weak or incapable', inferior 'for want of age, wisdom, courage, or any other qualities', while others were 'stoutest and bravest', 'ablest and most likely to rule well'.

44. Pufendorf 1934, 936 (VI.III.4).

45. Lawson 1659, 192 [governed]; Sir Thomas Pope Blunt writing in 1693, cited in Fletcher and Stevenson 1985, 14 [rabble]; Le Grand 1694, 397 [dulness]; Temple 1691, 69–70 [drudgery]; Nicholls 1701, II, 58–9, I. 84 [proprietors]. As Parry (1964, 174) states, in early-modern contract theory, government 'has no obligation to protect those who have come off badly in competitive existence'.

46. As Clark (1977, 706) has lucidly asserted, Locke's state of nature 'is simply seventeenth century England devoid of legitimate Lockean law and authority'.

47. Primrose 1655, 1 ('et infinitos praeterea alios, quibus viri non sunt obnoxii'); Goodman 1629, 322 [wonderful]; Riolan 1671, 85 [seed-plot]; Settala 1626, 25 ('tenero corpore'); Davis 1975, 71. On the lower wages of women, see also, for example, Bland et al. 1914, 348–9, 546–7. Compare Smith (in Carroll 1976, 99): 'To be a woman meant that one was subject to fevers and ill vapours arising from a

malfunctioning menstrual cycle, to hysteria resulting from a diseased womb, and to general bad health developing from a life of ease'.

48. Kelso 1956, 77 [unsuitable]; Stone 1977, 206 [deprived]; McIntosh 1991, 289 [craft-work].

49. E.,T. *The lawes resolution* 1632, 377 [ravishing]; Stone 1977, 93ff [violence]; Chartier 1596, 220 [shameful]; Topsell 1597, 150 [alone]; Temple 1691, 56 [master-buck]; Wiener 1975, 47 [crime].

50. *The batchelor's directory* 1694, sig. A7r [sad]; Stone 1990, 142 [deserted]; Ness 1695, II, 162 [misery]; Cochlaeus 1535, sig. G2r ('maxime virorum suorum ope per mortem destitutae'); McIntosh 1991, 295 [poverty]; Hufton 1984, 364 [allies]; Hanawalt 1986, 222 [wills]; Allestree 1673, 40, 145 [masterless]. On the economic problems that beset single women in early-modern Europe, see Hufton 1984, Prior 1985, 93–114; Willen 1988, especially 562; Earle 1989, 343; and Cullum in Goldberg 1992, 200. The arguments of Clark (1977, especially 711 and 722) often appear to be based on the assumption that early-modern contract theorists believed that women would thrive less well than males in a state of nature only because of the need to 'provide for their offspring'. In fact, all contemporary texts characterize women as weaker than and inferior to males, whether or not they were pregnant or nursing (see Chapter 2, The inferior sex). In Locke, children were a result of a marriage in which women have accepted subjection, not a precursor to it: 'The first society was between man and wife, which gave beginning to that between parents and children' (Locke 1988, 319, II.7). There is nothing in Locke to suggest that he viewed the state of nature as containing hosts of unmarried mothers.

51. Clark (1977, 711), for example, has argued that, in Locke, forced consent was no consent, but since women were unable to support themselves in the state of nature,

> how could he have believed them to be in a position to make the kind of contract which he himself believed to be necessary to create binding obligations? Such contracts, flowing from voluntary agreement, can be made only between equals, where both parties bargain from positions of equal strength.

Locke did not state that parties to a contract must be equally strongly placed, only that 'promises extorted by force, without right ... bind not at all' (1988, 392, II.16).

52. See Chapter 7, Coercion and consent.

53. Azor 1600, 27 (I.I.x 'condemnato ad mortem').

54. Huygens 1701, 33; Azor 1600, 27 (I.I.x) Covarruvias 1597, 166–7 (I.II.3–4.15). Maldonado used the case of a woman who refused to conceal a fugitive from some enemy unless he agreed to marry her; because the enemy did not inflict the fear in order to make him marry, 'the consent is deemed free, and the marriage valid' (1677, 446 'consensus censetur esse liber, et matrimonium esse verum'). See also Biderman 1621, 32.

55. Huygens 1696, 65. Huygens (ibid., 5) also argued that a drowning woman who promised a man marriage if he would save her, was bound by her promise; and Maldonado (1677, 446) argued the same in the case of a woman threatened by attack from a wild animal. See also Major 1509, f. 154v.

56. Kettlewell 1691, 31.

57. Locke 1988, 322 (II.85). Clark (1977, 703) has stated that in Locke's thought:

'The natural differences between the sexes ... override any presupposition of an equal right to autonomy as between men and women. Here and only here, a natural difference creates a justified domination of one person by another'. This statement seems to assume that Locke was in favour of universal adult male suffrage, rather than taking for granted that male servants (another group of the naturally inferior) should be excluded from active participation in the political process. In Locke's thought, the natural weakness of the majority of males might well lead them to consent to subjection, and hence to the just domination of their masters.

58. Nicholson (1986, 161), for example, has complained of the 'fallacies' in Locke's arguments that suggest that an individual woman must depend on an individual man. Pateman (in Shanley and Pateman 1991, 65–6) has stated,

> if there are free childless women in the first generation in the natural condition, there is no reason why they should not form protective confederations of their own by conquering men, or each other, and so obtaining servants. Women and men would then wage the war of all against all as masters of 'families' – and who knows who might win in the end?

59. Firth 1962, 68–93, 110–20; Parker 1988, 60–1. John Case, although a determined advocate of women's rule, admitted that women's 'frailty and weakness' prevented them waging war on males (1588, 134 'infirmae et imbecilles corpore').
60. Ness 1695, I, 110 [spared]; Turner 1683, 279.
61. Parker 1988, 59–60; Petit 1687, 113 [Amazons].
62. Hooper 1843, 381 [prone]; Barbaro 1677, 2 [congress]; Tully 1688, 15 [predominant]; *The batchelor's directory* 1694, 91 [second so numbered], 22 [mixture]; Larner 1981, 149 [uncomplicated].
63. *Aristotle's masterpiece* 1694, 3, 23 [gazing, eject]; Manzini 1587, 283 ('seminis qualitatem appetit'); Nifo 1641, De amore, 50 ('appetunt enim recipere semen'). On female desire for sexual relations and semen, and the illnesses that result from deprivation, see also Albertus Magnus 1596, sig. M4v–6r; Mercuriale 1591, 181–9; Chartier 1596, 3, 56; Dubois 1596, 55–64; Pineau 1598, 109–110; Rueff 1637, 27, 139; Riolan 1671, 87. See also Kelso 1956, 87. For the belief that women were less libidinous and more modest, see Wright 1621, 40; Durham 1675, 355.
64. King 1614, 21 [monastic]; Carter 1627, 66–7 [instructed]; Gataker 1623, 'A wife indeed', 37 [nuptial]; Primrose 1617, 228 [desire]; Stock 1865, 188 [meat]; Anderton 1642, 7 [chiefest]; Kellison 1605, 329 [sensuality]; Olizarovius 1651, 28 ('ad voluptatibus resistendum'). Lorenzo da Brindisi's discussion (1935, 293–5) shows how willingly Catholic priests conceded that virginity was particularly difficult for women. See also Kelso 1956, 121–2; Corbet 1685, 231; Monner 1561, 146; Whately 1624, 44, 67; Baxter 1830, 18.
65. Campanella 1968, 108 ('foeminae cum foemina'; 'natura ipsa'); *The batchelor's directory* 1694, 67 [confined]; Niccholes 1809, 275 [comfort]; Beard 1612, 419 [monsters]; *An essay towards* 1697, 140 [Amazons].
66. Bañez 1594, 18 (LVII.III, 'contra ordinem communem brutis et hominibus'); Vitoria 1991, 273 [pederasty]. Virtually all medieval and early-modern theorists believed that homosexual relations were contrary to the law of nature, because they did not result in offspring, the only legitimate object of sexual activity. See, for example, Abelard 1855b, col. 764; Guido [1500?], f. 91v; Wemyss 1633, II, 165; Preston 1635, 156–7; Elton 1637, 423–4; Reyner 1657, 79; Hughes 1672, 901; Turner

1685, 278. That fornication was illicit even in time of necessity, see Benincasa 1562, f. 105v–6r; Sordi 1602, f. 213r; Covarruvias 1597, 614, (II.II.1.5) and Chapter 5, Sex and marriage.

67. White 1656, III, 207 [abilities]; Bunyan 1860, 439 [ordinarily]; Poole 1696, II, sig. Nnn2r [weaker]; Rollock 1603, 344 [Rollock did concede that many women did not recognize their own craving for a superior]; Knewstubs 1584, 120 [benefit]; Tuvil 1635, 55–6 [puddle]; Allestree 1673, 39–40 [expediency].

68. Joannes de Sancto Geminiano 1585, f. 181r; Arelat, 1536 (Arelat) f. 18v ('quia uxores debent a maritis defendi'). English common law upheld the same principle, see Smith 1982, 133; Keble 1685, 446.

69. Gentili 1614, 21 ('sexus infirmitatem,'); Baynes 1866, 339 [jeopardy]; Byfield 1637, 641, 644 [shift]; Hopkins 1701, 165 [ready]; Lorenzo da Brindisi 1935, 292–3 ('de naturali desiderio', 'ad sustentationem, defensionem, protectionem, societatem, securitatem et delicias'). For the husband's duty of protection, see also Topsell 1597, 181; Allen 1600, 128–9; Pricke 1609, sig. K2r; King 1614, 26; Abbot 1653, 41; Trapp 1647, 100; Herbert 1648a, sig. V6v; Webb 1653, 79–80; White 1656, II, 96; Swinnock 1868, 507; Keble 1685, 446; *The house-keepers guide* 1706, 109.

70. Gouge 1630, 16, 50, 38–9 [submit]; Dickson 1659, 121–2 [expedient]; Poole 1696, II, sig. Rrr1v [grudge]. See also Bentley 1582, 15; Willet 1612, 7; Carter 1627, 12; Abbot 1653, 41, 43; Comber 1679, 77–8; *The house-keepers guide* 1706, 123.

71. Tuvil 1635, 3–4 [due]; Bernard 1845, 75 [loving]; Swinnock 1868, 503, 525; Tyrrell 1681, 112 [respect]; Pufendorf 1934, 860 (V.I.11) [each].

72. Nicholls 1701, 93 [all countries]; Stock 1865, 173 [companion].

73. Howe 1822, VII, 504–5 [strength]; Whately 1647, II, 97 [audaciousness]; White 1656, 72 [challenge]; Benincasa 1562, f. 22v ('sed paupertas ab ipso summo rerum conditore Deo Opt. Max. fuit inducta'); Ken 1838, 462 [Thou].

74. Hopkins 1701, 147–8 [fountain]; White 1656, 14, 70–2 [excellency]; Foreness 1683, 5 [superiority]; Allen 1600, 132, 126 [distinctions]; Kellison 1605, 168 [rule]; Lawson 1659, 186 [imparity]. See also Dering 1597, sig. A8r; Crooke 1625, 107; Tuvil 1635, 9–16; Maxwell 1644, 89; Leighton 1844, 397; Lathom 1683, 5.

75. Burgersdijck 1631, 165–6 ('imperfectiora perfectioribus inservire'; 'Praeterea mas natura sua praestantior est foemina: ideoque ille, imperare; haec parere debet'); Lagus 1566, 74–5 ('Non igitur violentia, aut arte humana, sed Deo'; 'Sicut nec femina aliqua Deo imputare potest, quod masculo sit inferior natura'); Fosset 1613, 51 [murmur]; Rollock 1603, 340–1 [estates]; Carter 1627, 63 [strangers]; Rogers 1642, 255 [otherwise].

76. B.,Ste. 1608, 50; Willet 1612, 4–5 [order]; Crooke 1615, 199; Heale 1609, 31 [pre-eminence]; White 1656, I, 110 [distinction].

77. Hooker 1977, 402, (V.73.2); Tuvil 1635, 33. Clark (1977, 723) has stated,

> the cornerstone of Locke's theory is the assumed natural disadvantage of women, and the ultimate objection to his theory is that it must convert a biological difference between the sexes into a socio-economic liability.

What Clark perceives as 'the ultimate objection' to Locke's theory, early-modern theorists regarded as its ultimate strength. Natural law theorists aimed at establishing social and political obligations in harmony with what they saw as man's natural, biological characteristics.

9

Conclusion

I opened by confessing that this was not a study of women's history. Some might suggest that – if the keynote of history is change rather than continuity – it is not a work of history at all. As one modern historian has acutely noted, many have argued that 'women's experience has no part in history, because everyone knows their role has been unchanging ... yet some would say change and the reasons for change is what history is all about'. But I have said little to suggest there were any significant changes in attitudes to women between 1500 and 1700. Certainly anyone looking for a steady progressive development from the belief that women were naturally inferior to the assertion that they are just as good as males in every respect will be sadly disappointed. The evidence does not even reveal a steady decline in status culminating in women's oppression and confinement in the home as capitalist shadow-labourers. For women the *'bon vieux temps'* are as hard to find in the history of ideas as they are in social history.[1]

In 1700, as in 1500, most theorists believed that women were physically, mentally, and psychologically inferior to males. In 1700, as in 1500, official theory maintained that wives should bear children, stay at home, concern themselves only with domestic – not public – affairs, and obey husbands who properly ruled married couples' affairs, chattels, and children. Contractarian theory offered the philosophical postulate of women's initial political equality with one hand and promptly took back any practical effect with the other. Protestant divines asserted the priesthood of all believers and excluded women from the ministry as completely as Catholic theologians paying lip service to the spiritual equality of the sexes. Married presbyters disparaged the celibacy that single priests exalted, but in virtually every other area of theory on relations between the sexes 'new presbyter' *was* 'but old priest writ large'. Initially, it appears that Renaissance and Reformation ruptured the strait-jacket of medieval orthodoxy in almost every area except that of attitudes to women. The student of early-modern history can easily gain the impression that everything changed (for males) and everything remained the same (for females). However, this characterization is not wholly accurate. Despite the significant degree of continuity in early-modern religious, moral, and political theory, attitudes to women did alter as the times changed. The shifts were gradual and far from universal; in 1700, some theorists look as like their forbears of 1500 as two peas in a pod. But eventually Renaissance and Reformation brought changes in other areas of thought that did have an impact on attitudes to women.

The Reformers were not feminists; Martin Luther, Jean Calvin, and their followers believed that males were naturally superior to females. Yet the Reformation brought about changes in attitudes towards women unintended by the Reformers. When they abandoned many dogmas enshrined in Catholic orthodoxy, Protestants called into question the authority of the Church, of the Fathers, and of generations of theologians. When they insisted on returning to the Bible and considering its words afresh, on finding Scriptural support for all the doctrines Christians must accept, they provoked a wholesale reconsideration of its true meaning. The newly critical attitude to orthodoxy did not immediately change attitudes to women, for Protestants were not protesting against doctrines of female subjection. Re-examining Scripture did not directly alter theory about the role of women, since the Bible is anything but unambiguously supportive of female equality. Instead, new approaches provoked changes in exegesis, little thought of by the early Reformers.

The way change actually occurred can be best illustrated by looking at the ramifications of a particular case – that of clerical celibacy. Martin Luther and other reformed divines accepted most of the traditional Catholic teaching on women and marriage, but with one significant exception: these clerics wanted to marry. Protestant rejection of a celibate ministry involved a radical break with the teaching on the supreme virtue of virginity that pervaded patristic and scholastic writings. To uphold their right to marry, protestant clergymen began to argue that Augustine, Jerome, and the countless theologians who had followed their teaching had got it all wrong; for clerics, they contended, virginity was not inherently superior to matrimony. Protestants attacked Jerome for being 'in his heat . . . so carried away' with the commendation of virginity that he 'wresteth Scripture to serve his turn; thus he counteth marriage filthiness, uncleanness, fleshliness'. Married protestant ministers could not endorse 'the rash and inconsiderate speeches of Jerome'. It was easy to progress from arguing that Jerome was 'an immoderate extoller of virginity', to dismissing his teaching as the product of psychological disturbance. It was simple to ascribe Jerome's 'rigor' to the fact that he 'more loved than favoured' the female sex. The Fathers' 'vehemency to Christian women' only stemmed from 'the severity of those times'. Jerome's and the Fathers' views were still being treated as authoritative in some sense, but the acid of historical relativism was being used to etch away their unacceptable features.[2]

The same historical analysis that neutered patristic teaching could equally well be applied to Scripture, and arguably needed to be. After

all, Saint Paul's enthusiasm for virginity was little less pronounced than
Jerome's. Eager to marry, protestant ministers set to the task with a will.
Paul's counsel to refrain from marriage, they declared, was provoked
only by particular historical circumstances; Christians were then perse-
cuted and ministers could perform their duties better if not tied down by
family responsibilities. His words, protestant divines proceeded, did not
apply in the sixteenth and seventeenth centuries. When Paul stated: 'It is
good for a man not to touch a woman' (I Corinthians 7:1) he had not
intended to dictate an immutable rule; he meant 'only more convenient
or better with respect to the troubled state of the church'. Paul's teaching
on issues other than marriage could be set aside with this convenient
device. Paul had asked rhetorically: 'Doth not even nature itself teach
you that if a man have long hair it is a shame unto him?' (I Corinthians
11:14). The influential French protestant divine, Claude de Saumaise,
argued that the answer was, No. Nature did not teach this, custom did.
Those reading Scripture must bear in mind that some precepts were perpetual, but others valid only in certain places, at certain times, for certain
people. For hundreds of pages, in enormous detail, Saumaise demonstrated that the appropriate length for male and female hair was purely a
matter of convention. Yet I Corinthians 11 was one of the key texts for
the subjection of women. Paul had insisted on the need for males to have
short hair as part of his argument that women could not pray in church
with their heads uncovered. Women (unlike males) must cover their
heads, because males not females were the glory of God; because woman
was created for the male, not the male for woman.[3]

If Saint Paul's conviction about the natural propriety of short hair for
males was just the product of a particular society's customs, why was
not his belief in the natural subordination of women equally historically
determined? Saumaise's arguments (unintentionally) implied that it
might be: In the East, he argued, both then and now males were powerful, women 'obsequious'. Dominant males made women cover their
heads as a sign of submission, so males growing their hair and covering
their heads 'in a feminine way' reduced themselves to women's status.
'Today these things are not so. Therefore Paul's decree does not pertain
to these times.' Of course, Saumaise did not wish to dismiss Paul's teaching on women's inferiority, only on hairdressing, but the logic of his
argument could be extended further than he intended. It was a small
step to arguing that in the East, in those times, wives were subject to
their husbands. Today these things are not so. Therefore Paul's decree
does not pertain to these times.[4]

Claude de Saumaise was produced as much by the Renaissance as the
Reformation; he was famous for his exceptional knowledge of classical
languages and texts. Much of the *Epistola* cited above was a brilliant display of his philological and historical skills. Without the revival of classical learning that was the keynote of humanism, it would have been

impossible for him to place Scriptural texts in their historical context as he did. The Renaissance, like the Reformation, was not directed at changing the role of women. Even humanists favouring women's education wanted only to equip women better for their existing social roles, not to alter them. They believed well-educated women would be more chaste and obedient wives, more caring mothers, more dutiful daughters. Humanists were not social revolutionaries; they were not even intellectual revolutionaries. They preached the need to develop new skills in Greek and Latin because they wanted better to understand the wisdom of Aristotle, Cicero, and other classical authors, not because they wanted to debunk them. Yet, just as the fervently religious Reformers instigated changes that led to the Bible being treated as one historical text among others, so humanist learning finally led to a loss of authority by classical texts. For centuries, Aristotle's philosophy, physiology, and sociology had been the point of departure for all Western learning. His attitude to women was enormously influential. The humanists of the early Renaissance were eager to understand him and developed new techniques to study his writings. But as their expertise became generally disseminated, the more Aristotle was understood the more he was dismissed.[5]

Aristotle's teaching on the physiology of human reproduction (along with other classical physicians') was gradually abandoned as human anatomy was considered anew. The Aristotelian hegemony was so undermined in philosophy that Thomas Hobbes could disparagingly talk of the 'contradictions and absurdities' that resulted 'from the bringing of philosophy and doctrine of Aristotle into religion'. It would be misleading to caricature medieval philosophy as a footnote to Aristotle. No one can read the scholastics without recognizing that they distorted or disagreed with Aristotle almost as often as they followed him. Nevertheless, the attitude of casual dismissal that Hobbes affected would have been impossible before the Renaissance. Aristotle's political thought suffered the same fate at Hobbes' hands. 'In these western parts of the world', he complained

> we are made to receive our opinions concerning the institution and rights of commonwealths from Aristotle, Cicero, and other men – Greeks and Romans – that living under popular states derived those rights not from the principles of nature, but transcribed them into their books out of the practice of their own commonwealths, which were popular.

Hobbes treated Aristotle as Claude de Saumaise treated Saint Paul. Aristotle's views on political rights did not stem from the principles of nature (any more than Paul's views on coiffure did), but from 'the practice' of his own commonwealth. Medieval and early-modern theory had generally insisted that Aristotle described what was natural. Christ and

his Apostles taught revealed truth; Aristotle, the truths of reason. The
Renaissance played a key part in destroying this belief.[6]

Professor Kelly's brilliant work has taught us to question whether
women had a Renaissance, and to note that the revival of 'classical culture with all its patriarchal and misogynous bias' was not an unmitigated good for women.[7] But we should be equally mistaken to characterize
it as an unmitigated evil. Humanists' views of women's role were as conservative as the scholastics' from whom they largely derived them. Yet
humanist learning enabled political and social theorists to conceive the
argument that in these Western parts of the world we are made to
receive our opinions concerning the rights of women from Aristotle and
other men who derived those rights not from the principles of nature,
but from the practice of their own commonwealths.

The Roman Empire and Western Europe adopted their religion from
the Eastern Mediterranean. The Bible was received as God's word. Old
Testament history showed woman in a nomadic, pastoral society where
she was little valued; Old Testament law gave woman status 'only in her
role as a mother' and decreed that she be 'always subject to the authority
of some male'. Saint Paul's Epistles attacked (what he regarded as) sexual libertinism stemming from the *mores* of the Roman Empire where
women 'had achieved a much higher legal status than women in Greece
or in the East'. Western European philosophers adopted the teachings of
Aristotle and other classical Greek thinkers who had written in a society
where women of their class were married at the onset of puberty and
thereafter strictly confined within the home. In the words of Professor
Woodbridge's seminal study of woman in early-modern English literature: 'The Bible and the classics were twin fountains of Renaissance
thought'. Early-modern thinkers regarded as authoritative, texts written
in societies where women's status was far lower and their freedom more
restricted. The consequence was tension and dissonance.[8]

Professor Woodbridge has perceptively and appositely noted that 'the
orthodox Renaissance view of woman-kind ... betrayed a radical discontinuity between theory and reality'. Theory taught that woman was inferior and subject; reality was 'the aggressive, liberty-minded
Englishwoman'.[9] Some historians have suggested that divines and
moralists preached female inferiority and wifely subjection to quell
women who were rebelling against imposed social roles. This explanation is inadequate not only because of the dearth of evidence for widespread early-modern feminism, but also because the teaching was so
often in wholly inappropriate media. Very few early-modern women
could read Latin, yet hundreds of early-modern Latin treatises contained
lengthy disquisitions and arguments about women's proper role. These
theorists were not trying to convince women; they were trying to convince themselves.[10]

Early-modern theorists spilt gallons of ink arguing about the (un)naturalness of polygamy, but not because there was any significant popular or academic movement in Europe to allow males many wives. Their problem was intellectual: culturally conditioned reflexes told them that polygamy was wrong, for their society abhorred it, but every day they read the Old Testament where it was treated as morally neutral, if not commendable. They endlessly debated whether adultery was worse in wife or husband, because social reality meant that the damage caused by a wife's adultery was far greater, but Augustine and the canonists told them that both sinned equally. Liberal contractarian theorists' admission that – naturally – women had equal political rights was provoked by the need to combat absolutist theory in a way consistent with the moral pre-suppositions and political axioms of their day, not by a novel conviction that women were physically, intellectually, and psychologically their equals. Chancery enforced equitable laws that allowed wives to control their own property, not because women were in a position to fight for legal and economic equality, but because the judges were convinced of the justice of 'the common way of proceeding' that kept property 'out of the power and reach of the husband'. Authoritative legal maxims taught that at marriage 'all manner of movable substance is presently by con-junction the husband's to sell, keep, or bequeath'; courts of law would not enforce the theory. Legal practitioners' own observation that upper-class women could and did control their property responsibly was more powerful than the inherited dictum: 'Every feme covert is *quodammodo* [in some way] an infant'. It was not coincidental that the development of equitable property rights for wives coincided with 'the emergence of women as independent adults'. Theoretical discussions attempted to resolve the tension between supposedly authoritative texts and wholly divergent cultural imperatives.[11]

Few early-modern theorists treated traditional teaching with the cava-lier contempt that Thomas Hobbes meted out to Aristotle. Instead, they reverently repeated the received wisdom that a wife 'must keep at home or in her house' and then fundamentally compromised the rule with numerous qualifications and exceptions: she could go abroad for reli-gious duties, works of mercy, 'for the health of her body, or the solace of her mind'. Still fewer theorists were willing to dismiss the teaching of Saint Paul. Instead, they interpreted his teaching in a way consonant with their own beliefs. Theologians glossed Paul's words when these suggested to the literally minded reader that only males were made in the image of God, or that females were saved by fecundity while males were saved by faith, or that a father could marry off his daughter with-out reference to her own inclinations. They did not themselves hold these beliefs and therefore they were certain that Paul's words could not have been intended in their literal sense. When Scripture or the classics clashed too jarringly with local customs and contemporary beliefs, they were explained away.[12]

256SEX AND SUBJECTION

This book has concentrated on official (not popular) culture and on theory (not practice). This is not because popular culture is unimportant, or because theory can be understood without reference to practice. Exactly the opposite is true. Early-modern Western thought is incomprehensible without reference to the customs and attitudes of society at large. The development of theory about sex and subjection in the early-modern period was intimately connected with the 'radical discontinuity between theory and reality' to which Professor Woodbridge has called our attention. Early-modern theorists debated the role of women in a bid to resolve the tensions that resulted from attempting to adapt ancient Eastern Mediterranean religion and philosophy to Western Europe, where the customary familial, social, and economic roles of women were wholly different.

In Western Europe between 1500 and 1700, views of the proper role of women did begin to change. Desultorily, and painfully slowly, a few theorists in these Western parts of the world ceased to believe that they must receive their opinions concerning the rights of women from Aristotle, Saint Paul, and other men who derived them, not from the principles of nature, but from the practice of their own commonwealths.

Notes

1. Hill 1993, 13; Hufton (1983, 126) in her stimulating survey of recent historiography on women has noted that early-modern historians are often given the 'unenviable task' of

> locating a *bon vieux temps* when women enjoyed a harmonious, if hardworking domestic role and social responsibility before they were down-graded into social parasites or factory fodder under the corrupting hand of capitalism. So far the location of this *bon vieux temps* has proved remarkably elusive.

Bennett (1993, 175) has noted 'the full historiography of women's work in the European past shows incontrovertibly that the idea of change-for-the-worse has a strong and powerful sway over the field'.
2. Willet 1634, 302 [filthiness]; Field 1847–52, I (1848), 299; Fulke 1601, 502 [immoderate]; Gauden 1656, 39, 113–14 [rigor]. Protestant theorists were slow to abandon the belief that virginity was inherently superior to marriage. William Fulke (1601, 3) complained that it was 'a mere slander' for Catholics to state of Protestants 'that we teach with Jovinian, that virginity and the continent life are not to be preferred before marriage'.
3. Poole 1696, II, sig. Fff3v [convenient]; Saumaise 1644, 9. The more conventional view that men should never wear their hair long was expounded at length in Prynne 1628, especially at 38–40, and Wall 1690.
4. Saumaise 1644, 665–6 ('obsequium'; 'muliebriter caput comit'; 'Hodie nulli tales sunt. Non igitur ad haec tempora pertinet Pauli dictum').
5. On the humanist view of the purpose of women's education, see Wayne in Hannay 1985, 17–19.

6. Hobbes 1991, 85 (I.12), 148–9 (II.21).

7. Kelly 1984, 35.

8. Bird in Ruether 1974, 57; Parvey in Ruether 1974, 119; Woodbridge 1984, 87.

9. Woodbridge 1984, 325. See Erickson (1993, 19–20) on 'the disjuncture between theory and practice' in English property law. Wiesner (in Friedman 1987, 169) has noted a similar tension between the reality and the theory of women's legal status in Germany.

10. As Howell has argued, historians must 'move beyond interpretation of the text itself to an analysis of reception. Who made this text? When? With what resources? Who read/heard/saw/otherwise experienced it?' (1991, 145).

11. Keck 1697, 225; E.,T. *The lawes resolution* 1632, 130, 141; Larner 1981, 102.

12. Pricke 1609, sig. L3r–v; see Chapter 3, Spiritual equality, and Chapter 7, Parental consent.

Bibliography*

Collections of documents

Adams, G.B. and Stephens, H.M., eds (1908). *Select documents of English constitutional history*, New York.

Berry, L.E. and Crummey, R.O., eds (1968). *Rude and barbarous kingdom*, Madison, WI.

Bland, A.E., Brown, P.A. and Tawney, R.H., eds (1914). *English economic history: Select documents.*

Bretschneider, C.G., ed. (1835). *Corpus Reformatorum*, vol. II, Halle.

Firth, C.H. and Rait, R.S., eds (1911). *Acts and ordinances of the Interregnum, 1642–1660.*

Gardiner, S.R., ed. (1886). *Reports of cases in the courts of Star Chamber and High Commission*, Camden Society, 2nd series vol. XXXIX.

Hair, P., ed. (1972). *Before the bawdy court.*

Howell, T.B., ed. (1809–28). *Cobbett's complete collection of state trials*, 34 vols.

Jones, D.L., ed. (1988). *A Parliamentary history of the Glorious Revolution.*

Lefkowitz, M.F. and Fant, M. (1977). *Women in Greece and Rome*, Toronto.

Pocock, N., ed. (1870). *Records of the Reformation*, 2 vols, Oxford.

Pollard, A.W. and Redgrave, G.R., eds (1976–91). *A short title catalogue of books printed in England, Scotland & Ireland*, 2nd ed., 2 vols.

Sadler, Ralph (1809). *The State papers and letters of Sir Ralph Sadler*, ed. Arthur Clifford, Edinburgh.

Schroeder, H.J., ed. (1941). *The canons and decrees of the Council of Trent*, transl. H.J. Schroeder, Rockford H.J.

Spalding, J.C., ed. (1992). *Reformation of the ecclesiastical laws of England, 1552*, transl. J.C. Spalding, Sixteenth century essays and studies, vol. XIX.

Stephenson, C. and Marcham, F.G., eds (1937). *Sources of English constitutional history*, New York.

Thornton, J., ed. (1934). *Table talk: From Ben Jonson to Leigh Hunt.*

Woodhouse, A.S.P., ed. (1938). *Puritanism and liberty.*

Zurich Letters (1845). transl. and ed. H. Robinson, 2nd series, Cambridge.

Primary sources

Abbot, Robert (1608). *A wedding sermon preached . . . 1607.*

Abbot, Robert (1653). *A Christian family builded by God.*

Abelard, Peter (1855a). Epitome theologiae Christianae. In *Patrologiae Cursus Completus*, ed. J.P. Migne, vol. CLXXVIII, cols 1685–758, Paris.

Abelard, Peter (1855b). Expositio in Hexameron. In *Patrologiae Cursus Completus*, ed. J.P. Migne, vol. CLXXVIII, cols 731–84, Paris.

*Place of publication is London except where stated

Agrippa, Henry Cornelius (1540). *The commendation of matrimony,* transl. D. Clapham.

Agrippa, Henry Cornelius (1980). Female pre-eminence. In *The feminist controversy of the Renaissance,* ed. D. Bornstein, New York.

Ainsworth, Henry (1627). *Annotations upon the five books of Moses.*

Albertus Magnus (1890–9). *Opera omnia,* ed. C.A. Borgnet, 38 vols, Paris.

Albertus Magnus, Pseudo- (1596). *De secretis mulierum libellus,* Lyons.

Alexander of Hales, Pseudo- (1948). *Summa theologica fratris Alexandri,* 4 vols Quaracchi.

Allen, Robert (1600). *A treasurie of catechisme, or Christian instruction.*

Allestree, Richard (1673). *The ladies calling in two parts,* Oxford.

Allestree, Richard (1683). *The whole duty of man.*

Ambrosiaster (1845). Commentary on Paul's epistles. In *Patrologiae Cursus Completus,* ed. J.P. Migne, vol. XVII, cols 45–508, Paris.

Ames, William (1639). *Conscience with the power and cases thereof,* Amsterdam (repr. Amsterdam, 1975).

Ames, William (1968). *The marrow of theology,* Boston.

[Anderton, Laurence] (1642). *The English nunne* [Saint Omer].

Andrewes, Lancelot (1854). *Two answers to Cardinal Perron,* Oxford.

Antoninus, Florentinus (1474). *Tractatus de censuris,* Venice.

Aquinas, Thomas (1872–82). *Summa theologica,* ed. J.P. Migne, 3 vols, Paris.

Archdekin, Richard (1679). *Theologia tripartita,* Cologne.

Arelat, Joannis Nicolat (1536). *Iurisprudentiae alumnis frugiferum de secundis nuptijs opus,* Lyons.

Aristotle (1984). *The complete works of,* ed. J. Barnes, 2 vols, Princeton, NJ.

Aristotle's masterpiece: or the secrets of generation (1694).

Astell, Mary (1730). *Some reflections upon marriage,* 4th ed. (repr. New York, 1970).

Atterbury, Francis (1735). *Sermons and discourses on several subjects and occasions,* 4th ed.

Attersoll, William (1618). *A commentary upon the forth booke of Moses, called Numbers.*

Atwood, William (1690). *The fundamental constitution of the English government.*

Augustine (1955). *Treatises on marriage and other subjects,* in The Fathers of the Church, ed. R.J. Deferrari, vol. 27, New York.

Augustine (1977). *City of God,* in Library of the Nicene and Post-Nicene Fathers, ed. P. Schaff, Michigan.

Augustine (1982). *The literal meaning of Genesis,* transl. J.H. Taylor, 2 vols, New York.

Aylmer, John (1559). *An harborowe for faithfull and trewe subiectes,* Strasbourg.

Ayrault, Pierre (1614). *A discourse for parents honour and authority over their children,* transl. J. Budden.

Azor, Juan (1600). *Institutionum moralium,* 2 vols, Rome.

B., Ste. (1608). *Counsel to the husband: to the wife instruction.*

Babington, Gervase (1615). *The workes of the right reverend father in God.*

Baconthorp, John (1526). *. . .Super quatuor sententiarum libros,* Venice.

Bale, John (1849). *Select works,* Cambridge.

Bañez, Domingo (1594). *. . .De iure and iustitia decisiones,* Salmantica.

Barbaro, Francis (1677). *Directions for love and marriage* (1st ed. 1513).

Barker, Peter (1624). *A iudicious and painefull exposition upon the ten commandements.*

Barlow, John (1632). *An exposition of the first and second chapters of the latter epistle of the Apostle Paul to Timothie.*

Barlow, Thomas (1692). *Several miscellaneous and weighty cases of conscience.*

Barlow, William (1690). *A treatise of fornication,* Oxford.

Baron and feme (1738). 3rd ed., Anon.

Bartolus of Sassoferato (1570). *In primiam ff partem,* 10 vols, Venice.

Bastingius, Jeremias [1593]. *An exposition or commentary upon the catechisme of Christian religion,* Cambridge.

The batchelor's directory: being a treatise on the excellence of marriage (1694). Anon.

Batty, Bartholomew (1581). *The Christian mans closet.*

Baudius, Dominic (1638). *Amores,* Amsterdam.

Baxter, Richard (1830). A Christian directory. In *The Practical works,* vol. IV.

Baynes, Paul (1866). *An entire commentary upon ... Paul to the Ephesians,* Edinburgh.

Beard, Thomas (1612). *The theatre of Gods iudgements.*

Becan, Martin (1625). *Compendium manualis controversiam huius temporii,* Douai.

Becon, Thomas (1844). *Prayers and other pieces,* Cambridge.

Bellarmine, Robert (1602–3). *Disputationem ... de controversiis Christianae fidei,* 4 vols, Venice.

Benincasa, Cornelius (1562). *Tractatus de paupertate ac eius privilegiis,* Persustae.

Bentley, Thomas (1582). *The monument of matrons: the sixt lamp of virginitie.*

Bernard, Francis (1684). *The Christian duty composed,* Aire.

Bernard, Richard (1845). *Ruth's recompence,* Edinburgh.

Besold, Christopher (1623). *Discursus politici,* Strasbourg.

Beust, Joachim von (1597). *Tractatus de jure connubiorum,* Leipzig.

Beverland, Adrian (1698). *De fornicatione cavenda admonitio.*

Beza, Theodore (1587). *Tractatio de polygamia,* Geneva.

Beza, Theodore (1591). *Tractatio de repudii et divortiis,* Geneva.

Biderman, Jacob (1621). *Matrimonii impedimenta,* Dillingen.

Bingham, Joseph (1855). The antiquities of the Christian church. In *Works,* vol. VIII, Oxford.

Blackwood, Christopher (1659). *An exposition upon the ten first chapters of ... Matthew.*

Blegny, Nicolas de (1676). *New and curious observations on the art of curing the venereal disease,* transl. W. Harris.

[Blount, Charles] (1693). *King William and Queen Mary conquerors.*

Blount, Charles (1911). A discourse of marriage. In *Penelope Rich and her circle,* ed. M.S. Rawson, pp. 320–42.

Blount, Henry (1638). *A voyage into the Levant,* 3rd ed.

Bodin, Jean (1962). *The six bookes of a commonweale,* ed. K.D. McRae, Cambridge, MA.

Bonaventure (1889). Commentaria in quatuor libros. In *Opera Omnia,* vol. IV, Quaracchi.

Bornstein, D. (1980). *The feminist controversy of the Renaissance,* New York.

Boughen, Edward (1646). *The principles of religion,* Oxford.

Bourgeois, Louyse (1626). *Observations diverses sur la sterilité,* Paris.

Bovet, Richard (1684). *Pandaemonium* (repr. Aldington, 1951).

Brenz, Johann (1566). *In scriptum Apostoli et Evangelistae Matthaei,* Tuebingen.

Brenz, Johann (1584). *A right godly and learned discourse upon the book of Esther*, transl. J. Stockwood.

Bridges, John (1573). *The supremacie of Christian princes*.

Brownlow, Richard (1651). *Reports: (A second part) of divers famous cases in law*.

Brydall, John (1680). *The white rose: or a word for the House of York*.

Brydall, John (1699). *Jus primogenti, or the dignity, right, and priviledge of the firstborn*.

Brydall, John (1703). *Lex spuriorum: or the law relating to bastardy*.

Bucer, Martin (1955). De regno Christi. In *Opera Latina*, vol. XV, Paris.

Bucer, Martin (1972). *Commonplaces*.

Bullinger, Henry (1541). *The Christen state of matrimonye*, n.p. (repr. Amsterdam, 1974).

Bullinger, Henry (1572). *A confutation of the Popes bull*.

Bullinger, Henry (1849). *Decades*, Cambridge.

Bunny, Edmund (1610). *Of divorce for adulterie*, Oxford.

Bunyan, John (1860). *The works*, vol. II, Glasgow.

Burgersdijck, Franco (1631). *Idea philosophiae tam naturalis tam moralis*, Oxford.

Burn, Richard (1788). *Ecclesiastical law*, 5th ed., 4 vols.

Busenbaum, Hermann (1663). *Medulla theologiae moralis*, Lyons.

B[utler], J[ohn] (1697). *The true state of the case of*.

Butler, John (1698). *Explanatory notes upon a mendacious libel*.

Byfield, Nicholas (1637). *A commentary upon the first three chapters of the first Epistle general of St. Peter*.

Cajetan, Cardinal (Tommaso de Vio) (1581). *Summa S. Thomae Aqyunatis ... praedicatorum ... Cum commentariis Thomae de Vio Caietani*, Lyons.

Calvin, Jean (1948). *Commentaries on the first book of Moses called Genesis*, transl. J. King, vol. I, Grand Rapids.

Camerarius, Joachim (1581). *Politicorum et Oeconomicorum Aristotelis interpretationes et explicationes accuratae*, Frankfurt.

Campanella, Tommaso (1968). *I sacri segni: Theologicorum liber XXIV de sacramentis*, vol. VI, Rome.

Canisius, Peter [1592–6] *A summe of Christian doctrine*, n.p.

Capreolus, John (1967). *Defensiones theologiae divi Thomae Aquinatis*, Frankfurt.

Cardwell, Edward, ed. (1841). *Postils on the epistles and gospels compiled and published by Richard Taverner in the year 1540*, Oxford.

Carter, Thomas (1627). *Carters Christian commonwealth; or domesticall dutyes deciphered*.

Caryll, Joseph (1647). *An exposition ... upon ... Job*.

Case, John (1588). *Sphaerae civitatis*, Oxford.

Catharinus, Ambrosius (Lancelotto Politi) (1551–2). *Enarrationes assertationes disputationes* (repr. Ridgewood, NJ, 1964).

Caussin, Nicholas (1634). *The holy court in three tomes* (repr. 1977).

Cawdry, Daniel (1656). *Family reformation promoted*.

Certain sermons or homilies (1756). Anon.

Chamberlayne, Edward (1673). *Anglia notitia*, 7th ed.

Chambers, David (1579). *Discours de la legitime succession des femmes*, Paris.

Charnock, Stephen (1866). *The complete works*, vol. V, Edinburgh.

[Chartier, Alain] (1596). *Detectable demaundes and pleasant questions, with their sev-*

eral answers, in matters of love.

Chemnitz, Martin (1978). *Examinations of the Council of Trent*, transl. F. Kramer, vol. II, St Louis.

Christian parents, The office of (1616). Anon., Cambridge.

Chrysostom, Saint John (1888–9). *Homilies*, in Library of the Nicene and Post-Nicene Fathers, ed. P. Schaff, vols X–XII, New York.

Cicero, Marcus Tullius (1534). *The three bookes of Tullyes offyces*, transl. R. Whittington.

Cipolla, Bartholomew (1547). *Varii tractatus D. Bartholemaei Caepollae Veronensis*, Lyons.

Clarke, Samuel (1659). *Medulla theologiae.*

Cleaver, Robert (1598). *A godly form of householde government.*

Cochlaeus, John (1535). *De matrimonio serenisssimi regis Angliae Henrici Octavi*, Leipzig.

Cock, Gijsbert (1668). ...*Vindiciae pro lege, imperio, and religione contra tractatus Thomae Hobbesii*, Utrecht.

Coke, Roger (1662). *A survey of the politicks of Mr Thomas White, Mr Thomas Hobbes and Hugo Grotius.*

Colet, John (1867). *A treatise on the sacraments of the Church.*

Colet, John (1985). *Commentary on First Corinthians*, ed. and transl. B. O'Kelly and C.A.L. Jarrott, Binghampton, NY.

Comber, Thomas (1679). *The occasional offices of matrimony.*

Concubinage and poligamy disprov'd (1698). Anon.

Conjugium languens (1700). Anon.

Conset, Henry (1685). *The practice of spiritual or ecclesiatical courts.*

Coras, Jean de (1605). *Paraphrase sur l'edict des mariages clandestinement contractex par les enfants de famille*, Lyons.

Corbet, John (1685). Matrimonial purity. In *The remains of John Corbet.*

Cosin, John (1843–51). *The works*, 4 vols, Oxford.

Coste, Hilarion de (1647). *Les eloges et les vies ... des dames illustres*, Paris.

Cottesford, Samuel (1622). *A very soveraigne oyle to restore debtors.*

The court of good counsell (1607). Anon.

Covarruvias, Diego (1597). De sponsalibus and de matrimonio. In *Opera omnia*, vol. I, Venice.

Craig, Thomas (1703). *The right of succession to the crown of England.*

Crompton, William (1632). *A wedding ring.*

Crooke, Helkiah (1615). *A description of the body of man.*

Crooke, Samuel (1625). *The guide unto true blessedness*, 4th ed.

Cucchi, Marco Antonio (1565). *Institutiones iuris canonici*, Pavia.

Cumberland, Richard (1727). *A treatise of the laws of nature*, transl. J. Maxwell.

D., L. (1692). *A check to debauchery.*

Davenant, John (1627). *Expositio epistolae ad Colossenses*, Cambridge.

Dawson, George (1694). *Origo legum: or a treatise of the origin of laws.*

Delfino, Giovanni (1553) ....*De matrimonii et caelibatu*, Camerino.

Dering, Edward (1597). *M. Derings workes.*

Diana, Anthony (1660). *Practicae resolutiones lectissimorum casuum*, Antwerp.

Dickson, David (1659). *An exposition of all St Pauls epistles.*

Dillingham, Francis (1609). *Christian oeconomy or household government.*

A discourse of the married and single life (1621). Anon.

Disputatio perjucunda ... mulieres homines non esse (1641). Anon. The Hague (1st ed. 1594).

Dod, John and Robert Cleaver (1629). *A treatise or exposition upon the ten commandements.*

Dod, John and William Hinde (1614). *Bathshebaes instructions to her son Lemuel.*

Domas, Jean (1705). *A treatise of the first principles of law in general*, transl. T. Wood.

Dominis, Marc' Antonio de (1617–20). *De republica ecclesiastica libri X*, 2 vols, n.p.

Doneau, Hugues (1828–33). *Commentariorum de jure civili*, 12 vols, Rome.

Donne, John (1953–62). *The sermons of John Donne*, ed. G.R. Potter and E.M. Simpson, 10 vols, Berkeley.

Donne, John (1971). *The complete English poems*, ed. A.J. Smith.

Dove, John (1601). *Of divorcement.*

Drexel, P. Hieremia (1642). *Tobias morali doctrina illustratus*, Antwerp.

Dubois, Jacques (1596). *De mensibus mulierum et hominis generatione*, Venice.

Duck, Arthur (1724). Treatise of the use and authority of Civil Law in England. In *The history of the Roman or Civil Law*, ed. C.J. de Ferriere.

Dugard, Samuel (1673). *The marriages of cousin germans*, Oxford.

Dugard, Samuel (1695). *Peri polutaisias or a discourse concerning the having many children.*

Duns Scotus, Johannes (1639). *Opera omnia*, Lyons (repr. Hildesheim, 1968).

Durandus of Saint Pourcain (1571). *In Petri Lombardi sententius theologicas commentarium*, 2 vols, Venice (repr. New Jersey, 1964).

Durham, James (1675). *A practical exposition of the X. commandments.*

E., T. (1632). *The lawes resolution of womens rights.*

Elton, Edward (1625). *God's holy mind touching matters moral.*

Elton, Edward (1637). *An exposition of the epistle to the Colossians*, 3rd ed.

Elyot, Thomas (1980). The defence of good women. In *The feminist controversy of the Renaissance*, ed. D. Bornstein, New York,.

Erasmus, Desiderius [1550?] *The censure and iudgement of*, transl. N. Less.

An essay towards a ... history of whoring (1697). Anon.

The Fall of Adam and Eve (1702). Anon.

Felden, Johann a (1653). *Annotata in Hug. Grotium de iure belli ac paci*, Amsterdam.

Feltham, Owen (1631). *Resolves, a duple century*, 4th ed.

Ferguson, Robert [1695]. *Whether the Parliament be not in law dissolved by the death of the Princess of Orange?*

Ferriere, Claude Joseph de (1724). *The history of the Roman or Civil Law.*

Field, Richard (1847–52). *Of the Church*, 4 vols, Cambridge.

Fifteen real comforts of matrimony (1683). Anon.

Filmer Robert (1991). *Patriarcha and other writings*, ed. J.P. Sommerville, Cambridge.

Finch, Henry (1627). *Law or a discourse thereof.*

Fisher, John (1597). *Opera omnia*, Wirceburgi (Gregg Press reprint 1977).

Fleetwood, William (1716). *The relative duties of parents and children*, 2nd ed.

Fonseca, Osorio da (1568). *A learned and very eloquent treatise* (repr. 1976).

Foreness, E. (1683). *A sermon preached at Manchester.*

Fosset, Thomas (1613). *The servants dutie.*

Frarinus, Petrus (1566). *An oration against the unlawfull insurrections* (repr. 1975).

Freig, John (1578). *Quaestiones oeconomicae et politicae*, Basle.

Freig, John [1591]. *Quaestiones Iustinianae in Institutiones Iuris Civilis*, Basle.

Fulbecke, William (1602). *The pandectes of the law of nations.*

Fulke, William (1601). *The text of the New Testament of Iesus Christ.*

Fuller, Thomas (1845). *A comment on Ruth*, Edinburgh.

Gataker, Thomas (1623). *A good wife Gods gift: and a wife indeed.*

[Gauden, John] (1656). *A discourse of auxiliary beauty.*

Gaule, John (1630). *Practique theories; or votive speculations.*

Gentili, Alberico (1614). *Disputationum de nuptiis libri septem*, Hanoviae.

Gentili, Scipio (1606). *De bonis maternis et de secundis nuptiis libri duo*, Hanoviae.

Geree, Stephen (1639). *The onely ornament of women.*

Gerson, Jean (1606). *Opera*, 4 vols, Paris.

Gibbon, Charles (1591). *A work worth the reading.*

A Glasse of the truth [1532?]. Anon.

Godolphin, John (1680). *Repertorium canonicum or an abridgement of the ecclesiastical laws.*

Gomersall, Robert (1634). *Sermons on St Peter.*

Goodman, Godfrey (1629). *The fall of man or the corruption of nature.*

[Gott, Samuel] (1670). *The divine history of the genesis of the world.*

Gouge, William (1622). *Of domesticall duties: Eight treatises.*

Gouge, William (1630). *An exposition on the whole fifth chapter of S. Iohns gospel.*

Grantham, Thomas (1641). *A marriage sermon.*

Grenaille, Francois de (1640). *L'honneste mariage*, Paris.

Grey, Richard (1730). *A system of English ecclesiastical law.*

Griffith, Matthew (1633). *Bethel: or a forme for families.*

Grotius, Hugo (1689). *De jure belli ac pacis libri tres*, Amsterdam.

Grotius, Hugo (1952). The exile of Adam. In *The celestial cycle*, ed. W. Kirkconnell, Toronto.

Gualther, Rudolph (1570). *D. Lucas Evangelista . . . homiliae CCXV*, Zurich.

Guido [1500?]. *Manipulus curatorum*, n.p.

Guillaume, Jacquette (1665). *Les dames illustres*, Paris.

Hacket, Roger (1607). *Two fruitful sermons needful for these times.*

Halkett, Anne (1979). *The memoirs of Anne, Lady Halkett and Ann, Lady Fanshawe*, ed. J. Loftis, Oxford.

Hall, Joseph (1837). *The works*, 10 vols, Oxford.

Hammond, Henry (1847). *Miscellaneous theological works*, vol. I, Oxford.

Harpsfield, Nicholas (1878). *A treatise on the pretended divorce*, ed, N. Pocock, Camden Society, 2. vol. XXI.

Harrington, William (1528). *Commendation of matrimony.*

Hay, William (1967). *William Hay's lectures on marriage*, transl. and ed. J.C. Barry, Edinburgh, The Stair Society.

Hayward, John (1603). *An answer to the first part of a certaine conference concerning succession.*

Heale, William (1609). *An apology for women*, Oxford.

Hemmingsen, Niels (1566). *De lege naturae apodictica methodus*, Viterbergae.

Henry VIII (1521). *Assertio septem sacramentorum adversus Martin Luther* (repr.

Ridgewood, NJ, 1966).

Herbert, William (1648a). *Herberts careful father and pious child.*

Herbert, William (1648b). *Herberts child-bearing woman.*

Hervet, Gentian (1593). *Oratio ad concilium qua suadetur . . .*, Venice.

Heydon, John (1658). *Advice to a daughter.*

Heywood, Thomas (1640). *The exemplary lives and memorable acts of nine the most worthy women of the world.*

Hickes, George (1682). *A discourse of the soveraign power.*

Hieron, Samuel (1613). *The bridegroome.*

Highmore, Nathaniel (1651). *The history of generation.*

Hincmar of Rheims (1852). De divortio Lothari et Tetbergae. In *Patrologiae Cursus Completus*, ed. J.P. Migne, vol. CXXV, cols 619–772, Paris.

Hobart, Henry (1671). *The reports of Sir Henry Hobart*, 3rd ed.

Hobbes, Thomas (1991). *Leviathan*, ed. R. Tuck, Cambridge.

Hooker, Richard (1977). *Of the laws of ecclesiastical polity*, Cambridge, MA.

Hooper, John (1843). *Early writings*, Cambridge.

Hopkins, Ezekiel (1701). *The works.*

Hotman, Antoine (1585). *Observationum quae ad veterum nuptiarum ritum pertinent*, Geneva.

Hotman, Antoine (1595). *Traicte de la dissolution du mariage*, 2nd ed., Paris.

Hotman, Francois (1594). *De castis incetisve nuptiis*, Lyons.

The house-keepers guide, in the prudent managing of their affairs (1706). Anon.

Howe, John (1822). *The whole works.*

Howson, John (1606). *Uxore dismissa propter fornicationem*, 2nd ed., Oxford.

Hugh of Saint Victor (1951). *On the sacraments of the Christian faith*, transl. R.J. Deferrari, Cambridge, MA.

Hughes, George (1672). *An analytical exposition of the whole first book of Moses.*

Hunnius, Aegidius (1604). *Praelectiones viginti and unum priora capita Geneseos*, Marburg.

Hutchinson, Roger (1842). *Works*, Cambridge.

Huygens, Gommarus (1696). *Breves observationes de sacramento in genere*, Liege.

Huygens, Gommarus (1697). *Breves observationes de prudentia, jurejustitia et restitutione*, Liege.

Huygens, Gommarus (1698). *Breves observationes de superstitione*, Liege.

Huygens, Gommarus (1701). *Breves observationes de contractu in genere*, Liege.

Jackson, Thomas (1617). *Nazareth and Bethlehem or Israel's portion in the son of Jesse*, Oxford.

Jansen, Cornelius (1641). *. . .Pentateuchus, sive commentarius in quinque libris Moysis*, Louvain.

Jermin, Michael (1639). *A commentary upon the whole booke of Ecclesiastes.*

Jerome, Stephen (1614). *Moses his sight of Canaan.*

Jewel, John (1850). Defence of the Apology. In *Works*, vol. IV, Cambridge.

Joannes de Sancto Geminiano (Fra Giovanni di Coppo) (1585). *Summa de exemplis et rerum similitudinibus locupletissima*, Lyons.

Joye, George [1541?]. *A contrarye (to a certayne manis) consultacion.*

Justinian (1913). *The Institutes of Justinian*, ed. J.B. Moyle, 5th ed., Oxford.

Keble, Joseph (1685). *Reports in the Court of King's Bench at Westminster*, 2 vols.

[Keck, Anthony] (1697). *Cases argued and decreed in the high Court of Chancery*.

Kellison, Matthew (1605). *A survey of the new religion* (repr. 1977).

Ken, Thomas (1682). *A sermon preached at the funeral of the Right Honourable the Lady Margaret Mainard*.

Ken, Thomas (1838). *The prose works*, ed. J.T. Round.

[Kettlewell, John] (1691). *The duty of allegiance settled*.

Kidder, Richard (1694). *A commentary on the five books of Moses*.

King, John (1614). *Vitis Palatina*.

King, William (1702). *De origine mali*.

Kipping, Heinrich (1664). *Exercitationes sacrae de creationis operibus et statu primis hominis*, Frankfurt.

Knewstubs, John (1584). *The lectures ... upon the twentieth chapter of Exodus*.

Knox, John (1949). *History of the Reformation in Scotland*, ed. W.C. Dickinson.

Knox, John (1985). *The political writings of John Knox*, ed. M.A. Breslow, Washington.

Kornmann, Heinrich (1610). *Sibylla trygandriana seu de virginitate*, Frankfurt.

Kraze, Joachim (1619). *Viridarium politicae, sive Reipublicae*, Venice.

Lagus, Conradus (1566). *Methodica iuris utrisque traditio*, Lyons.

Lamzweerde, Jan Baptist (1686). *Historia naturalis molarum uteri*, Leyden.

Lancelottus, Johannes (1564). *Institutionum iuris canonici libri quatuor*, Venice.

LaPeyrere, Isaac (1656). *Men before Adam or a discourse*.

Lapide, Cornelius a (1854–9). *Commentaria in sacram scripturam*, ed. X. Riario Sfortiae, 10 vols, Naples.

Lathom, Paul (1683). *The power of kings from God*.

Latimer, Hugh (1844–5). *Sermons*, 2 vols, Cambridge.

Laud, William (1847–60). *The works*, 7 vols, Oxford.

Lawrence, William (1680). *Marriage by the morall law of God*.

Lawson, George (1659). *Theo-politica or a body of divinity*.

Laymann, Paul (1625). *Theologiae moralis*, Monachii.

Le Grand, Anthony (1694). *An entire body of philosophy* (repr. New York, 1972).

Leighton, Robert (1844). *The works of*, Edinburgh.

Leslie, John (1569). *A defence of the honour of Marie Quene of Scotland* (repr. 1970).

Lessius, Leonard (1606). *De iustitia et iure caeterisque virtutibus*, Paris.

Lessius, Leonard (1621). *The treasure of vowed chastity in secular persons*, transl. I.W., n.p.

A letter to a member of Parliament (1675). Anon.

La Liberté des dames (1685). Anon., Paris.

Liguori, St Alphonso (1954). *Opera moralia*, 4 vols, Graz.

Littleton, Adam (1680). *Sixty-one sermons*.

Lloyd, William (1691). *A discourse of God's ways of disposing of kingdoms*.

Locke, John (1987). *A paraphrase and notes on the epistles of St Paul*, vol. I, Oxford.

Locke, John (1988). *Two treatises of government*, Cambridge (Student edition).

Loftus, Dudley (1677). *Difamiae adikia or the first marriage of Katherine FitzGerald*.

Lorenzo da Brindisi (1935). Explanatio in Genesim. In *Opera Omnia*, vol. III, Patavii.

Lotichius, Joannes (1630). *Gynaiecologia: id est: de nobilitate and perfectione sexus femineii*, Rinteln.

Lowde, James (1694). *A discourse concerning the nature of man* (repr. 1979).

Ludena, Joannis de (1563). *Disputatio theologica de coelibatu sacerdotum*, Padua.

Luther, Martin (1958). Lectures on Genesis. In *Works*, vol. 1, Saint Louis.

Luther, Martin (1959). The Babylonian captivity of the church. In *Works*, vol. 36, Philadelphia.

[Lyser, Johannes] (1682). *Polygamia triumphatrix: id est discursus politicus de polygamia;* Lund.

Mackenzie, George (1684). *Jus regium: or the just and solid foundation of monarchy.*

Mainwaring, Thomas (1673). *A reply to an answer to the defence.*

Major, John [1509]. *Quartus sententiarum*, Paris.

Maldonado, Juan (1677). *Opera varia theologica*, Paris (repr. Ridgewood, NJ, 1965).

Manzini, Celso (1587). *De cognitione quae lumine naturali haberi potest*, Venice.

Marca, Pierre de (1669). *Dissertationes posthumae, sacrae and ecclesiasticae*, Paris.

Marriage asserted in answer to a booke (1674). Anon.

Marriage promoted. In a discourse (1690). Anon.

Martin, T. (1676). *Mary Magdalen's tears wipt off.*

Massie, William (1586). *A sermon preached*, Oxford.

Mauser, Conrad (1597). Explicatio erudita ... de nuptiis. In *Tractatus de jure connubiorum*, ed. J. von Beust, Leipzig.

Maxwell, John (1644). *Sacro-sancta regum majestas: or; the sacred and royall prerogative of Christian kings*, Oxford.

Mayer, John (1622). *A treasury of ecclesiastical expositions.*

Mayer, John (1631). *A commentary upon the New Testament.*

Maynard, John (1668). *The beauty and order of the creation.*

Melanchthon, Philip (1834–60). *Operae quae supersunt omnia*, ed. C.G. Bretschneider and H.E. Bindseil, 28 vols, Halle.

Menochio, Giovanni (1678). *Commentarii totus S. Scripturae*, Antwerp.

Mercuriale, Girolamo (1591). *De morbis mulieribus praelectiones*, Venice.

Michel, Humfrey (1702). *Sovereignty subject unto duty.*

Milton, John (1959). *Complete prose works*, ed. E. Sirluck, vol. II.

Monner, Basil (1561). *De matrimonio*, Frankfurt.

More, Thomas (1989). *Utopia*, ed. G.M. Logan and R.M. Adams, Cambridge.

Musculus, Wolfgang (1548). A right godlye treatise of matrimony. In Hermann of Wied (Hermann V, Archbishop of Cologne) *The right institution of baptisme*, transl. R. Ryce, Ipswich.

Needler, Benjamin (1655). *Expository notes with practical observations.*

Ness, Christopher (1695). *The second volume of the sacred history and mystery.*

Ness, Christopher (1696). *A compleat history and mystery of the Old and New Testament.*

Nevizanus, Joannes (1570). *Sylvae nuptialis libri sex*, Venice.

Niccholes, Alexander (1809). A discourse of marriage and wiving. In *Harleian Miscellany* vol. III, pp. 251–88, ed. W. Oldys, 12 vols (1808–11).

Nicholls, William (1701). *The duty of inferiours towards their superiours.*

Nifo, Agostino (1641). *De pulchro liber*, Leyden.

Ochino, Bernardino (1657). *A dialogue of polygamy.*

Olizarovius, Aaron (1651). *De politica hominum societate*, Danzig.

Paget, Thomas (1650). *A religious scrutiny concerning unequal marriage*, 2nd ed.

Pagitt, Ephraim (1646). *Heresiography*, 3rd ed.

Pareus, David (1612). *Quaestiones controversiae theologicae de iure regnum et principum*, Hamburg.

Pareus, David (1631). ....*In S. Matthaei evangelium commentarius*, Oxford.

Parsons, Robert (1606). *An answere to the fifth part of Reportes* (repr. 1975).

Pascale, Filippo (1618). *Tractatus amplissimus de viribus patriae potestatis*, Naples.

Patrick, Simon (1695). *A commentary upon the first book of Moses called Genesis*.

Peleus, Julian (1602). *Quaestionum publice tractarum*, Paris.

Perkins, William (1617). *Commentary on Galatians*, ed. R. Cudworth (repr. New York, 1989).

Perkins, William (1618). *Works*, vol. III, Cambridge.

Perkins, William (1628). *The whole treatise of the cases of conscience*.

Petit, Pierre (1687). *De Amazonibus dissertatio*, 2nd ed., Amsterdam.

Pettus, John (1674). *Volatiles from the history of Adam and Eve*.

Pineau, Severin (1598). ....*Opusculum physiologum and anatomicum*, Paris.

Piscator, Johannes (n.d.) Appendix ad observationes, Unpublished Folger manuscript.

Plowden, Edmund (1816). *The commentaries or reports of*, 2 vols.

Plutarch (1603). *The Philosophie commonlie called The Morals*, transl. P. Holland.

Poole, Matthew (1696). *Annotations upon the Holy Bible*, ed. and completed by Samuel Clark and Edward Veale, 3rd ed., 2 vols.

Portius, Simon (1537). *De celibatu*, n.p.

Poullain de la Barre, François (1989). *The equality of the two sexes*, transl. A.D. Frankforter and P.J. Morman, Lewiston, New York.

Prandoni, Paulo (1649). *De parentum in pueros disciplina libri V*, Milan.

Preston, John (1635). *Sins overthrow*.

Pricke, Robert (1609). *The doctrine of superioritie and of subjection*.

Prideaux, John (1664). *Fasiculus controversiarum theologicarum*, Oxford.

Primrose, Gilbert (1617). *Iacobs vow opposed to the vows of monks and friers*.

Primrose, James (1655). *De mulierum morbis et symptomatis*, Rotterdam.

Pritchard, Thomas [1579] *The schoole of honest and vertuous lyfe*.

Prynne, William (1628). *The unlovelinesse of love-lockes*.

Pufendorf, Samuel (1934). *De jure naturae et gentium libri octo*, ed. H. Milford, transl. C.H. and W.A. Oldfather, Oxford.

R., R. (1615). *The house-holders helpe for domesticall discipline*.

Rainolds, John (1609). *A defence of the judgment of the Reformed churches*, n.p.

Raulin, Jean (1512). *Itenerarium pradiis . . . sermones*, Paris.

Raymond, Thomas (1696). *The reports of divers special cases*.

Reasons for the passing of the bill . . . suppressing vice (1699). Anon.

Reflections upon the opinions of some modern divines (1689). Anon.

Reflexions on marriage and the poetick discipline (1673). Anon.

Reyner, Edward (1657). *Considerations concerning marriage*.

Richardson, John (1655). *Choice observations and explanations upon the Old Testament*.

[Rider, Matthew] (1680). *The power of parliaments in the case of succession*.

Ridley, Lancelot (1548). *An exposicion in Englishe*.

Ridley, Thomas (1607). *A view of the civil and ecclesiastical law*.

Riolan, Jean (1671). *A sure guide; or the best and nearest way to physick and chyrurgery*, transl. N. Culpepper and W.R., 3rd ed.

Rogers, Daniel (1642). *Matrimonial honour*.

Rollock, Robert (1603). *Lectures upon the epistle of Paul to the Colossians*.

Rubeis, Flaminio de (1599). *Tractatus de adulteriis iuribus tum divinis, tum canonicis, and civilibus*, Venice.

Rueff, Jacob (1637). *The expert midwife*.

Rutherford, Samuel (1843). *Lex Rex or the law and the prince*, Edinburgh.

Salkeld, John (1617). *A treatise of paradise*.

Sanchez, Thomas (1654). *. . . De sancto matrimonii sacramento disputationum tomi tres*, Lyons.

Sanderson, Robert (1671). *XXXIV sermons*, 5th ed.

Sanderson, Robert (1685). *Nine cases of conscience*.

Sanderson, Robert (1851). *De obligatione conscientiae praelectiones decem*, ed. W. Whewell, Cambridge.

Saravia, Hadrian (1593). *De imperandi authoritate et Christiana obedientia*.

Saumaise, Claude de (1644). *Epistola ad Andream Colvium: super cap. XI. primae ad Corinth. Epist.*, Leyden.

Schickard, Wilhelm (1625). *Jus regium Hebraeorum*, Strasbourg.

Schneidewein, Johann (1585). *. . . in Institutionum imperialium titulum X de nuptii*, Jena.

Sclater, William (1633). *Utriusque epistolae ad Corinthios explicatio analytica*, Oxford.

Sclater, William (1650). *A brief and plain commentary*.

[Sedgwick, Richard] (1624). *A short summe of the principall things*.

Selden, John (1991). *On Jewish marriage law*, transl. J.R. Ziskind, Leiden.

Seneca, Lucius Annaeus (1614). *The workes both morrall and natural*, transl. T. Lodge.

Settala, Ludovico (1626). *. . . De ratione instituenda and gubernandae familiae*, Milan.

Sharp, John (1702). *A sermon preach'd at the coronation of Queen Anne*.

Sherlock, Thomas (1714). *A sermon preached . . . on March 8 . . . 1713/4*.

Shower, John (1694). *Family religion in three letters to a friend*.

Sidney, Algernon (1990). *Discourses concerning government*, Indianapolis.

Smith, Henry (1866). *Works*, 2 vols, Edinburgh.

Smith, Thomas (1982). *De republica Anglorum*, ed. M. Dewar, Cambridge.

Snawsel, Robert (1631). *A looking glasse for married folkes*.

Sordi, Giovanni (1602). *Tractatus de alimentis*, Venice.

Soto, Domingo de (1619). *De iustitia et iure libri decem*, Salmanticae.

Spinoza, Benedict de (1958). *The political works*, transl. A.G. Wernham, Oxford.

Stock, Richard (1865). *A commentary upon the prophecy of Malachi*, Edinburgh.

Stockwood, John (1589). *A Bartholomew fairing for parentes*.

Stubbs, John (1968). *John Stubbs's Gaping gulf*, ed. L.E. Berry, Charlottesville, VA.

Suarez, Francisco (1872). *Tractatus de legibus ac Deo legislatore*, Naples.

Subertus, Petrus (1508). *De cultu vinee dni. liber*, Paris.

Swan, John (1635). *Speculum mundi Or a glasse representing the face of the world*, Cambridge.

Swinburne, Henry (1686). *A treatise of spousals or matrimonial contracts*.

Swinnock, George (1868). *The works*, vol. I, Edinburgh.

Sydenham, Thomas (1844). *Opera omnia*.

Tapper, Ruard (1582). . . . *Omnia, quae haberi poterunt, opera*, Cologne.

Tasso, Ercole (1599). *Of marriage and wiving.*

Taylor, Jeremy (1828). *The whole works*, 15 vols.

Taylor, Thomas (1625). *A good husband and a good wife.*

Taylor, Thomas (1633). *Three treatises.*

Temple, William (1691). *Miscellanea. The first part*, 3rd ed.

Thomas, William (1661). *Christian and conjugall counsell.*

Thorndike, Herbert (1844–56). *Theological works*, 6 vols, Oxford.

Tilney, Edmund (1568). *A brief and pleasant discourse of duties in marriage.*

Tiraqueau, André (1555). *Commentarii in l. boves*, Venice.

Tiraqueau, André (1574). *Commentarii de nobilitate et iure primigeniorum*, Venice.

Topsell, Edward (1597). *The reward of religion.*

Torshell, Samuel (1650). *The womans glorie*, 2nd ed.

Towerson, Gabriel (1676). *An explication of the decalogue.*

Trapp, John (1647). *A commentary or exposition upon all the epistles.*

Trapp, John (1656). *A commentary or exposition upon all the books of the New Testament.*

Trapp, John (1662). *Annotations upon the Old and New Testaments in five distinct volumes.*

Treason, popery &c. brought to a publique test (1680). Anon.

Tuke, Thomas (1616). *A treatise against painting and tincturing of men and women.*

[Tully, George] (1688). *An answer to a discourse concerning the celibacy of the clergy*, Oxford.

Tunstall, Cuthbert (1518). *Cuthberti Tonstalli in laudem matrimonii oratio.*

Turner, James (1683). *Pallas armata. Military essays.*

Turner, John (1685). *Boaz and Ruth.*

Turner, John (1686a). *Addenda and mutanda.*

Turner, John (1686b). *An argument in defence of the marriage of an uncle.*

Turner, John (1698). *A discourse on fornication.*

Turriani, Francisci (1563). *De matrimoni clandestinis explicatio*, Venice.

Turturetti, Vincenzo (1629). *Parallella ethica et juridica*, Paris.

Tuvil, Daniel (1616). *Asylum veneris, or a sanctuary for ladies.*

Tuvil, Daniel (1635). *St Pauls threefold cord.*

Tyrrell, James (1681). *Patriarcha non monarcha.*

Tyrrell, James (1691/2). *Bibliotheca politica: or a discourse.*

Vaughan, John (1706). *Reports of . . . Sir John Vaughan.*

Vaughan, William (1600). *The golden grove.*

Vauts, Moses à (1650). *The husband's authority unvail'd.*

Vaux, Laurence (1599). *A catechisme or Christian doctrine, necessary for children and ignorant people*, [Douai?].

Vegio, Maffeo (1613). . . . *Opera quae hactenus haberi potuerunt*, Lodi.

Venette, Nicholas [1688]. *Tableau de l'amour considere dans l'estat du mariage*, n.p.

Veron, John [1562]. *A strong defence of the marriage of priests.*

Vettori, Peter (1584). *Comentarii in X libros Aristotelis*, Florence.

Vincent of Beauvais (1624). *Speculum quadruplex: naturale, doctrinale, morale, historiale*, 2 vols, Douai (repr. Graz, 1964–5).

Vitoria, Francisco de (1991). *Political writings*, ed. and transl. A. Pagden and J. Lawrence, Cambridge.

Waker, Nathaniel (1664). *A sermon preached at the funeral of Mr. Lucas Lucie*.

Wall, Thomas (1690). *God's holy order in nature*.

Watson, Thomas (1558). *Holsome and catholyke doctrine concerninge the seven sacraments*.

Webb, George (1653). *The practice of quietnes*, 8th ed.

Wemyss, John (1633). *The workes of Mr. J. Weemes*, 3 vols.

Weyer, Johann (1991). De praestigiis daemonum. In *Witches, devils, and doctors in the Renaissance*, ed. G. Mora and B. Kohl, transl. J. Shea, Medieval and Renaissance texts and studies, Binghampton, New York.

Whaley, Nathaniel (1698). *Two sermons: One against adultery*.

Whately William (1617). *A bride-bush or a wedding sermon*.

Whately, William (1624). *A care-cloth*.

Whately, William (1647). *Prototypes, or, the primary precedent presidents out of the Book of Genesis*.

Whetstone, George (1586). *The English myrrour*.

White, John (1656). *A commentary upon the three first chapters of . . . Genesis*.

Wilkinson, Robert (1607). *The merchant royall: A sermon preached at the nuptials of Lord Hay*.

Willet, Andrew (1605). *Hexapla in Genesin*, Cambridge.

Willet, Andrew (1608). *Hexapla in Exodum*.

Willet, Andrew (1612). *A treatise of Solomons marriage*.

Willet, Andrew (1614). *An harmonie upon the second booke of Samuel*, Cambridge.

Willet, Andrew (1634). *Synopsis papismi*.

Wilson, Thomas (1615). *Theological rules*.

Wiseman, Robert (1657). *The law of laws*.

Wood, Thomas (1712). *A new institute of the imperial or Civil Law*, 2nd ed.

[Woodhead, Abraham] (1687). *Two discourses, the first, concerning the spirit of Martin Luther*, Oxford.

Wright, Thomas (1621). *The passions of the mind*.

Xenophon (1923). *Memorabilia and Oeconomicus*, transl. E.C. Marchant.

Zouche, Richard (1682). *Quaestionum juris civilis centuria*, 3rd ed.

Secondary sources

Alleman, G.S. (1942). *Matrimonial law and the materials of Restoration comedy*, Philadelphia.

Amussen, S.D. (1988). *An ordered society*, Oxford.

Anderson, M. (1980). *Approaches to the history of the Western family 1500–1914*.

Bailey, D.S. (1959). *Sexual relation in Christian thought*, New York.

Baker, F. *et al.*, eds (1981). *1642: Literature and power in the seventeenth century*, Essex.

Baker, J.H. (1971). *An introduction to English legal history*.

Baker, J.H., ed. (1978). *Legal records and the historian*.

Bald, R.C. (1970). *John Donne: A life*, Oxford.

Bay, J.C. (1934). Women not considered human beings. *The Library Quarterly*, IV: 156–64.

Bennett, J.M. (1993). Women's history: a study in continuity and change. *Women's History Review*, 2: 173–84.

Boxer, M.J. and Quataert, J.H. (1987). *Connecting spheres: Women in the Western world 1500 to the present*, Oxford.

Brennan, T. and Pateman, C. (1979). Mere auxiliaries to the commonwealth: Women and the origins of liberalism. *Political Studies*, XXVII: 183–200.

Brundage, J.A. (1987). *Law, sex and Christian society in Medieval Europe*, Chicago.

Butler, M.A. (1978). Early liberal roots of feminism: John Locke and the attack on patriarchy. *American Political Science Quarterly*, 72: 135–50.

Cahn, S. (1987). *Industry of devotion: The transformation of women's work in England, 1500–1660*, New York.

Cairncross, J. (1974). *After polygamy was made a sin.*

Carlson, E.J. (1990). Marriage reform and the Elizabethan High Commission. *Sixteenth Century Journal*, XXI: 437–51.

Carroll, B.A., ed. (1976). *Liberating women's history: Theoretical and critical essays*, Urbana.

Chrimes, S.B. (1966). *English constitutional ideas in the fifteenth century*, New York.

Cioni, M.L. (1985). *Women and law in Elizabethan England with particular reference to the Court of Chancery.*

Clark, A. (1919). *Working life of women in the seventeenth century*, New York.

Clark, L.M.G. (1977). Women and John Locke: or who owns the apples in the garden of Eden. *Canadian Journal of Philosophy*, VII: 699–724.

Clark, P. (1983). *The English alehouse: A social history.*

Collinson, P. (1988). *The birthpangs of protestant England*, New York.

Corbett, P.E. (1930). *The Roman law of marriage*, Oxford.

Davis, N.Z. (1975). *Society and culture in early modern France*, Stanford.

Durston, C. (1988). Unhallowed wedlocks: The regulation of marriage during the English Revolution. *The Historical Journal*, 31: 45–59.

Durston, C. (1989). *The family in the English Revolution*, Oxford.

Dusinberre, J. (1975). *Shakespeare and the nature of women.*

Earle, P. (1989). The female labour market in London in the late seventeenth and early eighteenth centuries. *Economic History Review*, 2nd series XLII: 328–53.

Eells, H. (1924). *The attitude of Martin Bucer toward the bigamy of Philip of Hesse*, New Haven.

Erickson, A.L. (1990). Common law versus common practice: the use of marriage settlements in early modern England. *Economic History Review*, 2nd series XLII: 21–39.

Erickson, A.L. (1993). *Women and property in early modern England.*

Falk, Z.W. (1966). *Jewish matrimonial law in the Middle Ages*, Oxford.

Firth, C.H. (1962). *Cromwell's army.*

Fitzmaurice-Kelly, J. (1927). Woman in sixteenth-century Spain. *Revue Hispanique*, 70: 557–632.

Fletcher, A. and Stevenson, J., eds (1985). *Order and disorder in early modern*

England, Cambridge.

Friedman, J. ed. (1987). *Regnum, religio et ratio: Essays presented to Robert M. Kingdon*, Ann Arbor, MI.

Gagen, J.E. (1954). *The New Woman: Her emergence in English drama 1600–1730*, New York.

Gillis, J.R. (1985). *For better for worse: British marriages 1600 to the present*, Oxford.

Goldberg, P.J.P., ed. (1992). *Woman is a worthy wight: women in English society c. 1200–1500*, Stroud.

Goody, J. (1983). *The development of the family and marriage in Europe*, Cambridge.

Greaves, R.L., ed. (1985). *Triumph over silence: Women in protestant history*.

Halkett, J. (1970). *Milton and the idea of matrimony*, New Haven.

Hanawalt, B.A. (1986). *The ties that bound: Peasant families in Medieval England*, Oxford.

Hannay, M.P., ed. (1985). *Silent but for the word: Tudor women as patrons, translators and writers of religious works*, Ohio.

Harris, B.J. (1990a). Property, power, and personal relations: Elite mothers and sons in Yorkist and early Tudor England. *Signs*, 15: 606–32.

Harris, B.J. (1990b). Women and politics in Tudor England. *The Historical Journal*, 33: 259–81.

Harris, F.R. (1912). *The life of Edward Mountagu, K.G.*, London.

Harrison, W. (1992). The role of women in Anabaptist thought and practice: The Hutterite experience of the sixteenth and seventeenth centuries. *Sixteenth Century Journal*, XXIII: 49–69.

Haw, R. (1952). *The state of matrimony*, London.

Healy, E.T. (1956). *Woman according to Saint Bonaventure*, Erie, PA.

Helmholz, R.H. (1987). *Canon law and the law of England*.

Helmholz, R.H. (1990). *Roman canon law in Reformation England*, Cambridge.

Hill, B. (1993). Women's history: a study in change, continuity or standing still. *Women's History Review*, 2: 5–22.

Hirst, D. (1975). *The representative of the people?* Cambridge.

Hogrefe, P. (1972). Legal rights of Tudor women and their circumvention by men and women. *Sixteenth Century Journal*, III: 97–105.

Honeyman, K. and Goodman, J. (1991). Women's work, gender conflict, and labour markets in Europe, 1500–1900. *Economic History Review*, 2nd series XLIV: 608–28.

Houlbrooke, R. (1984). *The English family: 1450–1700*.

Houlbrooke, R. (1985). The making of marriage in Mid-Tudor England: Evidence from the records of matrimonial contract litigation. *Journal of Family History*, 10: 339–52.

Howell, C. (1983). *Land, family and inheritance in transition: Kibworth Harcourt 1280–1700*, Cambridge.

Howell, M.C. (1991). A feminist historian looks at the new historicism: What's so historical about it? *Women's Studies*, 19: 139–47.

Hufton, O. (1983). Women in history: Early modern Europe. *Past and Present*, 101: 125–41.

Hufton, O. (1984). Women without men: widows and spinsters in Britain and France in the eighteenth century. *Journal of Family History*, 9: 355–76.

Hulliung, M. (1974). Patriarchalism and its early enemies. *Political Theory*, 2: 410–19.

Ingram, M. (1987). *Church courts, sex and marriage in England, 1570–1640*, Cambridge.

Johnson, J.T. (1971). The covenant idea and the puritan view of marriage. *Journal of the History of Ideas*, XXXII: 107–18.
Jordan, C. (1987). Women's rule in sixteenth-century British political thought. *Renaissance Quarterly*, 40: 421–51.
Jordan, C. (1990). *Renaissance Feminism: Literary texts and political models*, Ithaca.
Joyce, G.H. (1948). *Christian marriage*.

Kelly, H.A. (1976). *The matrimonial trials of Henry VIII*, Stanford.
Kelly, J. (1984). *Women, history and theory*, Chicago.
Kelso, R. (1956). *Doctrine for the lady of the Renaissance*, Urbana.
Kent, J. (1973). Attitudes of Members of the House of Commons to the regulation of Personal Conduct. *Past and Present*, XLVI: 41–71.
Koerner, K.F. (1985). *Liberalism and its critics*.

Lake, P. (1987). Feminine piety and personal potency: The emancipation of Mrs Jane Ratcliffe. *The Seventeenth Century*, 11: 143–65.
Larner, C. (1981). *Enemies of God: The witch-hunt in Scotland*.
Laslett, P. (1977). *Family life and illicit love in earlier generations*, Cambridge.
Lee, P.A. (1990). A Bodye politique to govern: Aylmer, Knox and the debate on queenship. *The Historian*, LII: 242–61.
Leites, E. (1986a). *The puritan conscience and modern sexuality*, New Haven.
Leites, E. (1986b). The family as history. *Partisan Review*, 53: 111–25.
Lemaire, A. (1975). *Les lois fondamentales de la monarchie française*, Geneva.
Levine, D. (1977). *Family formation in an age of nascent capitalism*, New York.
Levine, M. (1973). *Tudor dynastic problems 1460–1571*.

Macaulay, T.B. (1913–15). *The history of England*, ed. C.H. Firth., 6 vols.
Macfarlane, A. (1978). *The origins of English individualism: The family, property and social transition*, Cambridge.
Macfarlane, A. (1986). *Marriage and love in England: Modes of reproduction 1300–1840*, Oxford.
McIntosh, M.K. (1984). Servants and the household unit in an Elizabethan English community. *Journal of Family History*, 9: 3–23.
McIntosh, M.K. (1991). *A community transformed: The manor and Liberty of Havering, 1500–1620*, Cambridge.
Mack, P. (1992). *Visionary women: Ecstatic prophecy in seventeenth-century England*, Berkeley.
Maclean, I. (1980). *The Renaissance notion of woman*, Cambridge.
Macpherson, C.B. (1964). *The political theory of possessive individualism*, Oxford.
Maine, H.S. (1875). *Lectures on the early history of institutions*.
Mendelson, S.H. (1987). *The mental world of Stuart women*, Brighton.
Mitterauer, M. and Sieder, R. (1982). *The European family: patriarchy to partnership from the Middle Ages to the present*, Chicago.

Morgan, E.S. (1942). The puritans and sex. *New England Quarterly*, XV: 591–607.

Morgan, E.S. (1944). *The puritan family*, Boston.

Muldoon, J. (1979). *Popes, lawyers, and infidels*, Liverpool.

Nadelhaft, J. (1982). The Englishwoman's sexual civil war: feminist attitudes towards men, women and marriage 1650–1740. *Journal of the History of Ideas*, XLIII: 555–79.

Nicholson, L.J. (1986). *Gender and history*, New York.

Noonan, J.T. (1986). *Contraception*, Cambridge, MA.

Nozick, R. (1974). *Anarchy, state, and Utopia*, New York.

Okin, S.M. (1979). *Women in Western political thought*, New Jersey.

Okin, S.M. (1983–4). Patriarchy and married women's property in England: Questions on some current views. *Eighteenth Century Studies*, 17: 121–38.

Outhwaite, R.B., ed. (1981). *Marriage and society: Studies in the social history of marriage*, New York.

Ozment, S. (1983). *When fathers ruled: Family life in Reformation Europe*, Cambridge, MA.

Parker, G. (1988). *The military revolution*, Cambridge.

Parry, G. (1964). Individuality, politics and the critique of paternalism in John Locke. *Political Studies*, 12: 163–77.

Pateman, C. (1988). *The sexual contract*, Stanford.

Pattison, M. (1892). *Isaac Casaubon: 1559–1614*, Oxford.

Phillips, J.E. (1941–2). The background of Spenser's attitude towards women rulers. *Huntington Library Quarterly*, 5: 5–32.

Pollock, L. (1993). *With faith and physic: The life of a Tudor gentlewoman*.

Potter, M. (1986). Gender equality and gender hierachy in Calvin's theology. *Signs*, 11: 725–39.

Prest, W.R. (1991). Law and women's rights in early modern England. *The Seventeenth Century*, VI: 169–87.

Prior, M., ed. (1985). *Women in English society: 1500–1800*.

Quaife, G.R. (1979). *Wanton wenches and wayward wives*, Rhode Island.

Rawls, J. (1971). *A theory of justice*, Cambridge, MA.

Roby, H.J. (1902). *Roman private law in the times of Cicero and of the Antonines*, Cambridge.

Rougement, D. de (1983). *Love in the Western world*, New York.

Ruether, R.R., ed. (1974). *Religion and sexism: Images of woman in the Jewish and Christian traditions*, New York.

Rushton, P. (1986). Property power and family networks: The problem of disputed marriage in early modern England. *Journal of Family History*, 11: 205–19.

Scalingi, P.L. (1978). The scepter or the distaff: The question of female sovereignty. *The Historian*, 41: 59–75.

Schlatter, R.B. (1940). *The social ideas of religious leaders 1660–1688*, Oxford.

Schuecking, L.L. (1969). *The puritan family: A social study from literary sources*.

Scott, J.W. (1986). Gender: a useful category of historical analysis. *American Historical Review*, 91: 1053–75.

Scott, J.W. (1988). *Gender and the politics of history*, New York.

Shanley, M.L. (1979). Marriage contract and social contract. *Western Political Quarterly*, XXXII: 79–91.

Shanley, M.L. and Pateman, C., eds (1991). *Feminist interpretations and political theory*, Pennsylvania.

Sharpe, J.A. (1980). *Defamation and sexual slander in early modern England*, York.

Skinner, Q. (1978). *The foundations of modern political thought*, 2 vols, Cambridge.

Smith, A.L. (1964). *Church and state in the Middle Ages*, New York.

Smith, H. (1982). *Reason's disciples: Seventeenth century English feminists*, Chicago.

Sommerville, J.P. (1986a). *Politics and ideology in England 1603–1640*.

Sommerville, J.P. (1986b). History and theory: the Norman Conquest in early Stuart political thought. *Political Studies*, XXXIV: 249–61.

Sommerville, J.P. (1992). *Thomas Hobbes: Political ideas in historical context*, New York.

Stone, L. (1977). *The family, sex and marriage in England 1500–1800*.

Stone, L. (1990). *Road to divorce*, Cambridge.

Strype, J. (1711). *The life and acts of Matthew Parker*.

Strype, J. (1824). *Annals of the Reformation*, Oxford.

Tavard, G.H. (1973). *Woman in Christian tradition*, Notre Dame.

Thomas, K. (1959). The double standard. *Journal of the History of Ideas*, XX: 195–216.

Thomas, K. (1958). Women and the Civil War sects. *Past and Present*, 13: 42–62.

Tilly, L.A. (1989). Gender, women's history and social history. *Social Science History*, 13: 439–62.

Todd, M. (1987). *Christian humanism and the puritan social order*, Cambridge.

Trumbach, R. (1978). *The rise of the egalitarian family*, New York.

Tully, J. (1988). *Meaning and context: Quentin Skinner and his critics*, Princeton, NJ.

Turner, J.G. (1987). *One Flesh: Paradisal marriage and sexual relations in the age of Milton*, Oxford.

Vacant, A. and Mangenot, E., eds (1903–46). *Dictionnaire de Théologie Catholique*, 15 vols, Paris.

Wall, A. (1990). Elizabethan precept and feminine practice: The Thynne family of Longleat. *History*, 75: 23–38.

Warnicke, R.M. (1989). Lady Mildmay's journal: A study in autobiography and meditation in Reformation England. *Sixteenth Century Journal*, XX: 55–68.

Weber, K.E.M. (1930). *The protestant ethic and the spirit of capitalism*.

Wiener, C.Z. (1975). Sex-roles and crime in late Elizabethan Hertfordshire. *Journal of Social History*, 8: 38–60.

Willen, D. (1988). Women in the public sphere in early modern England: The case of the urban working poor. *Sixteenth Century Journal*, XIX: 559–75.

Willen, D. (1992). Godly women in early modern England: puritanism and gender. *Journal of Ecclesiastical History*, 43: 561–80.

Woodbridge, L. (1984). *Women and the English Renaissance*, Urbana.

Wright, L.B. (1935). *Middle-class culture in Elizabethan England*, New York.

Index

THE CULT OF CHILDHOOD

Giovanni Francesco Caroto, *Fanciullo con pupazzetto*
(Museo di Castelvecchio, Verona).

GEORGE BOAS

THE CULT OF CHILDHOOD

SPRING PUBLICATIONS, INC.
DALLAS, TEXAS

DUNQUIN SERIES 18

© 1966 by The Warburg Institute. All rights reserved. Appendix © 1990 by
Spring Publications, Inc. All rights reserved. First Spring Publications printing
1990. Printed in the United States of America. Text printed on acidfree paper.
Cover designed by Margot McLean. Production by
Charlotte Milholland. Cover image is a mixed-media collage, untitled, by
Jean Klebs, 1985. Photo reproduction by Jonathan Lam.
Published by Spring Publications, Inc., P. O. Box 222069, Dallas, TX 75222,
by arrangement with The Warburg Institute, London, which has kindly granted
permission to use the original typography. *The Cult of Childhood* was first published
as volume 29 in the series Studies of the Warburg Institute (1966).

International distributors:
Spring; Postfach; 8803 Rüschlikon; Switzerland.
Japan Spring Sha, Inc.; 12–10, 2-Chome,
Nigawa Takamaru; Takarazuka 665, Japan.
Element Books Ltd; Longmead Shaftesbury;
Dorset SP7 8PL; England.
Astam Books Pty. Ltd.; 162–168 Parramatta Road;
Stanmore N.S.W. 2048; Australia.
Libros e Imagenes; Apdo. Post 40–085;
México D.F. 06140; México.
Zipak Livraria Editora Ltda; Alameda Lorena 871;
01424 São Paulo SP; Brazil.

Library of Congress Cataloging-in-Publication Data

Boas, George, 1891–
The cult of childhood / George Boas.
p. cm. – (Dunquin Series ; 18)
Reprint. Originally published: London : Warburg Institute,
University of London, 1966. (Studies of the Warburg Institute ; vol. 29)
Includes bibliographical references.
ISBN 0–88214–218–6
1. Children in literature. 2. Children as artists. I. Title.
II. Series: Studies of the Warburg Institute ; v. 29.
PN56.5.C48B6 1990
809'.93352054 – dc20   90–33312
CIP

TABLE OF CONTENTS

PREFACE

In 1935 the writer of this essay, in collaboration with A.O. Lovejoy, published what was to have been the first of four volumes of a documentary history of primitivism. The second war and the failure of our colleagues to write the volumes assigned to them resulted in the appearance of *Primitivism and Related Ideas in Antiquity* only. In 1948, two years after demobilization, a series of essays on the same set of ideas in the Middle Ages was published, and it was hoped that work might be resumed on the original programme. This hope was not realized. Meanwhile, it became clear that whereas cultural anthropologists were steadily destroying what we had called chronological primitivism, a belief in cultural primitivism still persisted. But its forms and their exemplifications had both changed.

With the settlement of the two Americas there arose an understandable scepticism about the nobility of the savage. The place of the American Indian was then taken by the Polynesian. Thanks to the novels of Melville and the short stories of Somerset Maugham, thanks also to the paintings of Gauguin, life in Polynesia was imagined to be a dream of erotic bliss, of dance, song, and feasting. The gentleness, the beauty, the accessibility of the Polynesian women became a standard theme of lyric expression. But again the ethnologist stepped in and a more intimate knowledge of these people led to partial disenchantment. For though Europe might be blamed for the introduction of measles and venereal diseases, it was hardly to be held guilty for yaws and elephantiasis. Nevertheless, enough remained of a soft life to attract the soft primitivist. He who loved a life in which the goddess Nature took the place of Art, who furnished human beings with all terrestrial delights and did so gratuitously, could find nothing better than the islands of the South Seas.

When Art leaves the stage for good, only the hard primitivist

can be satisfied. One of the ancient promoters of this cult was
Juvenal with his acorn-eating, cave-dwelling anthropoids who
scorned the refinements of comfort. Both hard and soft primitivists agreed that civilization was but an accumulation of superfluities, superfluities of knowledge, of laws, of government, of
the various arts and crafts. But the soft primitivist maintained that
the simple and natural life was actually more pleasant and less
fatiguing than the complicated and artificial life of modern society.
The hard primitivist had nothing but contempt for such hedonism
and preferred to do without those things that can be dispensed
with. He took as his model Diogenes of Sinope, who threw away
his cup when he saw a dog lapping up water from a stream.

One might think that when it was proved that the only remaining examples of primitive man did not meet the requirements of the cultural primitivist, that idea would have disappeared.
Quite the contrary. A search was then made for a new exemplar
that would be, if not the chronological *Urmensch*, at least the cultural. The outstanding results of this search were Woman, the
Child, the Folk (rural), and later the Irrational or Neurotic, and
the Collective Unconscious. Oddly enough and in spite of the
clear differences both existential and qualitative amongst these
beings, they were all supposed to have some, if not all, of the
characteristics which had been atttibuted to the Noble Savage.
Above all there was a kind of intuitive wisdom in them as contrasted with learning; second, a keener appreciation of beauty, not
of course the beauties of the academy but of something called
Nature; third, there was a greater sensitivity to moral values.
But the praise of all these innate propensities was part and parcel
of a general anti-intellectualism which had been steadily growing
since the sixteenth century. It will surprise no cultural historian
that this should have happened synchronously with the advancement of the natural sciences.

A complete study of the history of cultural primitivism since
the sixteenth century would require a very fat volume. If I have
selected one strand in this history, the cult of childhood, it is

because in the United States, where I have done most of my studies, it has reached amazing proportions and because to see before one's eyes the exemplification of an idea in which one has been interested is always stimulating. One finds this idea in the *mores* of North American societies: in the love of 'cuteness', the passion for joining and organizing secret societies in which special rites and costumes are required and strange titles given to the members, in the adolescent rough and tumble of the American farce, in the rapid innovations in slang, in the love of comic strips which go to the length of translating the Bible and the great classic narratives into pictures so that nothing remains but moments of action. But one also finds it in programmes of education, in what is sometimes called a philosophy of life, and in the criticism of works of art. If adults are urged to retain their youth, to 'think young', to act and dress like youngsters, it is because the Child has been held up to them as a paradigm of the ideal man. It has often been pointed out that in the United States children are indulged in all their desires. No one, I suspect, enjoys having his desires repressed and, were it possible to think of a life in which all were gratified and the one law of Thélème was never violated, that might be thought of as the ideal. American children have almost reached that limit of bliss. From the days of Mrs. Trollope to our own, they have been represented as spoiled darlings, allowed to do whatever they wish and never be reprimanded or punished. The picture is of course, vastly exaggerated, but there must be a good bit of truth in it for so many observers to have noted it. Discipline in schools has become greatly relaxed—indeed there was room for relaxation—and it has been suggested that the child's interests should alone determine the curriculum that he would pursue. Self-expression was the aim sought by all, though few ever raised the question of what the self was or whether there were good and bad ways of expressing it. In the arts the teaching of any technique was deplored, for the very fact that what a child made was childlike sufficed to justify him.

I speak of this as if it were peculiar to the United States. Belief in its implications is probably more widespread there than elsewhere, but as a matter of historical fact the roots of the ideas involved are European. The religious roots are to be found in St. Mark and St. Matthew. The epistemological are in Montaigne and Agrippa von Nettesheim. One finds the complex as an appraisal of life in seventeenth-century England. Its aesthetics began probably in Vienna and the prophecy that the twentieth century would be the 'Century of the Child' was made in Sweden. That the child was a creature *sui generis* with rights and privileges of his own and not simply a duodecimo edition of a man was emphasized in eighteenth-century France. But all this will be sketched in the pages that follow.

I have followed the practice in this essay of quoting at length rather than paraphrasing and of leaving most of the quotations in the language in which they were written. This does not make for easy reading but it does leave the reader free to see for himself just what the authority quoted said. I have been aided in various ways by friends and colleagues to whom it is a pleasure to express my deep gratitude. Amongst these are Professors Grace Frank, Harold Cherniss, E. H. Gombrich, the staff of the Warburg Institute, and Allan Ludwig, Mrs. Bryson Burroughs and Miss Martha J. Hubbard. Without their help this study would be even more skeletal than it is. It is, I hope, unnecessary to say that the interpretation of the materials which they have been good enough to send me is my own, not theirs.

Ruxton (Maryland) — London 1964   G.B.

THE CULT OF CHILDHOOD

Before beginning the history of the Cult of Childhood, it may be well to point out that it is easy to substitute the Child for the chronologically primitive, for childhood is the obvious first stage in any individual's biography. If one is a chronological primitivist, one will find admirable traits in whatever is primordial or one will invent them if need be. In the Christian tradition this attitude was reinforced by the first chapter of Genesis, in which God saw that His work was good; that it is no longer quite so good was clear to anyone who had eyes to see. Consequently the degeneration of Creation was a problem and a problem that some people met by urging their fellows to return to the primitive condition of the race. The case of the Greek and Roman primitivists was different. For reasons which can no longer be discovered, they had at an early period of their cultural history a legend, that of the Golden Age or the Age of Kronos, paralleled in Latin literature by the *Saturnia regna*, from which history had declined. When it was possible for men to see in the Child a replica of Adam before the Fall or of Man in the Golden Age, it was also possible to praise him as the possessor of all the virtues that had been in the original.

The Child is also a good model for the cultural primitivist. For it is equally obvious that as he emerges into our adult world he is innocent of all the arts and sciences, unspoiled by the artifices of civilization. This in itself would not prove him to be better than the adult, but if one is a cultural primitivist, one is committed to the thesis of Rousseau's *Discourse on the Sciences and the Arts*. The theme of innate wisdom v. acquired knowledge is an old one, suggested, if not elaborated, as early as Theognis and Pindar.[1] If there is such a thing as congenital wisdom and other virtues,

[1] See A. O. Lovejoy and G. Boas, *Primitivism... in Antiquity*, Baltimore, 1935, p. 196, n. 8.

then it ought to appear in childhood. But though there are hints of the cult of childhood in classical antiquity, it does not seem to have had much popularity. In general the Ancients had a low opinion of children if they appraised them at all.

I

One might think that the myth of Meno would have led men to seek in infancy a repository of all knowledge and goodness, but when all is said and done the slave-boy would have recalled nothing if he had not been teased along the path by Socrates. In the *Laws* (653 A-B) children are spoken of as creatures whose first sensations are pleasure and pain. Virtue and vice enter their soul in this form. 'But as to wisdom and settled true opinions, a man is lucky if they come to him even in old age'.[1] Children are incapable of reasoning. And later (*Laws*, 760 A) Plato says that even though a man be born with a favourable nature, if his education is bad, he will turn into the 'wildest creature on earth'. The artistic tastes of little children rise no higher than the puppet show (*ibid.*, 658 C) and those of youth are satisfied with comedies. If one should say that this is simply the opinion of a grumpy disillusioned old man, let one turn to *Phaedo* (77 E) where Cebes says to Socrates, 'There is within us a sort of child who fears such things' — as the dissipation of the soul at death. It is the philosopher's task to rid men of such childish fears. In spite of the adumbration of recent psychogenetic theories of the retention of childhood in the adult, there is no anticipation of our main theme here.

In Aristotle the situation is similar. It is the opinion of antihedonists, Aristotle says (*Nicomachean Ethics*, 1152b19) that children and brutes pursue pleasure, but, though he is tinged with hedonism himself, he is far from being opposed to this opinion. When he deals with family relations, he maintains that the father is the monarch of the family (*ibid.*, 1160 b 25). The child's psyche is so vacuous that he has not even any innate love for his parents (*ibid.*, 1161 b 16) and acquires filial affection only when he also

---

[1] Translation by R. G. Bury, Loeb Classical Library, vol. I, p. 89.

acquires the power of understanding and discrimination. As for his intellect, we read (*ibid.*, 1174 a 2): 'No one would choose to live with the intellect of a child throughout his life, however much he were to be pleased at the things that children are pleased at, nor to get enjoyment by doing some disgraceful deed, though he were never to feel any pain in consequence'.[1] In the *Historia animalium* (588 a 38) we discover that children are dwarflike in their bodily structure and, like dwarfs, have abnormally weak memories. Their memories are weak because of the weight of their upper parts in relation to the rest of their bodies. Finally, in Pseudo-Aristotle's *Rhetorica ad Alexandrum* (1441 a) where the technique of a eulogy is outlined, the eulogist is warned that 'in children it is generally considered that orderliness and self-control are due not to themselves but to those who have charge of them, and so they must be dealt with briefly'.[2]

The only child in classical literature who is presented to us at length is, as far as I know, Ion in Euripides's tragedy of that name. For neither Astyanax in Homer nor Ascanius in Vergil, not to mention Dionysius of Halicarnassus and Strabo, is portrayed as a precocious sage. We are not given Ion's age, but he was at least old enough to serve in the temple. His one 'noble thought' begins at line 625 when he says that he would rather live a 'demotic' life than a royal life. 'I would prefer the life of the people', he says, 'rather than that of a king whose pleasure it is to have low friends and who hates nobles because he fears to die'. It may be assumed that his priestly education has not been entirely without influence here and that what he says does not spring from innate childlike wisdom. Nor is there anything germane to our story in Creusa's account of her rape by Apollo and her subsequent exposure of the fruit of that experience (lines 961ff.), though the child stretches out his arms to her as if he were pleading with her to do him no harm. This might suggest to some readers that he had a dim intuition of what was about to happen, but

[1] Translation by W. D. Ross, Oxford, 1924.
[2] Translation by E. S. Forster, Oxford, 1924.

it is slim evidence of any exaltation of childhood. Like Racine's Joas, whom he prefigures, he has always been sheltered from evil and, unlike some modern child-heroes, has no good influence on any other character in the play.

Similar remarks may be made about Pre-Socratic opinion. When Heraclitus speaks of time as a child playing with draughts or counters, for no one can be sure what the game is, it is surely not to praise childhood. And when Democritus says that if children are allowed to do as they please, they will not learn letters, music, or that which comprehends all virtue: to have respect (Frg. B 179, Diels), and adds to this that if you have to have a child, you had best adopt one (Frg. B 277), for then you can choose the one you want, he is surely expressing no high opinion of them. But when reading authors whose works exist only in scattered fragments preserved by writers who often post-date them by centuries, one must beware of attributing to them sentiments which may not have been expressed in the context from which their words were lifted. Similarly one has also to be wary of treating them as fair examples of 'the' Greeks or Romans. In fact, we do not know what Greek or Roman opinion as a whole was on anything or even if there was a Greek or Roman opinion as a whole.

There is one phrase in Cicero which might have been expected to be handed on and frequently quoted, but which seems to have had few echoes. It occurs three times in *De finibus*, once where he attributes to his *bête noire*, Epicurus, the idea that children and beasts are *specula naturae* (Bk. II, x, 32), once when Piso is made to say that perpetual repose is unendurable and refers to infants as evidence (Bk. V, xx, 55), and once (Bk. V, xxii, 61) to prove that moral excellence is an end in itself. That Piso does not believe that we should stop with Nature is clear enough from an earlier passage in which he describes the ages of man as beginning with the helplessness of the newly born, *tamquam omnino sine animo sint* (Bk. V, xv, 42). 'Nature' was, to be sure, a sacred word to both Stoics and Epicureans and though children are *specula naturae*, yet they are dark mirrors, the *prima...commendatio quae a natura*

nostri facta est nobis incerta et obscura est (Bk. V, xv, 41). Children do not then seem to be norms for adult behaviour, but when one thinks of the popularity of Cicero in the Renaissance and the Enlightenment, one wonders why the Child as the *speculum naturae* did not have more vogue.[1]

With the coming of Christianity new literary material was available for the admirer of childhood, though it was seldom used. This material was, of course, scriptural. It was of two sorts, prophecies and similar texts extolling the wisdom of a given child and mentioning the wisdom of children as something wonderful, and New Testament texts praising the childlike nature. Of the former, Psalm VIII, 2 was the most obvious: 'Out of the mouths of babes and sucklings hast thou ordained strength, because of thine enemies; that thou mightest still the enemy and the avenger'. Even more impressive was the so-called Messianic prophecy of Isaiah (VII, 14 ff.). But in the latter the child Immanuel apparently must eat 'butter and honey' in order 'to know to refuse the evil and choose the good', and, it is added by the Prophet, 'Before the child shall know to refuse the evil and choose the good, the land that thou abhorrest shall be forsaken of both her kings'. No one, as far as I know, has maintained that this refers to children in general, but rather to an exceptional child, though biblical exegesis has been used in strange ways. It is nevertheless unlikely that anyone has ever contradicted the plain meaning of the words, '*before* the child shall know how to refuse the evil and choose the good...' In short, even if Immanuel is the Messiah, he requires a certain experience before knowing good and evil. Similarly the

---

[1] The idea of the Child as a *speculum naturae* must be much older than the time of Cicero for we find in Herodotus, Book II, the story of King Psammetichus who, wishing to know whether the Phrygians or the Egyptians were the older race, made the experiment of isolating two newly born infants and waiting to learn what their first words would be. Their first words, or sounds, were reported to be *beccos* which was the Phrygian word for *bread*. Phrygian, he inferred, must be the older language, the natural language of mankind. Dr. Harold Cherniss has called my attention to two passages in late Greek literature which speak of the special 'purity' of children, purity that comes of their being 'unsullied' by the world of the body, making them beloved of the gods: Iamblichus, *Vita Pythagorica*, 51-52; *Corpus Hermeticum*, X, 15 (ed. Festugière and Nock I, p. 120, 13-21).

story of Jesus preaching to the Doctors in the Temple (Luke, II, 42-49), pertains not to children as such but to a supernaturally gifted child, a boy of twelve. Milton understood it in this way and described the event in *Paradise Regained* (Bk. I, lines 201ff.).

> When I was yet a child, no childish play
> To me was pleasing; all my mind was set
> Serious to learn and know, and thence to do
> What might be public good: myself I thought
> Born to that end — born to promote all truth,
> All righteous things; therefore above my years,
> The law of God I read, and found it sweet,
> Made it my whole delight; and in it grew
> To such perfection, that, ere yet my age
> Had measured twice six years, at our great feast
> I went into the Temple, there to hear
> The teachers of our law...[1]

If now one adduces the story of the Adoration of the Magi, told only in Matthew (II, 1-2, 9-12) or that of the Shepherds (Luke, II, 15-17) as the elevation of childhood to a position of eminence, it must be once more insisted that this is the adoration of one child, not of childhood. The details of Jesus's boyhood which were to be related later in paintings come from the Protevangelium of James (chapter 21), Pseudo-Matthew (chapter 16), and the Arabic Gospel of the Infancy (chapter 7). But iconographically and verbally both Adorations are historical and thus unique events having no reference to the general superiority of childhood over maturity.

[1] Cf. Book IV, 214-20, where it is clear that Milton was no anti-intellectualist. Lest someone wonder at the need of an incarnate omniscient Deity's having to study before He knew the Law which He Himself had revealed to man, it should be pointed out that if God was to be made man, He must undergo the whole fortune of a man. For all Milton's admiration for science, he was sometimes antiphilosophic. In *Paradise Regained*, Book IV, 285 ff., he writes of the futility of Greek philosophy, e.g., lines 288 ff.,

> ..... He who receives
> Light from above, from the Fountain of Light,
> No other doctrines needs, though granted true...

And in line 295 he says of Plato that he 'to fabling fell and smooth conceits'.

It would be tempting to dwell upon such themes as the adoration of the Infant Jesus by his Mother, but that theme goes back only to the fourteenth century. The earliest representation of it which I have been able to discover is a detail in the Tree of Life by Pacino di Bonaguida in the Accademia in Florence, based upon the Vision of St. Birgitta.[1] But unless one interprets this as a symbol of childhood and not the representation of a particular child, one can infer nothing about the natural endowment of the infantile mind from it. The same would be true of statues of the Christ-child as king, crowned and in robes of state, which date from the seventeenth century, according to Emile Mâle,[2] of pictures of the Child in an attitude of benediction, or even of pictures of St. Catherine's mystic marriage with the Infant Jesus. In all such cases the emphasis is upon one specific child. If the comparison is not out of place, one could make a similar remark about the various child-prodigies who have been recorded from time to time: Cicero writing his *De ratione dicendi* at the age of twelve or thirteen, Tiberius delivering a funeral oration for his maternal grand-mother when he was nine, Marcus Aurelius beginning the study of philosophy at twelve.[3] But again, at the risk

[1] I owe this to Professor Millard Meiss. See his *Painting in Florence and Siena after the Black Death*, Princeton, 1951, p. 149, n. 73. The theme, as everyone knows, became very popular. Among the best known representations of it in the fourteenth and fifteenth centuries are those of Gentile da Fabriano and Paolo Veneziano. But it would be sheer pedantry to make a long list of them. For a reproduction of the Pacino, see Plate II, 6, in Richard Offner's *Corpus of Florentine Painting*, Section III, vol. ii, part i, New York, 1930.

[2] Emile Mâle, *L'Art religieux après le Concile de Trente*, Paris, 1932, p. 328.

[3] In that mine of curious information, Adrien Baillet's *Jugemens des Savans*, one will find half a volume given over to *Les Enfans célèbres*. I use the edition of 1722, vol. IV. Cf. Jacob Burckhardt, *Civilization of the Renaissance in Italy*, translated by S. G. C. Middlemore, N. Y. and London, (1931 ?), p. 273, n. 2 for the mention of several prodigies in the Renaissance. Vasari provides numerous stories of the exceptional skill of some artists as children, beginning with the best-known, that of the child Giotto. See the lives of Brunelleschi, Giuliano da Maiano, Baldovinetti, Lippo Lippi, Andrea dal Castagno, Mantegna, Leonardo, Bramante, and Raphael, amongst others. And although he urges artists to follow their 'natural bent,' rather than attempt performances of which they are not capable, he does not praise the immature artist because he expresses some childlike trait but on the contrary because he is as good as a well trained painter or sculptor. Similarly when praising, for instance, Raphael for abandoning the idea of equalling Michelangelo, an artist whose type of genius was foreign to Raphael's nature, he never deprecates study.

of labouring the obvious, these are all exceptional cases and they
are recorded because they are exceptional. On the other hand there
are certain New Testament verses which might well have been
used by the extreme anti-intellectualist. I refer of course to Matthew XVIII, 1 ff., particularly to verse 31: 'Except ye be converted and become as little children, ye shall not enter into the kingdom
of heaven'. But the Evangelist does not say in what respect one
should become as little children and it may well be that he was
referring to a re-birth. Or again, Mark X, 14: 'Suffer the little
children to come unto me, and forbid them not: for of such is the
kingdom of heaven'. To which should be added Luke IX, 47.[1]
In contrast to these lessons is the famous statement of St. Paul
(I Corinthians, xiii, 11): 'When I was a child, I spake as a child, I
understood as a child, I thought as a child; but when I became a
man, I put away childish things'. Surely no comment is required.

The early Christian Fathers and ecclesiastical writers made
little use of these New Testament verses which might induce
them into an adoration of childhood as such. Adam before the
Fall, says Theophilus, 'was in age still an infant, wherefore he was not
able to receive knowledge worthily. For even now, when a child
is born, it is not able to eat bread, but first is fed on milk and then
with its advance in years it proceeds to solid food'. So it was with
Adam.[2] But Theophilus adds that this was merely the first step
towards a better state, in spite of the fact that Adam was created
in the image and likeness of God. Tertullian, whose anti-intellectualism might have led him into admiration of innate wisdom,
says that both Adam and Eve were sexually immature, children
both physically and emotionally, yet rational.[3] And Clement of

---

[1] The first of these three references oddly resembles a passage of the *Tao-te Ching*,
chapter 55. See the translation by Wing-tsit Chan, *The Way of Lao Tsu*, in the Library
of Liberal Arts, Indianapolis and N.Y., 1963, p. 197. In a comment on this passage
Mr. Chan refers to a saying of Mencius: 'The great man is one who does not lose his
child's heart'. But these passages, as I suspect is also true of Matthew XVIII, urge
the purgation by the Sage of his accumulated prejudices and sins and the return to
'innocence'.

[2] *Ad Autolycum*, II, 25. See *Patrologia Graeca*, vol. VI, col. 1091.

[3] See G. Boas, *Essays on Primitivism...in the Middle Ages*, Baltimore, 1948, p. 18.

Alexandria, like Theophilus, speaks of Adam as a child, playing freely in the Garden of Eden and, though he was a child of God, he succumbed to the charms of pleasure. Thus he was not yet an exemplar.[1] Clement seems to be the first of the Fathers to depreciate learning and his theory of levels of meaning might have led to a fervent anti-intellectualism which would stituate the child's mind at the apex of wisdom. But actually there is next to nothing in his writings which adds to the *topos* which we are discussing.

As one proceeds through mediaeval Latin literature, one again fails to find any use being made of Matthew XVIII or similar passages. There are, to be sure, poems in which children are spoken of in eulogistic terms, but The Child as a model for adults never occurs in any of the writings that I have been able to examine. On the contrary, when a specially gifted child is mentioned, it is because he is unlike other children. Thus Ausonius writes of a young scribe who can take dictation faster than his master can talk.[2] Dracontius in his *Aegritudo Perdiccae* writes of a boy who had an incestuous love of his mother, Castalia, and hanged himself when he discovered that he could not be cured of this passion. But Perdicca was obviously neither a baby nor a young child.[3] Modoinus' *Eclogue* has as speakers an old man and a boy, but the boy is neither prematurely wise nor even a hero. Sedulius, who wrote in the Carolingian period, has a poem to Eberhard on the birth of a son; but it contains merely the usual congratulatory phrases. Another poem written *Ad quendam impubem* simply urges the boy to study.[4] Similarly a poem on the Holy Innocents, though it refers to them as *innocuos Christi testes nondum loquentes*, does not embroider on the theme.[5] Marbod of Rennes (11th century) writes to one of his pupils a set of instructions on how to spend his day, but this would have been unnecessary if the

[1] *Ibid.*, p. 25.
[2] Epigram CXIV. *Quis ordo rerum tam novus/veniat in aures ut tuas/quod lingua nondum absolverit?* Quoted in F.J.E. Raby, *A History of Secular Latin Poetry in the Middle Ages*, 2nd ed., Oxford, 1957, vol. I, p. 58.
[3] Raby, *op. cit.*, vol. I, p. 204.
[4] See *Poetae Latini Aevi Carolini*, vol. III, pp. 202 and 355.
[5] *Ibid.*, vol. IV, p. 351.

boy was born with wisdom innately implanted in him.[1] And though Baudry of Bourgueil wrote a poem addressed *virginibus puerisque*, he wrote it down to their level of understanding and pointed out that it would appeal to them both presumably for that reason.[2] Even when a man like Peter Damian writes a diatribe against Plato, Aristotle, and other philosophers, he ends with the words, 'Let the simplicity of Christ be my teacher; let the true homeliness of the wise loosen the bonds of my doubts' (*Christi me simplicitas doceat, vera sapientium rusticitas ambiguitatis meae vinculum solvat*), and exclaims, *Mea igitur grammatica Christus est*.[3] He might have introduced to good purpose here some of the New Testament verses referred to above, but he fails to do so.[4]

When one preaches anti-intellectualism, one has to find some human state in which there is happiness or well-being without the gifts of reason, unless one is satisfied with despair. One may reject learning in favour of intuition, the mystic vision, instinct, or even momentary sensual pleasure, but one can hardly construct a philosophy of life without pointing to some human being who exemplifies it. Otherwise one is preaching a futile programme. The anti-intellectualist has gone for his exemplar to primitive man, to the peasant, to woman, to the Unconscious, and, as this essay is attempting to show, to the child, each of whom is supposed to possess untaught wisdom. Hence, though the mediaeval writers are not explicit paidolaters, if I may be permitted this neologism, the sources from which they might draw inspiration,

---

[1] Raby, *op. cit.*, vol. I, p. 337.

[2] *Ibid.*, vol. I, p. 342.

[3] *Dominus vobiscum* in *Patrologia Latina*, vol. 145, col. 231, and Epistle VIII, 8. Cf. the famous attack on learning by Gregory the Great and John XIII, anticipated even in St. Anselm. See G. Boas, *op. cit.*, pp. 121 ff., on the Vanity of the Arts and Sciences.

[4] Just as the Adoration of the Infant Jesus in painting is almost post-mediaeval, so, according to Coulton, the theme of Christ Blessing the Children, is 'seldom or never' portrayed. See G.G. Coulton, *Mediaeval Faith and Symbolism*, New York (Harper Torchbooks), 1958, p. 295. Emile Mâle, *op. cit.*, chapter VII, section 5, pp. 325 ff., dates the cult of the Child Jesus back to St. Francis of Assisi, following Borély's *La Dévotion du saint Enfant Jésus au berceau*, Paris, 1664. Hans Wentzel, however, in his article 'Christkind-Bilder aus alter Zeit', in *Die Kunst und das schöne Heim*, III, 1961, pushes the cult itself back into the fourteenth century, though his illustrations date no earlier than the fifteenth century.

if any was needed, are clear enough. Added to the hints which we have listed, and they are only hints, are the various stories of wonderful children, which proved that on certain occasions babes and sucklings did indeed speak truth without instruction. There was, for instance, the recognition of the unborn Christ by the unborn John the Baptist during the Visitation. There was the case of St. Brice, who succeeded St. Martin of Tours in the fourth century, who was accused of having seduced a nun; her child of thirty days calls out, *Briccius non est pater meus*.[1] There is the story of St. Augustine meditating on the Trinity by the seashore and seeing a child trying to empty the sea with a seashell. When he tries to point out to the child the vanity of such an enterprise, the child replies that it is equally vain to try to explain the dogma of the Trinity. And there are saints such as Christopher, Simpert of Augsburg, Nicholas, Anthony of Padua, Vincent de Paul, who are associated with the Infant Jesus or with human babies. It is hard to resist the impression that such stories kept alive the feeling that Jesus as an infant and ordinary children too could on certain occasions perform miracles and speak like prophets.

But the full development of our theme does not come until our own times. As scientific progress advanced, men seemed to despair of finding terrestrial happiness in its gifts. There emerges almost step by step with the growth of natural science a fear of reason. We turn now to the beginnings of the cult of childhood in the scepticism of the sixteenth century.

II

No proof is needed that the sixteenth century saw several movements that weakened the principle of authority. The recovery of the ancient classical texts which had been lost or forgotten, the explorations of the navigators, the new astronomy and physics, the many inventions, the Protestant Reformation, the consolidation of the new nations, the rise of vernacular literatures, all combined to destroy the notion that everything worth

---

[1] The scene is depicted in the crypt of St. Ours de Loches.

knowing was already known. Amongst these influences was philosophical scepticism. It found popular expression in the paradoxical literature which, like the *dissoi logoi* of antiquity, must have suggested that traditional beliefs about morals and customs could not only be criticized but also rejected. Then there was that group called by Guy de Brués the *Nouveaux Académiciens* of whom Brués made the poet Baïf the spokesman. Such scepticism was directed towards free thought in scientific matters. Though on the suface the scepticism was nothing more than criticism of the principles of scientific investigation, when extended it could be turned towards religion and theology as well. There was also a kind of scepticism which was the handmaid of religion, the kind that one finds in Agrippa von Nettesheim (1487-1535) and which in subtler form reappeared in such seventeenth century writers as Pascal and Huet of Avranches. When one calls the roll of men who might properly be termed sceptics, one comes upon the names of Rabelais, Montaigne, Vanini, Sánchez, and later of Charron, La Mothele Vayer, Sorbière, Glanvill, and Bayle. These men obviously were not in agreement on all points, but their combined influence upon some minds must have been anti--authoritarian. If the principle of authority was rejected, the thinker must either throw up the sponge or discover some source of truth within himself to take its place. For the word "authority" at that time, as throughout the Middle Ages, almost always referred to tradition and tradition was by its very nature something social rather than individual. To the Church, tradition began with the Covenant of Abraham, passed on through Moses to his brother, Aaron the High Priest, through the succeeding High Priests to St. Peter, and from him down through the ages from Pope to Pope. But to some Protestant sects tradition was not so important as Scripture and to others, such as the members of the Religious Society of Friends, its place was taken by what amounted to special revelations given to individuals by the Spirit. Meanwhile there were always mystics of one sort or another, for mysticism, as is well known, has appeared at all times and in all places. The

direct intuition that comes in the mystic vision has no authority other than itself; its truths are self-substantiated, analogously to the supposed truths of immediate sensory experience. The mystic can be an anti-intellectualist, though he need not be, but he is always an anti-authoritarian as far as religious information is concerned.

The book of Agrippa's which is of the most importance for our present purpose is his *De incertitudine et vanitate omnium scientiarum et artium* which appeared in 1531, though written a few years earlier.[1] The author makes it perfectly clear at the outset that he is writing this book in the interests of revealed religion. He planned to take up every science and art in turn and show how it conflicted with the Word of God and turned men towards heresy. *Non enim in lingua sed in corde veritatis sedes est* (p. 2), a sentiment anticipated by St. Augustine and echoed by Pascal. It would be perhaps superfluous to point out that the heart refers to the *lumen naturale*, common sense, or what has sometimes been called instinct. Its precise meaning need not concern us here for it becomes clear, as one reads Agrippa, that he was thinking of such faculties as conscience or good will, the mechanism of which is seldom, if ever, conscious. Thus we find him saying (p. 4), 'True happiness does not consist in the knowledge of good things, but in a good life'.

> Vera enim beatitudo non consistit in bonorum cognitione, sed in vita bona: non intelligere, sed in intellectu vivere; neque enim bona intelligentia, sed bona voluntas conjungit homines Deo, nec aliud efficiunt disciplinae fortis adhibitae, nisi quia conditionem nobis quandam purgatoriam adhibent, ad beatitudinem aliquid conducentem, non tamen rationem ipsam, qua nobis beatitudo compleatur, nisi eis adsit et vita, in ipsam bonorum translata naturam; saepissime enim est compertum, ut ait Cicero pro Archia, ad laudem atque virtutem [plus] naturam sine doctrina quam doctrinam sine ratione valuisse.

[1] I use the edition of 1643, Lugduni Batavorum.

Presumably the good life can be lived without reason and the good will needs no doctrinal skill to guide it. Nature without doctrine is superior to doctrine without nature. So far this might mean simply that one's beliefs may be ineffective as far as our character is concerned; a man innocent of learning may be more virtuous than a man of great learning. This is one of those commonplaces which have been repeated over and over again, a platitude. But Agrippa feels that there is more to the story than that, for he believes that learning is an evil in itself. It is inherently bad. Adam (p. 5) would never have been expelled from Eden if he had not sought knowledge of good and evil; Paul wanted to expel from the Church those who demanded more knowledge than was proper; Socrates was called the wisest of men by the Oracle because he alone knew that he knew nothing. 'Nothing', says Agrippa (pp. 5-6), 'can happen to a man more pestilential than knowledge (*scientia*)':

> Haec est vera illa pestis, quae totum et omne hominum genus ob unum subvertit, quae omnem innocentiam expulit, et nos tot peccatorum generibus, mortique fecit obnoxios, quae fidei lumen extinxit, animas nostras in profundas conjiciens tenebras, quae veritatem damnans, errores in altissimo throno collocavit.

But it is obvious that it is not merely knowledge which is pestilential. It is a kind of knowledge. For that which comes from the heart or from faith is to be sought. It is *scientia* which expels all innocence, which extinguishes the light of faith, and casts our souls into the depths of darkness and sets error on the loftiest throne.

Agrippa thus distinguishes between the apprehension of truth through a non-rational faculty and the reason. What truth is apprehended is of course the dogmas of the Church, which he realizes are contradictory to the conclusions of *scientia*. In maintaining this he is in agreement with the sentiments, if not the words, of those Popes and Fathers whom I have mentioned above.

For though the mediaeval Catholic philosophers were, as everyone knows, determined to give a rational account and demonstration of their theological and ethical beliefs, there did exist a persistent anti-rational tradition running alongside of scholasticism. It is doubtful whether any mediaeval philosopher wrote so complete a denunciation of learning as Agrippa's, for he takes up one science after another, one art after another, and shows to his own satisfaction that they all build up human pride and arrogance and weaken faith. And he concludes with the old exemplum of the ass as the model of wisdom and virtue. His *encomium asini* comes at the very end of the book. It is based on the historical fact that Jesus chose as His disciples neither scientists not philosophers, but humble fishermen and carpenters. These disciples, says Agrippa (p. 306), are Christ's asses. The ass lives on little and is satisfied, bears heavy burdens, suffers poverty, hunger, labour, blows, neglect, ill treatment, and does not complain. He is very patient, of simple and innocent heart, and was honoured by God Himself under the Old Law, for though sheep could be sacrificed, it was forbidden to sacrifice the ass. He was, moreover, a witness of the birth of Christ and, having borne Christ's body, is marked with a cross on his back. Balaam's ass was given a voice and Samson killed the Philistines with the jawbone of an ass:

> Nonne Christus in bucca asinorum suorum simplicium et rudium idiotarum, apostolorum et discipulorum suorum vicit et percussit omnes Philosophos gentium et legisperitos Iudaeorum, omnemque humanam sapientium prostravit atque confecit, propinans nobis ex illorum suorum asinorum maxilla aquae vitae et sapientiae aeternae? (p. 308).

In fact, Agrippa goes so far as to say that the early Christians were called *asinati* 'and were accustomed even to depict Christ Himself with the ears of an ass'.[1]

---

[1] Agrippa could not have known of the *graffito* from the walls of the Imperial Palace on the Palatine of the crucified ass with the inscription, 'Alexamenos worships God'. He refers to Tertullian as authority for the opinion that the early Christians were called *asinati*. This may be based on Tertullian's *Ad nationes*, Bk. I, chap. xi,

I have described elsewhere how this type of argument led into praise of animals who know good and evil without possessing reason, are virtuous without instruction, and have no need of the intellect.[1] In some cases, as in Gelli's *Circe*, this may have been simply a paradox, but in others it was probably sincere dispraise of human reason as well as admiration, like Walt Whitman's, for the beasts. In extreme form we find authors maintaining that the beasts could reason and reason better than men can, but on the whole the 'theriophiles' were satisfied to show that instinct was superior to intellect. Yet in spite of the popularity of this notion, one finds little evidence that the admirable traits of animals were transferred to children. That was to come later.

This is also true of Montaigne. Montaigne was not a consistent anti-intellectualist. He had indeed little regard for logical consistency. He was painting a portrait in his *Essays* and not writing a thesis. And though he could praise primitive man in the *Cannibals* and the animals in his *Apology for Raimond Sebond*, he had firm ideas about the training of children which did not include allowing them to express themselves to the point where their innate wisdom would shine forth and obscure that of adults. On the contrary,

> Ie treuve que nos plus grands vices prennent leur ply dez nostre plus tendre enfance, et que nostre principal gouvernement est entre les mains des nourrices.[2]

Even in his attack on pedantry (Bk. I, xxiv), he defends merely good actions as superior to learning but does not go so far as to

---

but that is not certain. In a French version of Agrippa's *De incertitudine*, 1582, we find that truth is to be found only in the poor in spirit, children who are 'desnués de tous Thresors des sciences, qui sont purs du coeur, nects de toute ordure des sciences, et l'esprit desquels est ainsi qu'un beau papier blanc, auquel n'a esté encor escrit aucune chose des traditions humaines' (p. 539). The cross on the donkey's back appears frequently in English verse. See *inter alia* Wordsworth's *Peter Bell* (971-75). But above all see Frances A. Yates, *Giordano Bruno and the Hermetic Tradition*, London and Chicago, 1964, p. 259 and n. 4.

[1] See G. Boas, *The Happy Beast in French Literature of the Seventeenth Century*, Baltimore, 1933.

[2] *De la coustume, et de ne changer ayseement une loy receue*, Bk. I, chap. xxii. I use the edition of LeClerc-Prevost-Paradol, Paris, 1865. The reference is to vol. I, p. 131.

locate them in a childlike heart. His main point of attack is both our failure to absorb our learning so that our lives may be guided by the wisdom of sages and our use of it as decoration.

> Si nostre ame n'en va un meilleur bransle, si nous n'en avons le iugement plus sain, i'aymerois aussi cher que mon escholier eust passé le temps à iouer à la pausme; au moins le corps en seroit plus alaigre.[1]

He makes a sharp distinction between ideas and practices and has little use for the former unless they lead to an amelioration of the latter. He does not maintain that goodness is the criterion of truth but simply that truth which does not support goodness is not worth possessing.

> Tel a la veue claire qui ne l'a pas droicte; et par consequent veoid le bien, et ne le suyt pas; et veoid la science, et ne s'en sert pas.[2]

As far as children are concerned, Montaigne in the same essay refers to Xenophon's account of the Persians who teach their young virtue before letters. For the first seven years the child is given physical training to *rendre le corps beau et sain* and at fourteen he is put into the hands of four tutors, the wisest, the justest, the most temperate, and the bravest, learning from them respectively religion, truthfulness, self-restraint, and fearlessness. No further argument is needed to show that for Montaigne, in spite of his scepticism and his subordination of learning to right action, the child was not to be free of instruction. In fact in his essay *De l'institution des enfants*, dedicated to Diane de Foix (Bk. I, xxv), we find him contrasting the immature animal and the immature man and pointing out that the former show their innate character at once, whereas the latter soon disguise it.[3]

[1] Vol. I, p. 176.
[2] *Ibid.*, p. 181.
[3] Cf. Cicero, *De finibus* V, xv, 41.

Les petits des ours et des chiens montrent leur inclination naturelle; mais les hommes, se iectant incontinent en des accoustumances, en des opinions, en des loys, se changent ou se desguisent facilement: si est il difficile de forcer les propensions naturelles.[1]

The child then does have a congenital 'character', but it is not to be given free rein. The pupil is not to remain silent before his tutor and absorb what he is told without question. But the value of this procedure is not that it permits the child to develop his innate personality but rather to permit the tutor to see what manner of problem he has to solve. For truth transcends personality and is common to all.[2] It is that common truth that the teacher has to instil in his pupils. And when Montaigne comes to lay out the curriculum, one sees how full it is.[3] The sceptic is willing to prescribe lessons in spite of his *que sçais-je*.[4]

Montaigne, then, retained much of the traditional curriculum in his programme and had no illusions about the possibility of a child's developing all knowledge or even that which would be profitable to him as he grew up out of his congenital potentialities. He was, as is well known, against violence in teaching the young and may have helped to bring about the kindlier treatment of children, which in turn may have bolstered a new idea of what childhood was. If he did have this influence, it took some time to make itself felt, for caning continued to be a practice in some schools down to our own days. Judging from the comments on British schools, made by such men as Sydney Smith, Macready, Southey, Crabbe, Cowper, and even Lowes Dickinson, to say nothing of the impression made, one would imagine, upon foreigners by such a book as *Tom Brown's School Days*, such institutions were anything but agreeable to the more sensitive youngsters who were sent to

[1] Vol. I, p. 193.
[2] *Ibid.*, p. 197.
[3] *Ibid.*, pp. 208 ff.
[4] For suggestions of Montaigne's influence on pedagogy in the seventeenth century, see Alan H. Boase, *The Fortunes of Montaigne*, London, 1935, Index, *s.v. Education*.

them.[1] It was Rousseau's *Emile* (1762) that introduced a more radical appraisal of childhood into European thought. For the seventeenth century in France, though one given to speculating upon the childhood of Christ, saw no hope of terrestrial happiness in it. Emile Mâle quotes Bérulle as speaking of childhood as *l'état le plus vil et le plus abject de la nature humaine, après celui de la mort*. Condren is cited as the author of the thought that in childhood Adam lives and rules in us while Jesus is in captivity. And Pascal went to the length of saying that the child was close to the animal.[2] In distinction to this we find Lord Morley saying that at the time when *Emile* was published, everyone was discussing education[3] and judging from the success of the book, popular opinion must have shifted considerably. In the group which composes 'everybody' Morley included the Abbé de Saint-Pierre, Mme d'Epinay, Mme de Grafigny, Grimm, and Helvétius. For that matter, everybody was talking about pretty nearly everything else too after the *Encyclopédie* began to appear. But regardless of that, a kindlier feeling even towards animals was being expressed.[4] But Rousseau's thoughts were bound to take a new orientation and *Emile*, like the *Contrat Social*, was directed towards the correction

[1] See Smith's 'Professional Education', originally published in the *Edinburgh Review* in 1809; Macready's *Reminiscences*, New York, 1875, pp. 9 and 14; *The Life and Correspondence of Robert Southey*, ed. by C. C. Southey (his son), London, 1849, vol. I, pp. 48, 77, and 133; Crabbe's *The Borough*, Letter 24, especially the part on Leonard, the teacher, and the bully; Cowper's *Tirocinium*; Dickinson in E. M. Forster, *Goldsworthy Lowes Dickinson*, London (reprint), 1947, p. 24. These are only a few scattered testimonies which could be paralleled by contrasting opinions. Dr. Arnold, for instance, was of course an admirer of the public school and had, moreover, a low opinion of boyhood in general. He said that he hoped to produce Christian men from his boys, 'for Christian boys I can scarcely hope to make; I mean that, from the natural imperfect state of boyhood, they are not susceptible of Christian principles in their full development upon their practice, and I suspect that a low standard of morals in many respects must be tolerated amongst them, as it was on a larger scale of what I consider the boyhood of the human race'. See A. P. Stanley, *Thomas Arnold*, quoted in Arnold Whitridge, *Dr. Arnold of Rugby*, New York, 1928, p. 94. It will be observed that Dr. Arnold apparently accepted that form of the Law of Recapitulation which became very popular in the generation after his. Cf. his article, 'Discipline of Public Schools', *Quarterly Review of Education*, 1835.

[2] See Emile Mâle, *L'Art religieux après le Concile de Trente*, chapter VII, section v, particularly p. 326.

[3] John Morley, *Rousseau*, London, 1896, vol. II, p. 199.

[4] See Hester Hastings, *Man and Beast in French Thought of the Eighteenth Century*, Baltimore, 1936.

of what its author believed to be 'unnatural'. He knew that children require some kind of instruction and his main innovation in the theory of pedagogy was in the technique of teaching. The work began, it will be recalled, as the *Contrat Social* began, by pointing out that everything is good as it leaves the hand of the Creator and that everything degenerates in the hands of men.[1] The child therefore is born good but becomes vitiated under the direction of schoolmasters who would deform him to suit their own ends. Prejudices, authority, necessity, bad examples, social institutions, all stifle a child's natural aptitudes. At the same time we are weak, helpless, and ignorant at birth and hence require force, help, and judgment. These can be gained only through education. But education, to be successful and not warp the child's congenital character or frustrate his potentialities, must rely on the methods that nature uses. These methods are primarily the warnings of pleasure and pain. 'Les premières sensations des enfans', he says, echoing Plato and Aristotle (Bk. I, p. 419), 'sont purement affectives; ils n'appercoivent que le plaisir et la douleur'. But by means of these affections one can guide a child towards those ends which he ought to seek and away from those which he ought to avoid. There is therefore no innate sense of right and wrong in the child, an idea which was to be contradicted even by some of Rousseau's disciples. Second, it is worth noting that Rousseau wishes to introduce a sense of responsibility into the child's mind at an early age by avoiding the acquisition of habits. He must be prepared for the exercise of his free will as soon as he possesses one, and this can be done by varying his regimen from day to day. Contemporarily Rousseau's Calvinistic colleagues in colonial America were saying just the opposite, as we shall see. By building up the child's strength and his self-reliance, one will build up his desire for goodness. 'Toute méchanceté (p. 422) vient de foiblesse; l'enfant n'est méchant que parce qu'il est foible; rendez-le fort, il sera bon; celui qui pourroit tout ne feroit jamais

---

[1] My page-references will be to the *Oeuvres complètes de J. J. Rousseau*, Paris, 1883, Vol. II.

de mal': surely one of the strongest apologies for autocrats ever penned. The bully, one suspects, is the boy who has been deprived of freedom in his childhood. The sense of pleasure and pain will suffice, when one is very young, to teach one what to shun, what one can accomplish, and what lies beyond one's powers. Nature (Bk. II, p. 430) sets up a balance between a child's abilities and his desires and when these are balanced, one is happy. The child is not a man. This principle, which might have seemed self-evident, is one of Rousseau's cardinal principles, and his main point of attack on education, as he believed it to be practised at that time, is that it treats the child as if he were an adult. He is, on the contrary, closer to the animal and should be allowed to live as his animal nature demands. He is equipped with the means of self-preservation and the rest lies dormant within him to mature at a later date. To be close to the animals was not depravity to Rousseau as it was to Pascal. It is simply a stage through which we must inevitably pass; depravity would consist in our remaining in it. The child then is neither a beast nor a man; he is a child. He can be made to feel his weakness and at the same time not to suffer from it.

The simple truism that the child is a child and not a man is one of those obvious sentences the importance of which is usually overlooked. Whatever Rousseau's faults, he saw clearly that it is futile to neglect the individuality of what you are dealing with in any practical matter. The limitations of a given 'nature' may and usually do exist alongside of its special aptitudes and by pointing out what a thing is not, one may also be pointing out what it is. Rousseau seems to have initiated the idea that childhood is something inherently different from manhood and—but this did not necessarily follow—it has its rights and privileges. The chief of these was the right to be itself.

Midway then between animality and humanity, the child cannot be taught to reason. Reason is a faculty which evolves with the greatest difficulty and at the latest date. It cannot be used to develop those faculties upon which it depends. It is the fruit of our

other propensities, not their seed. Since the child is, so to speak, *sui generis*, it must be recognized that he has his own ways of seeing, thinking, feeling, and nothing is more foolish than to attempt a substitution of our ways for his. Rousseau, let me insist, does not ask men to become as little children or revert to a condition from which they inevitably emerge. He says that children have no need to reason and is himself no great admirer of scientific powers, but nevertheless he has enough sense to realize the impossibility of replacing the adult's kind of life by the child's. He is pleading rather against forcing a child to behave in a manner that is foreign to his nature and natural equipment. Lacking the power of reason he cannot profit from sermons. Sermons must be replaced by experience (p. 439). Rational concepts of good and evil are alien to him and hence in the child's world there is no morality which transcends experience: 'Dépourvu de toute moralité dans ses actions, il ne peut rien faire qui soit moralement mal et qui mérite ni châtiment ni reprimande' (*ibid.*) Punishment is meaningless to a child, though it may be effective in forcing him to act or to refrain from acting as an adult demands. But the harm which is done is the harm always done by imposing upon a human being a way of behaving that he is incapable of understanding. A child is guided by his *amour-propre*. What that demands is what is good in his cosmos. What it forbids is evil. He will thus become self-sufficient and ready for the rational life at the proper time. But it must be remembered that the little Emile was to be brought up in the country and by a tutor and beyond the influence of society.

So far there was little to induce a reader to admire childhood. And Rousseau remained sober until he felt compelled to draw a picture of the child as contrasted with the adult. At that point he turned to metaphor and instead of painting a man as the goal of childhood, as the fruit into which the child was to develop, he permitted his love of children — a love which he failed to nourish in his own life — to run away with him. The child, he then says, is to the man as spring is to autumn. The child is 'bouillant, vif, animé, sans souci rongeant, sans longue et pénible prévoyance;

tout entier à son être actuel, et jouissant d'une plénitude de vie qui semble vouloir s'étendre hors de lui' (p. 489). One suspects that a man might be found to whom the same description might apply, but when Rousseau comes to sketch the adult, we find quite an antithetical picture. When we look at a man as he is, or as we imagine him to be in his old age, 'l'idée de la nature dé-clinante efface tout notre plaisir...l'image de la mort enlaidit tout' (ibid.).[1] Consequently Rousseau could not contemplate his pupil's being led by a man with any satisfaction. The schoolmaster approaches, takes Emile by the hand, and the boy's gaze fades and his gaiety is wiped out: 'Adieu la joie, adieu les folâtres jeux' (ibid.). The boy is led into a room lined with books: 'Des livres! quel triste ameublement pour son âge! Le pauvre enfant se laisse entrainer, tourne un oeil de regret sur tout ce qui l'environne, se tait, et part les yeux gonflés de pleurs qu'il n'ose répandre, et le coeur gros de soupirs qu'il n'ose exhaler' (ibid.). Whereupon Rousseau exclaims that he hates books (p. 507). But by now the boy is fifteen years old and must enter upon his life as a man.

Rousseau's dislike of society led him to express an admiration for childhood which was not inherent in his theory of the child's nature. Nor did he pay much attention to the inevitability of social life if one was to survive. We are here concerned not with criticizing Emile but rather with indicating its place in the history of an idea. In that history it is clear that subsequent corollaries were implicit. If the child is of a special and unique character, and if that character is good, then it might follow that a philosopher should try to preserve it as long as possible. But if a child either dies or turns into a man, as it seems reasonable to believe, then the question is whether the retention of childhood in maturity is either possible or desirable. On Rousseau's premises it would seem to be impossible and hence, whether desirable or not, it was not to be advocated. Other educationalists, however, prepared

---

[1] Earlier, in Book II (p. 432), Rousseau had praised the savage for his resignation before the fact of death: 'Les sauvages, ainsi que les bêtes, se débattent fort peu contre la mort et l'endurent sans se plaindre'.

the way for the retention of childhood or, if one prefers, its prolongation.

Of Rousseau's many followers there are two whose writings had special popularity and perhaps influence, Bernardin de Saint-Pierre and Pestalozzi. These two men, moreover, directed their thoughts to two different ends. Bernardin de Saint-Pierre thought of himself as a philosopher, indeed as a metaphysician; Pestalozzi was content to be an educator. Furthermore he had the opportunity, rare in his time, of actually putting his pedagogical ideas into practice.

The work of Bernardin de Saint-Pierre which is of interest to us is his *Paul et Virginie*, for it was here that the prolongation of childhood under the guidance of Nature was most clearly propounded. The theme of the story might be called the conflict between nature and custom (or civilization). Of the two children who are the main characters in the story, born within the sphere of natural influences and little else, one, Virginie, is shipped off to Paris where she is made unhappy by the corrupting effects of an artificial life. She is forced to learn reading and writing, to dress like an urban woman, to assume her title of *Comtesse*, to go to school, and when she proves recalcitrant to all this, she is sent back to her island home where she was once again to fall into the arms of Mother Nature. Unfortunately a violent storm breaks out as her ship nears the coast and her innate modesty prevents her from undressing when the vessel is about to break up on the rocks and, since her would-be rescuer cannot get her to land fully clothed, she loses the chance of being saved. Hence she is drowned, modesty and all. *Nature* in this book is not distinct from *God*. And apparently Bernardin de Saint-Pierre did not think of religion as part of civilization, but rather as an expression of a congenital piety on man's part. Both the children and their mentors go to church and trust in God. Their morality is not that of the animals nor hedonists, but rather that of the simple unquestioning child who 'instinctively' believes in and practises charity and the simple life. Their main occupation is garden-

ing, as Adam's was, at least according to the second chapter of Genesis, if not the first. Both are of course beautiful, friendly, and good in every respect. But at the same time Virginie's mother, Mme de la Tour, is uneasy lest Paul's manhood assert itself and that he unite with Virginie before she is of the proper age.[1] For Virginie begins to feel the stirrings of sexuality before Paul does. The analogy to the story of Adam and Eve is clear enough to be passed over, but maybe the death of Virginie is payment for her mother's distrust of Nature.

Nature-God (*Deus sive Natura*) is thus set over against Society. The author was not so ferocious an opponent of society as his master was, for the two children do live in a social group and assume duties towards their fellow men. But their society is 'natural' and not artificial and the sacred adjective atones for all the faults of the substantive. The moral then turns out to be: Whoever follows Nature will be happy and virtuous; whoever departs from her rule will be miserable. Should one object that a storm at sea is as natural as a sunny day, the answer, I suppose, would be that the word 'natural' includes the meaning 'beneficent'. That is, in reading Bernardin one should not forget the lessons of, for instance, his *Etudes de la Nature*, in which the presence of God and His purposes must both be thought of when ruminating about historical events, their causes and goals. This comes out fairly clearly in the epilogue to the story where the *Vieillard*, who takes the place of the *Vicaire Savoyard*, delivers the moral.

Paul, pretty badly shaken by the death of Virginie, wanders about disconsolate. The *Vieillard* preaches a long sermon to the boy which includes the following passage:

> Mon fils, la bienfaisance est le bonheur de la vertu; il n'en a point de plus assuré et de plus grand sur la terre. Les projets de plaisir, de repos, de délices, d'abondance,

[1] One is tempted to say that there is a psychological connection between the suspicions of Mme de la Tour and her social rank. She is of the nobility, whereas Paul's mother is of the lower orders. Like a good Rousseauist, Bernardin professes a contempt for titles.

de gloire, ne sont point faits pour l'homme, faible, voyageur et passager. Voyez comme un pas vers la fortune nous a précipités tous d'abîme en abîme. Vous vous y êtes opposé, il est vrai; mais qui n'eût pas cru que le voyage de Virginie devait se terminer par son bonheur et par le vôtre? Les invitations d'une parente riche et âgée, les conseils d'un sage gouverneur, les applaudissements d'une colonie, les exhortations et l'autorité d'un prêtre, ont décidé du malheur de Virginie. Ainsi nous courons à notre perte, trompés par la prudence même de ceux qui nous gouvernent.[1]

But even in this story where all the artifices of society are absent, where there are no clocks, no history books, no philosophic texts (p. 101), and where the children live as brother and sister, Virginie like Eve, 'douce, modeste, confiante', and Paul like Adam, 'ayant la taille d'un homme avec la simplicité d'un enfant' (p. 104), it is the influence of Nature and not the innate goodness of childhood that brings into being what happiness they have. Bernardin was not one of the strongest intellects in European letters, but at least he understood that children had to grow up, to live in the environment of other human beings, and to carry out the social duties of such a life. The book, aside from other qualities, is a defence of the simple life, not of extreme cultural primitivism.

With Pestalozzi the situation is somewhat different. For in him we have at least a glimmer of the cult of childhood. It has not yet reached the proportions which it attained in more recent times, but is nevertheless clearly discernible. The child in his opinion requires training, just as everyone does. But there are certain highly desirable traits of childhood which are innate. One of these is belief in God.[2] Faith in God, says Pestalozzi, 'is *not* the consequence of training and education; it is the consciousness of the pure and simple, who with innocent ear listen

[1] Edition of 1838 (Curmer), p. 279.
[2] I quote from *Pestalozzi's Educational Writings*, ed. J. A. Green with the assistance of Frances A. Collie, New York and London, 1912. I use this edition because it is compendious and avoids the prolixity of the author. My reference here is from 'Evening Hours of a Hermit', pp. 24 and 28.

to Nature's voice and know that God is their Father'. Historians
of philosophy will hear in these words an echo — or coincidental
repetition — of the seventeenth-century doctrine of innate ideas,
for the existence of God was listed by both the English Platonists
and Descartes as one of the primary unlearned ideas of the soul.
But none of them, as far as I know, ever said that innate ideas
were clear to babies. Just what Pestalozzi's faith in God asserts
is nebulous. One is not told whether it is faith that there exists a
God whose characteristics are undefined or whether it is the faith
that there exists a power in the universe which is protective of
mankind. But he does say that 'Nature's way of establishing faith
in God' is the discovery of joy in our days, fortification in suffering, and the certainty of the 'overwhelming preponderance of
good'. Out of this faith arise 'the feelings of fatherhood and broth-
erhood among men — the source of all righteousness' (p. 28).
He does not mean of course that the child is capable of putting
his faith into words. He has little use for words, as is clear from
his essay, 'Christopher and Elizabeth' (p. 42). The child is inherently a theist and philanthropist, in the etymological sense of
that term.

In the second place there is a moment when the child is 'completely pure'.[1] Unfortunately it is only that moment in which he
comes into the world. 'But', says Pestalozzi, 'no sooner is it there
than it is passed. It is gone at his first cry. From that time onwards
the child passes farther and farther away from this condition'.[2]
Consequently one can hardly think of a child as one comes upon
him in real life who retains his primordial innocence. The same is
true of the race. It too came into the world in a state of purity,
but Pestalozzi says that there is no record of such a state of affairs.
(One wonders why so religious an author should have overlooked
the record in Genesis). Regardless of the lack of a record, racial

[1] 'Enquiries concerning the Course of Nature in the Development of the Human
Race', p. 62.
[2] Cf. the selection from John Earle, p. 42 below and also Wordsworth's 'Ode on
Intimations of Immortality' where the poet speaks of the 'vision splendid' fading
away in manhood.

innocence lasted 'to the point at which [man] first knew of evil; his first wrong action, his first deception, began his degradation' (p. 63). This, like the child's immediate loss of purity, came very early in the history of mankind. Nevertheless, in his famous 'How Gertrude Teaches her Children' (p. 91), he points out that 'God has, happily, made children very resistant to these destroying agencies' — the errors and follies of grown-up men in stifling Nature in the heart and mind of a child — 'and this resistance is backed up by the realities of Nature in the midst of which they live'. Hence we may conclude, on the assumption that Pestalozzi is systematic in his thoughts, that the loss of purity and innocence does not entail the weakening of resistance to evil and that the purity in question lies in the factor of resistance. The educator, we are told, should remember this and attach more importance 'to the immeasurable influence of the child's own nature upon his education' (p. 91). One who does remember this will draw out of the child abilities which are potentially within him instead of trying to cram things that are alien to his nature into his immature mind. Learning, says Pestalozzi (p. 86), begins at birth. As soon as the child becomes sensitive to external impressions he begins to learn. 'To instruct men is nothing more than to help human nature to develop in its own way, and the art of instruction depends primarily on harmonizing our message and the demands we make upon the child with his powers at the moment' (p. 87). This is obviously an extraordinary extension of the lessons of Plato's *Meno*, but an extension which seemed more and more reasonable as the years went by, until in our own time all the arts became a form of self-expression, though neither 'the self' nor 'expression' were ever clearly defined. Since the powers in question are at first those of observation and communication (*ibid.*), these must be developed before the child is taught to read or spell. But since Nature cares only for the type and nothing for the individual, 'she cannot therefore come into harmony with the intellectual and moral nature of man' (p. 133). The individual, as an individual, then is non-natural, and probably supernatural, though

Pestalozzi is not explicit on that point. At any rate he distinguishes, as many others have done, between what he calls the heart and the head, and in this place he makes a parallel distinction between 'our sensory nature' and our moral and intellectual life. It is no secret that the greatest influence on Pestalozzi was Rousseau, both in *Emile* and the *Contrat Social*, and we see here in this insistance on man's dual nature the same distinction that was made by the *Vicaire Savoyard*. In short, in spite of his constantly repeated emphasis upon the necessity of following Nature, he recognizes that human nature too has to be followed. We happen to have an animal nature as well as a human nature and at moments Pestalozzi seems to be saying that the former is our sensory life. For he maintains that man's sensory nature must be subordinated to his 'spiritual' nature which is alone truly human. Our spiritual nature, though potentially present in everyone, has to be developed. And its development is in the long run the aim of education. But it begins to develop while the child is under the care of its mother, and in passages which deal with this period, as in the 13th Letter of 'How Gertrude Teaches her Children' (pp. 142f.), we find that love, confidence, gratitude, and then brotherly love are all awakened by maternal care. These are clearly emotional states and are directed towards another person. Whether they would be stimulated by anybody who was not one's mother or a substitute for one's mother, is not argued. Pestalozzi seems to believe that once actualized, these potential virtues would be exercised whenever appropriate. In his letters to the English educator, J. P. Greaves (Letter 11, p. 213), we are told that this is the difference between human beings and animals. Animals can go only so far as their instincts take them and the limit of their development is the frontiers of morality. Men, on the contrary, can transcend their instincts. The danger, however, is always present that when the child is no longer dependent on its mother's care, it will become self-assertive and indulge in 'sensuous enjoyment and self-will' (p. 146). At this point 'we can no longer trust Nature' (p. 147). The child's capacity for faith and

love are as the roots of a tree; it produces nourishment for the higher life.[1]

Not even the most fervent admirer of Pestalozzi would say that he was a precise expounder of his doctrines. Like Froebel he was a man of deep sympathy with children and tended to invest them with traits that would awaken a similar sympathy of the part of his readers. But since it was granted that these traits were but promises in the very young, an easy way out of the problem was to speak of them as potential. Both Pestalozzi and Froebel had suffered during their primary schooling. The former was left fatherless from the age of six; the latter motherless from that of nine months.[2] Both looked for a clue to the proper principles of education in what they called Nature and both thought of Nature as discoverable only in the rural districts. Both were interested in the observation of flowers and birds and the things which as a whole were later to be called 'Nature-study', but according to Froebel, Pestalozzi was too empirical and seemed 'in no wise to recognize the Divine element in science'.[3] His disciple, Barop, felt that Pestalozzi thought of children as 'receptive', whereas his master believed them to be 'creative', a word which by now has taken on an almost sacred character.[4] We are happily not concerned here with the educational theories and practices of these two men but rather with their ideas of the nature of childhood and in that they were in fundamental agreement. Both, without referring to Cicero, as far as I have discovered, nevertheless conceived of the child as the *speculum naturae*, and since they both admired nature

[1] 'Address to my House' (1818), p. 189. The botanical metaphor became popular in Germany during the eighteenth century. It appears with special force in Herder. The idea seems to have been that a plant has a power of growth 'within itself', much like Bergson's *élan vital*. Why this should not have been true of animals and men also, I do not know. One sees the influence of the metaphor in pedagogical theories which treat of learning as growth, propelled, so to speak, from within.

[2] See the *Life, Educational Principles, and Methods of John Henry Pestalozzi*, ed. Henry Barnard, New York, 1862, 3rd ed., p. 49, and the *Autobiography of Friedrich Froebel*, translated by Michaelis and Moore, Syracuse, N. Y., 1889, p. 3.

[3] *Autobiography*, p. 117.

[4] *Ibid.*, p. 129. For a contrast between Pestalozzi and Froebel, see the latter's autobiography, p. 111, note.

above art and urban civilization, they must be held partly responsible for the cult of childhood.

But meanwhile the cult had received reinforcement from *belles-
lettres* and theology. We shall now turn to those fields.

III

The poets of the seventeenth and eighteenth centuries, in so
far as they enter into our discussion, were men who had absorbed
religious ideas into their very being and when they wrote, they
utilized them as if they were part and parcel of the general stock
of knowledge. They had no need to argue about the innocence
of childhood, for everyone probably believed in it without defining it precisely. We find, for instance, in Racine's *Athalie* (Act
II, scene vii) the usual comments on innocence, on the intimacy
between the child and God, and Racine did not hesitate to put
into the mouth of little Joas, who was nine years old,[1] lines whose
truth needed no demonstration:

> Je suis, dit-on, un orphelin
> Entre les bras de Dieu jeté dès ma naissance.
> Et qui de mes parents n'eus jamais connaissance.

And when Athalie asks him who has taken care of him, he replies:

> Dieu laissa-t-il jamais ses enfants au besoin?
> Aux petits des oiseaux il donne leur pâture,
> Et sa bonté s'étend sur toute la nature.

These famous verses thwart the aims of the wicked queen. But
the very fact that they are spoken without any deliberate purpose

---

[1] Racine in the preface to this play says that he gave Joas nine years 'pour le mettre
en état de répondre aux questions qu'on lui fait'. He goes on to say, 'Je crois ne lui
avoir rien fait dire qui soit au-dessus de la portée d'un enfant de cet âge qui a de
l'esprit et de la mémoire'. But, as if he were a bit uneasy about this, he adds that 'il
faut considérer que c'est ici un enfant tout extraordinaire'. For another precociously
religious child, see the mediaeval *Filius Getronis* in which the boy, Adeodatus, reviles
the pagan god, Apollo:

> Deus tuus mendax et malus est;
> stultus, cecus, surdus et mutus est.

Quoted from Grace Frank, *The Medieval French Drama*, Oxford, 1954, p. 50.

to thwart her may have given them greater poignancy. And though Joas is indeed an *enfant tout extraordinaire*, yet there is enough similarity between him and an ordinary child in his very sincerity and lack of guile to make him a prefiguration of the twentieth-century child, all innocence and innate wisdom. The seventeenth century in France was not enthusiastic about childhood, as we have seen. But by the time of Rousseau the idea that childhood was more 'natural' than maturity reached even into the writings of Diderot. He, for instance, thought that children were closer than adults to *la poésie naturelle*, but he did not advocate that adults imitate them.[1] In fact it will be found, I think, that when French writers dealt with children, they were more likely to write about the sorrows of the child than about his happiness.

In English *belles-lettres* the situation is somewhat different. One of the earliest expressions of the cult of childhood that I have come across is in John Earle's *Microcosmographie* (1628). The opening sketch in this delightful work is called *The Child* and it runs as follows:

> [The Child] is the best copy of *Adam* before he tasted of Eve or the apple; and he is happy whose small practice in the world can only write this Character. He is nature's fresh picture newly drawn in oil, which time, and much handling, dims and defaces. His Soul is yet a white paper unscribbled with observations of the world, wherewith, at length, it becomes a blurred notebook. He is purely happy, because he knows no evil, nor hath made means by sin to be acquainted with misery. He arrives not at the mischief of being wise, nor endures evils to come, by foreseeing them. He kisses and loves all, and, when the smart of the rod is past, smiles on his beater. Nature and his Parents alike dandle him, and 'tice him on with a bait of sugar

---

[1] I owe this reference to Yvon Belaval, *L'Esthétique sans paradoxe de Diderot*, Paris, 1950, p. 146. Of course Diderot always was a great admirer of genius and genius is innate. Art v. nature, instruction v. genius, the primitive v. the civilized: it is easy to move on to maturity v. childhood. But Diderot did not take that step.

to a draught of wormwood... His hardest labour is his tongue, as if he were loath to use so deceitful an Organ; and he is best company with it when he can but prattle... The older he grows, he is a stair lower from God; and, like his first father, much worse in his breeches. He is the Christian's example, and the old man's relapse; the one imitates his pureness, and the other falls into his simplicity. Could he put off his body with his little coat, he had got eternity without a burthen, and exchanged but one Heaven for another.

This is an epitome of the whole story. At the risk of pointing out the obvious, I shall emphasize certain details in this sketch. First, the young child is a close copy of Adam before the Fall — one of Pelagius's errors by the way. Here the cultural and chronological primitivism of the writer coalesce. But it should also be noted that this anticipates what was to be said a century or so later: that the life of the individual recapitulates the history of the race. Thus when a man like Earle saw in the child prelapsarian Adam, nineteenth-century writers would see in him primitive man. Second, growing up is degeneration, at least morally. As in Hesiod's story of the Ages there is steady progress from good to bad, with the exception of the Age of Heroes, so in the life of the human being each degree of age brings new evil in its train. The baby is closest to God, the adult farthest from Him. Third, experience, instead of being a gradual fruition of the child's potentialities, is a defacement of his original purity. The metaphor of the white paper was to be repeated and popularized by John Locke (*Essay*, Bk. II, i. 2), but in him what was written on the paper was so much added to the stock of our ideas and a good bit of it was advantageous. To Earle, however, experience blackens the white paper of the soul and gives to the growing child nothing of moral value. Fourth, the child is both pure and simple and his soul, could it be decorporealized, would be like that of the inhabitants of Heaven. The purity of the child must

of course be freedom from sin; just what his simplicity consists in is more questionable. But one suspects that it is the simplicity of the Wise Fool or the Simpleton in some of Grimm's Fairy Tales, the youngest of three brothers who always turns out to be more intelligent than his elders. He is the child who appears in Andersen's *The Emperor's Clothes*.

Now there is nothing here that is very clear and Earle apparently thought that any reader would understand what he was talking about. He was not an historian of ideas, but an essayist. What he had to say was in all probability based on commonly accepted assumptions. He would believe, as his contemporaries no doubt did too, that newly born babies would go straight to Heaven if baptized; that they would remain in Limbo until presumably the Last Judgement, a special Limbo, the *Limbus puerorum* or *parvulorum* as it was sometimes styled. But there was also some reason to believe that they had inherited the sin of Adam and that, unless they had received grace from God bestowed for reasons which no human being could conjecture, they would be punished just as other sinners were punished. Hence when Calvinism became part of the Puritan tradition, as in the Massachusetts Bay Colony, the inherited wickedness, rather than the innocence, of childhood was emphasized. There was no question of a child's innocence of any sin committed by him personally. But nevertheless the New England ministers chose to overlook that in favour of man's guilty heritage. John Hersey, for instance, lowered the babe down to a position close to that of Satan. 'Break their wills', he says, 'betimes. Begin this work before they can run alone, before they can speak plainly, or speak at all. Whatever pain it costs, conquer their stubbornness; break their wills if you would not damn the child... Therefore let a child from a year old be taught to fear the rod, and cry softly'. No explanation was given of the adverb. 'At all events at that age make him do as he is bid, if you whip him ten times running to do it; let none persuade you it is cruel to do this'. This may be said to be written from the Augustinian point of view, the point of view from which man could be said

to have lost his *posse non peccare*. On the opposite side of the dispute one could take only the Pelagian point of view and that had been condemned as early as 418 at the Council of Carthage. The condemnation of Pelagianism does not seem to have worried some writers and it is noticeable that the cult of childhood flourished more in Protestant than in Catholic communities. Even the formidable Calvinism of the Massachusetts Bay Colony died out to all intents and purposes when the Colonies gained their freedom and at the same time easier living conditions.

No account of the literary history of this idea would be complete if some mention were not made of Thomas Traherne (ca. 1637-74). In his *Centuries of Meditations*, which was not, however, published until 1908, he drew a bright picture of a happy childhood, his own. Since this remained so long in manuscript, it serves here only as an example of the criteria of infantile happiness. The one item that fits in with our idea most neatly is in his sentence, 'I knew by intuition those things which since my Apostasy, I collected again by the highest reason', for there is no question that Traherne thought it better to know intuitively than rationally. At least it required less effort. 'I seemed', he continued, 'as one brought into the Estate of Innocence. All things were spotless and pure and glorious'. This resembles Earle's comment on the child as the image of the unfallen Adam. And in fact Traherne goes on to say, 'I saw all in the peace of Eden; Heaven and Earth did sing my Creator's praises, and could not make more melody to Adam than to me. All Time was Eternity, and a perpetual Sabbath. Is it not strange, that an infant should be heir to the whole world, and see those mysteries which the books of the learned never unfold?' This, if taken seriously, and Traherne surely did take it seriously, is an exaltation of the child's wisdom to a point which it attained only in the childhood of Jesus. Whether Traherne was suggesting, if not saying, that the child repeated the childhood of the Second Adam as well as of the First Adam, I do not know. What is certain is that as the child grows up, he loses his intuitive grasp of supernatural truths, just as Adam

by eating of the fruit of the Tree of Knowledge paradoxically enough lost the only knowledge that was of importance.[1]

At this point we must recall a curious bit of iconographical history, the details of which must be filled in by more expert hands. I refer to the representation of angels as little children. These children were originally the *Erotes* of Hellenistic art. They are dispersed from Antioch to Pompeii. In their primordial function they hardly served as symbols of purity or innocence and when they were resuscitated in the Italian Renaissance as the *putti*, they sometimes were attendants of Venus, but also fluttered about the Blessed Virgin. In time these *putti* became known as cherubs, though they were far removed from the Cherubs of the Bible. The Lord could hardly ride upon these cherubim as in David's song (II Samuel xxii, 11) nor could have they been placed on either side of the Ark as described in Exodus xxv, 18ff. The Biblical Cherubim were in fact rather terrifying beings, having in Chaucer's words 'a fyr-reed cherubinnes face'.[2] The mediaeval painters seem to have retained the notion that angels of any rank should be represented as adults, young but not babies. Fouquet, Mantegna, and Jean Bellegambe turn the angels into *putti* and this in no frivolous spirit.[3] Mantegna's 'Adoration of the Shepherds' in the Metropolitan Museum in New York, Fouquet's 'Madonna and Child' in Antwerp, Borgognone's 'Assumption' also in the Metropolitan, as well as scores of other paintings contain childlike angels. One of the most famous examples is the two *putti* at the foot of the Sistine Madonna of Raphael. Sometimes these cherubs are bodiless, which complicates the iconography, for the lack of a body, as in the dome of the Baptistery in Florence (thirteenth century), may, but I cannot say 'does', mean that spiritual beings either have no bodies, or that their glorified bodies cannot be paint-

---

[1] See Monica Kiefer, *American Children through their Books, 1700-1835*, Philadelphia, 1948, esp. chapters entitled 'War with the Devil' and 'The Art of Decent Behaviour'.

[2] For the representation of the Cherubim in art, see the article *s. v.* in *Reallexikon zur deutschen Kunstgeschichte*, vol. iii, 1953ff.

[3] For a short history of the *putto*, see the section *Kinderengel* of the article *Engel*, in *Reallexikon zur deutschen Kunstgeschichte*, vol. v, 1963ff., col. 464ff.

ed, or that the body is too closely associated with sin to be attached to an angel's head. The bodiless cherub found its last resting place on the tombstones in New England cemeteries, where wings frequently sprout from the sides of the head rather than from the shoulders as in Hellenistic genii.

There is a possible source for the use of children as angels in the mediaeval drama, but the evidence is too uncertain to prove anything definitely. Gustave Cohen in his *Histoire de la mise en scène* (ed. of 1951, p. 37), maintains that the role of angel was *souvent* given to children. Later in the same volume (p. 208), he is more forthright about this and says that children played such roles *presque toujours* and adds that 'dans les rubriques le mot "Angeli" est souvent remplacé par "pueri" '. But on the other hand Karl Young in his *Drama of the Mediaeval Church*, while indicating a few clear examples of boys playing the roles of angels, also shows that they played those of the Maries and Apostles. In short, one suspects that boys were used wherever it was desired to use them,[1] with due regard for propriety. Angels would have to be played by males; that much is certain. And if it was sometimes thought desirable to choose young boys to play their part, one suspects it may have been because the boy before puberty was also believed to be sexually pure. But all this is conjecture. When the bodiless 'cherub' was placed upon tombstones, it was to represent the soul of the deceased and not a member of the angelic hosts.[2] It is

---

[1] Passages referred to will be found in vol. I, p. 244; II, pp. 12 and 14. This should be supplemented by Grace Frank, *op. cit.*, p. 33 where a boy, 'representing an angel', announces the birth of Christ. But Mrs. Frank does not say that children were usually cast in angelic roles.

[2] Professor Allan Ludwig of Yale is preparing a work on New England tombstones. He writes me that the bodiless heads with wings 'in New England...probably stood for purity, innocence and the childlike character the soul was supposed to have'. Professor Ludwig was also kind enough to send me some lines written by Edward Taylor (1642-1729), a clergyman of Westfield, Massachusetts, which express the hope that he will be reborn as a 'babe in Christ':

> Make mee thy Babe, and him my Elder Brother.
> A Right, Lord Grant me in his Birth Right high.
> His Grace, my Treasure make above all other.

The first line recalls George Herbert's conclusion to *The Collar*. On the interpretation of this difficult poem, see Rosalie L. Colie, '*Logos* in the Temple', *Journal of the Warburg and Courtauld Institutes*, vol. XXVI (1963), p. 333.

impossible to tell whether they are all childlike in face since the carvings are often so crude that they give little information about the age of the head represented. The situation then is about as follows: in the fifteenth century or thereabouts painters began to use *putti* as angels and this may have been because they thought of little children as more symbolic of angelic purity than men would have been. At the same time, there is no need to insist, certain angels, Gabriel for instance, were almost always represented as adults.[1]

There is a curious parallel to the identification of children with angels in none other than the Swedish seer, Swedenborg. In his *Heaven and its Wonders and Hell*[2] he dwells, as so many others have done, on the innocence of childhood, an innocence whose nature 'is thus far unknown' (p. 213). It is, however, visible to the eyes, 'as seen in the face, speech, and movements, particularly of children'. Though it is not 'genuine innocence' (par. 277), 'nevertheless one may learn from it what innocence is'. Children do not know the difference between good and evil, truth and falsity. They have no prudence, no purposiveness, no power of deliberation. This is what Swedenborg calls external innocence, 'for their minds are not yet formed, the mind being understanding and will and thought and affection therefrom' (*ibid.*). In short the knowledge of good and evil, true and false, is based on intellectual powers of discrimination and a child is naturally or inherently a being who lives without the need of such powers. But once he

[1] Since completing this essay I have received, thanks to Professor E. H. Gombrich, a photograph of an onyx gem in the British Museum, published in Dalton's *Catalogue of Christian Antiquities*, pl. III, no. 104, dated 9th century. The gem shows a standing draped figure of a woman and a winged naked child beside her. The inscription, which is in Greek, identifies the women as the Mother of God and the winged child as the angel Gabriel. Above the figures is the opening of the *Ave Maria*. It is also said to be an ancient gem with an inscription added in the 8th or 9th century. See the *Dictionnaire d'Archéologie Chrétienne et de Liturgie*, t.I, 2e partie, art. *Anges*, fig. 610. This is the only case that I have found of an Annunciation in which Gabriel is represented by a *putto*. But if the gem is ancient and the inscription 9th century or thereabouts, then clearly the substitution of a winged Cupid for the Archangel could not have been so strange as it might seem.

[2] I use the translation issued by the Swedenborg Foundation, New York, n.d. The first edition, in Latin, was dated 1758.

dies in childhood, regardless of whether he is baptized or not, of whether parents are pious or not, he 'is received... by the Lord and trained up in heaven and taught in accordance with Divine order, and imbued with affections for what is good, and through these with a knowledge of what is true; and afterwards as he is perfected in intelligence and wisdom is introduced into heaven and becomes an angel' (p. 270, par. 329). Children consequently are not the highest form of human life, but the highest form this side of Paradise. The sentiment is not unlike that expressed by Sir Aubrey de Vere (1788-1846) in the opening lines of his sonnet on the Children's Crusade:

> All holy influences dwell within
>   The breast of Childhood: instinct fresh from God
>   Inspire it, ere the heart beneath the rod
> Of Grief hath bled, or caught the plague of sin.

One might imagine that the next step forward in this story would have been taken by William Blake. But actually, except for the distinction made by him between innocence and experience in the titles of two sets of poems, a distinction which was to be utilized later by Péguy, there is very little that he contributed to the development of the cult of childhood. To begin with, his influence was negligible until another generation had come into prominence and in the second place his vision of childhood was somewhat like a nightmare. He seemed more interested in the sorrows of childhood, lost children, chimney sweeps, little black boys, the child Orc 'bathed in springs of sorrow', than either in the happiness of innocent babes or in their wisdom. It is rather to Wordsworth to whom we must turn for the next development of our idea.

One would of course immediately think of parts of the *Ode on Intimations of Immortality* as perfect expressions of the themes which are in question. For not only does Heaven lie about us in our infancy, but as the child grows, the shades of the prison house begin to close about him. So the claims of the body begin

to gain ascendancy in Earle's paragraph. That great stanza of
Wordsworth's is a history of the individual's moral degeneration
up to the time when we reach the stage of manhood at which the
'vision splendid' fades entirely. The supposed freedom of the child,
freedom presumably from the inevitable yoke of years, fits into
the picture of holy childhood which we have been sketching.

But Wordsworth's attitude towards childhood varied. At times,
as in the verses just referred to, he idealized that period of life,
but at others he expressed simply the usual delight which most
people are likely to feel in the sight of children. One can infer
little, for instance, of a philosophic nature from his modern
version of the *Prioress's Tale*, in spite of the child's abnormal
interest in religion and his zeal in learning the *Alma Redemptoris*.
Again, in his *Two Thieves*, which deals with two far from moral
youngsters, he expresses nothing more, as far as our subject is
concerned, than a vague anti-intellectualism, though he suggests
that children have an innate ability for poetry in the lines

> O now that the genius of Bewick were mine,
> And the skill which he learned on the banks of the Tyne!
> Then the Muses might deal with me just as they chose,
> For I' d take my last leave both of verse and of prose.
>
> What feats would I work with my magical hand!
> Book-learning and books should be banished the land...

There may be a thin echo of *Emile* here, but it is very thin and in
reality committed Wordsworth to next to nothing. As a matter of
cold fact the poem was included in 'Poems referring to the
Period of Old Age', and the notion that is suggested of innate
genius is more clearly expressed in the manuscript version of the
stanzas than in the printed version.[1]

In his lines on the 'Influence of Natural Objects in Calling
forth and Strengthening the Imagination in Boyhood and Early

---

[1] See *The Poetical Works of William Wordsworth*, ed. William Knight, London,
1896, vol. II, p. 60, n. 1, where the comparison is between 'the poet who lives on the
banks of the Tyne' and Sir Joshua Reynolds, to the latter's disadvantage.

Youth', there is a somewhat clearer suggestion of the diluted Neo-Platonism of the *Ode*, for one suspects that the 'Wisdom and Spirit of the Universe', the 'Soul' that is the 'Eternity of thought', is the *Anima mundi*. The poem is autobiographical, as most of Wordsworth's are, and in it he attributes to this Soul the power of giving 'forms and images a breath/And everlasting motion'. He continues, addressing this spirit,

> ... not in vain
> By day or star-light, thus from my first dawn
> Of childhood didst thou intertwine for me
> The passions that build up our human soul;
> Not with the mean and vulgar works of Man:
> But with high objects, with enduring things,
> With life and nature: purifying thus
> The elements of feeling and of thought,
> And sanctifying by such discipline
> Both pain and fear—until we recognize
> A grandeur in the beatings of the heart.

That this is not merely autobiographical but contains an idea which might be applied to other poets may be seen in his sonnet beginning

> A Poet! — He hath put his heart to school,
> Nor dares to move unpropped upon the staff
> Which Art hath lodged within his hand — must laugh
> By precept only, and shed tears by rule,
> Thy Art be Nature...

But it would be folly to attempt to reconstruct a consistent philosophy out of the works of William Wordsworth. In spite of his writing the famous line on the child's being father to the man, anticipated by Milton, Dryden, Chatterton, and even Pope, he yet could wonder what age would bring forth in the six-year-old Hartley Coleridge, as well he might in view of the boy's future history. Earth and heaven are fused in the child's experience and, if Wordsworth remembered these early verses in later years, he

must have wondered why the father of the man had given so few
signs of what he was to be. Yet in *The Prelude* he reiterates the idea
that the fusion of heaven and earth is characteristic of boyhood.
One has only to think of the lines in the first book (581 ff.):

> ...oft amid those fits of vulgar joy
> Which, through all seasons, on a child's pursuits
> Are prompt attendants, 'mid that giddy bliss
> Which, like a tempest, works along the blood
> And is forgotten; even then I felt
> Gleams like the flashing of a shield; — the earth
> And common face of Nature spake to me
> Rememberable things.

These, he says, might have come by chance,

> .... yet not in vain
> Nor profitless, if haply they impressed
> Collateral objects and appearances,
> Albeit lifeless then and doomed to sleep
> Until maturer seasons called them forth
> To impregnate and to elevate the mind.

Book V, which deals largely with schooling, recalls the lines referred to above in that it emphasizes the triviality of book-learning
and the profundity of that 'wise spirit' who is 'at work for us...
most prodigal of blessings, and most studious of our good,/Even
in what seem our most unfruitful hours' (lines 360-63).

Wordsworth's contemporary, Emerson,[1] had been subjected to
the same Platonistic-post-Kantian-mystical influences. He man-
aged to add a few fresh touches to the picture by attributing to children an ability to distinguish 'false reasons which their parents
give them in answer to their questions, whether touching natural
facts, or religion, or persons'. They are, he says (*Uses of Great
Men*, p. 33), secure 'from infusions of evil persons, from vulgarity

---

[1] I use the Riverside Edition of Emerson's works. To avoid the multiplication of
footnotes, I have put the name of the essay quoted and the page in parentheses after
each quotation. The quotation above is from 'Worship', p. 128.

and second thoughts'. His essay on 'Domestic Life' opens with a dithyramb on childhood, quoting Milton's lines from *Paradise Regained* (IV, 220) to the effect that 'childhood shows the man, as morning shows the day'. This praise of the child plays largely on the supposed pleasures of the child's imagination and on how the imagination invests all experience with magic. In 'Self Reliance' (pp. 49f.) we find that children's minds 'being whole, their eye is as yet unconquered and when we look in their faces we are disconcerted. Infancy conforms to nobody; all conform to it; so that one babe commonly makes four or five out of the adults who prattle and play to it'. Non-conformity was one of Emerson's pet slogans and in his essay on *Character* (p. 104) in which character is said to be 'nature in her highest form', he announces—for Emerson always announced and never argued his theses—that this masterpiece, as he called it, is always best seen in children upon whom no hands but nature's have ever been laid. The infant non-conformist is a paradigm for the adult who would become a masterpiece of character in his turn and there is even a hint of what was beginning to be a popular belief, that children represent the first stages of civilization in their sincerity and spontaneity, traits which are also found in the 'first lines of written prose and verse of a nation'. On the other hand, in his essay on 'Nature' he admits, but by oversight, that children do believe in an external world, but 'with culture' its status as mere appearance will be apprehended. Children, though oracles, can be fooled.

In nineteenth-century France the clearest and best known example of a poet's adoration of childhood comes from Victor Hugo's *L'Art d'être Grand-père*. Hugo was never restrained in either his loves or his hates and in this collection of verses he is his usual exuberant self. In his *Epopée du Lion*,[1] it will be recalled, a naked child, a girl, is able by the power of her mere presence to tame a lion which is carrying in its mouth her brother. 'Seule et nue', she was singing in her bed, 'car l'enfant chante même alors que tout se tait'. She sees the lion and cries, *Frère, mon frère*,

---

[1] My quotations are from the *Oeuvres Complètes*, Paris, 1914, vol. VIII.

> et debout, rose dans la lumière
> Qui la divinisait et qui la réchauffait,
> Regarda ce géant des bois, dont l'oeil eût fait
> Reculer les Typhons et fuir les Briarées.
> Qui sait ce qui se passe en ces têtes sacrées?
> Elle se dressa droite au bord du lit étroit,
> Et menaça le monstre avec son petit doigt.
> Alors, près du berceau de soie et de dentelle,
> Le grand lion posa son frère devant elle,
> Comme eût fait une mère en abaissant les bras,
> Et lui dit; Le voici. Là! ne te fâche pas!

Thus the innocence of the child conquers the ferocity of the beast. But Hugo was capable of going much farther than this. In *Le Syllabus* (p. 531 f.) we find that when he bends over to listen to the *âme pure* of a child, he seems to penetrate into heaven.

> Car vous étiez hier, ô doux parleurs étranges,
> Les interlocuteurs des astres et des anges,
>   En vous rien n'est mauvais;
> Vous m'apportez, à moi qui gronde la nue,
> On ne sait quel rayon de l'aurore inconnue;
>   Vous en venez, j'y vais.

This vague reminiscence of the doctrine of pre-existence is repeated in *Les Enfants Pauvres* (p. 538):

> Prenez garde à ce petit être;
> Il est bien grand, il contient Dieu.
> Les enfants sont, avant de naître,
> Des lumières dans le ciel bleu.
>
> Dieu nous les offre en sa largesse;
> Ils viennent; Dieu nous en fait don;
> Dans leur rire il met sa sagesse
> Et dans leur baiser son pardon...
>
> La misère de l'innocence
> Accuse l'homme vicieux.
> L'homme tient l'ange en sa puissance...

There was of course a trace of mysticism in Hugo, a mysticism which at times led him to praise the imperfection of the universe, as in *Contemplations* (VI), and at others induced a state of *docta ignorantia*. In his *Encore de l'Immaculée Conception* he is willing to go so far as to read into the gurgling of babies a profound sense which he realizes that he cannot articulate.

> Le babil des marmots est ma bibliothèque;
> J'ouvre chacun des mots qu'ils disent, comme on prend
> Un livre, et j'y découvre un sens profond et grand,
> Sévère quelquefois.

He ends with a contemptuous reference to the dogma of original sin, a dogma which could hardly be accepted by one who thought that babies came fresh from their conversations with stars and angels into this *bas-monde*. But perhaps one of the most interesting of his poems on childhood is *Les Griffonages de l'Ecolier* in which the boy Charle (*sic*) tired of the text of his school-book, improvises with his pen all over the margins and between the lines. I say that this is specially interesting since it is the only poem in the collection, as far as I have been able to discover, which deals with the artistic genius of the child. As in many of Hugo's poems, rhyme is the rudder of his verses, but, regardless of that, or possibly because of it, his alexandrines reveal something which two generations later was to become a source of admiration for infantile drawing.

> Dans un livre, partout, en haut, en bas, des fresques,
> Comme on en voit aux murs des alhambras moresques,
> Des taches d'encre, ayant des aspects d'animaux,
> Qui dévorent la phrase et qui rongent les mots,
> Et, le texte mangé, viennent mordre les marges...
> Ce chevreau, le caprice, a grimpé sur les vers.
> Le livre, c'est l'endroit; l'écolier, c'est l'envers.
> Sa gaîté s'est mêlée, espiègle, aux stigmates
> Du vengeur qui voulait s'enfuir chez les sarmates.

Les barbouillages sont étranges, profonds, drus.
Les monstres! Les voilà perchés, l'un sur Codrus,
L'autre sur Néron. L'autre égratigne un dactyle.
Un pâté fait son nid dans les branches du style...
Partout la main du rêve a tracé le dessin;
Et c'est ainsi qu'au gré de l'écolier, l'essaim
Des griffonages, horde hostile aux belles-lettres,
S'est envolé parmi les sombres hexamètres.
Jeu ! songe! on ne sait quoi d'enfantin, s'enlaçant
Au poëme, lui donne un ineffable accent,
Commente le chef-d'oeuvre, et l'on sent l'harmonie
D'une naïveté complétant un génie...
Grâce à lui, ce vieux texte est un lieu singulier
Où le hasard, l'ennui, le lazzi, la rature,
Dressent au second plan leur vague architecture.
Son encre a fait la nuit sur le livre étoilé.
Et pourtant, par instants, ce noir réseau brouillé,
A travers ses rameaux, ses porches, ses pilastres,
Laisse passer l'idée et laisse voir les astres...

Unfortunately for Charle, a monitor who does not share Hugo's
enthusiasm for dark hieroglyphics, arrives, sees the book, and
gives the boy a thousand verses to copy. Thanks to Hugo's fame,
Charle has won out and historians see in his scribbling an anticipation of what was to be the art of the Unconscious in a later century.

One has to wait until the twentieth century to find anything
more resembling this in French poetry. But in Charles Péguy's
Le Mystère des Saints Innocents we find something parallel in
sentiment if not in form. This poem is what the seventeenth century would have called a paradox in the sense that it denies the
common opinion of mankind:

Or c'est l'innocence qui est pleine et c'est l'expérience qui est
vide.
C'est l'innocence qui gagne et c'est l'expérience qui perd...

C'est l'innocence qui sait et c'est l'expérience qui ne
    sait pas...
Voilà, dit Dieu, ce que j'en fais de votre expérience.[1]

Written on the eve of the first World War, this poem should be
interpreted more as a protest against the complacency of adults
who were unaware of the misery that lay alongside their wealth
and comfort than as a hymn to childhood. It is an expression of
its author's disgust directed against men who in spite of experience were guided into folly. It is a cry of delight in the beauty of
innocence and candour rather than a statement of philosophic
import. Nevertheless it is also a plea for recognizing that in the
supposed innocence of childhood there is a direct and intimate
communication with God. And it may not be irrelevant to point
out that the poem ends with a paradox which has a profounder implication than appears in the simplicity of the words, spoken by God:

> ...je ne connais rien de si beau dans le monde
> Que cet enfant qui s'endort en faisant sa prière
> (Que ce petit être qui s'endort de confiance),
> Et qui mélange son *Notre Père* avec son *Je vous salue Marie*.
> Rien n'est aussi beau et c'est même un point
> Où la Sainte Vierge est de mon avis.
> Là-dessus.
> Et je peux bien dire que c'est le seul point où nous soyons du
> même avis. Car généralement nous sommes d'un avis contraire.
> Parce qu'elle est pour la miséricorde.
> Et moi il faut bien que je sois pour la justice.

The thought accords, as everyone knows, with one strain in the
Christian tradition, though perhaps not the dominant strain, and
I shall not labour it.[2]

---

[1] From *Cahiers de la Quinzaine*, XIII, 12 (1912), pp. 185 ff.

[2] The paradoxical nature of man's condition may be said to reduce to his loss of
innocence and his retention of the need for choosing the right. If we are doomed
to choose the worse, then only *misericordia*, not justice, will save us. Cf. Dante's
Purgatorio, Canto XVI; but Dante is not praising man in lines 85-93 where he speaks
of the 'fanciulla/che piangendo e ridendo pargoleggia', etc.

The most surprising admirer of childhood is Swinburne, for
who would have imagined that the author of *Laus Veneris*
would have had anything good to say about innocence? Yet his
poems on childhood, and indeed on babies, reach a point of ecstatic enthusiasm which becomes ludicrous. Children, as late as
eight years of age, are 'sinless as the spring' ('Eight Years Old');
their smiles gladden eyes and ears, 'yet sometimes sweeter than
their words or smiles are even their tears';[1] they are 'gods in
heavenly kindness' ('A Child's Thanks'); they cannot be deceived
or flattered, and they always 'know the truth' ('Cradle Songs', IV).
The genealogy of this *topos* is now fairly clear.

But in *belles-lettres* it was undoubtedly George Eliot's *Silas
Marner* which did most to spread the idea that the very presence
of a child was a beneficent influence. We have, to be sure, lines like
those of Landor

> Around the child bend all the three
> Sweet Graces: Faith, Hope, Charity.
> Around the man bend other faces:
> Pride, Envy, Malice, are his Graces.

And there are several well-known verses of Tennyson, such as

> O, the child too clothes the father (*Locksley Hall*, 91)

or the beginning of *Willow Water* which speaks of the 'child's heart
within the man's.' But *Silas Marner* (1861) is perhaps the first
novel written by a serious author in which the child becomes
a redeemer simply because of its childlike nature. There is no need
to spin out the details of this story, for almost all Americans, if
not Britons, were made to read it in school.[2] The point that
interests us is that little Eppie's mere presence in the home of the
crabbed and avaricious Silas was enough to convert him back to

---

[1] See 'A Child's Pity' which tells how a child wept over the death of a mother-
crocodile whose children were left orphaned.

[2] There are fifty-three editions of this book, many of which were published 'for
class-room use', listed in the catalogue of the Library of Congress for 1943. British
children may have escaped it.

the ways of his moral youth. She is described as 'a creature of endless claims and evergrowing desires, seeking and loving sunshine and living movements; making trial of everything, with trust in new joy, and stirring the human kindness in all eyes that looked on her'.[1] Whereas Silas had hoarded gold pieces before her coming and had taken his greatest delight in counting them over at night, now her golden curls replaced them. (It will be recalled that his sovereigns had been stolen). At the end of the fourteenth chapter we find the significant words, 'In the old days there were angels who came and took men by the hand and led them away from the city of destruction. We see no white-winged angels now. But yet men are led away from threatening destruction: a hand is put into theirs, which leads them forth gently towards a calm and bright land, so that they look no more backwards; and this hand may be a little child's' (p. 184). And a little child shall lead them. Unfortunately for those who seek Biblical authority, the antecedent of 'them' in Isaiah XI, 6 would appear to be the wolf and the lamb and the leopard and the young lion. George Eliot was not so frivolous as not to prepare the way for Silas's conversion from gold sovereigns to golden curls, for in the opening chapters of the book he is described as an upright Godfearing young man deceived in love. In short the change operated by Eppie was not so miraculous as it might seem. Eppie was not the revolting little prig who came into the world with Elsie Dinsmore and her kind, but then George Eliot was not Martha Finley either. The influence of a little child, however, was not confined to goodygoody books. Dickens invented David Copperfield, that angelic boy, along with a number of other saintly children, and Hans Andersen the unnamed child who saw that the Emperor was naked. There followed through the nineteenth and twentieth centuries a procession of child-heroes and heroines. We all remember characters like Poil de Carotte, Maisie in *What Maisie Knew*, Balzac's Louis Lambert, and lesser lights such as Little

---

[1] Quoted from *The Complete Works of George Eliot*, New York, n.d., (Harpers), vol. XVI, p. 175 f.

Lord Fauntleroy and his literary cousin, Sara Crewe, the children in *Little Men* and *Little Women*, the Prince in Mark Twain's *The Prince and the Pauper*, the family in Kenneth Grahame's *Dream Days* and *The Golden Age*, and of course, though far from being of the same kith, Mrs. Stowe's Little Eva. There were even a few bad boys, created by Mark Twain, Kipling, Booth Tarkington, and Thomas Bailey Aldrich, and recently children have appeared as downright wicked in the works of Richard Hughes and William Golding. But these authors are exceptions, though they may have started a new trend. It would require, however, a volume even to list the books that have appeared in English alone centring about children.

The climax of all this is to be found in Ellen Key. She made one of the few prophecies about cultural history which have come true, in her famous book, *The Century of the Child*. The English translation of that work came out in 1909 and went through three editions. It was translated into several languages at the time of its publication, perhaps because it attained a certain notoriety through its advocacy of 'free love'. But since Ellen Key based her conclusions on the Law of Recapitulation, we had best relate some of the early history of that law as it applied to human children before discussing Ellen Key.

IV

From the time of St. Augustine down, mediaeval writers conceived of human history under the metaphor of ages,[1] each age corresponding to one of the days of creation and also to a period of human life. In general such accounts list six ages, the first corresponding to infancy, the second to boyhood, the third to youth, the fourth to maturity, the fifth to middle age, and the sixth to senility, though there were occasionally slight variations. This metaphor gave thinkers of later date a background against which they could, if so disposed, appraise each age, seeing history either

[1] *Essays on Primitivism...in the Middle Ages* discusses this theme in some detail and gives a list of passages in which it is utilized, pp. 177 ff.

as growth and decline ending in death, or as an undulatory process swinging from good to bad and back again. To see human
history as if it were the biography of a single individual became a
commonplace. But it is interesting to observe that St. Augustine,
for instance, did not think of the first age, that of infancy, as the
best. On the contrary, it is the fourth age, that of youth, which is
the best, for as on the fourth day of creation the sun and the moon
were made, so in the fourth age there arose King David and the
splendour of his reign. If any period of an individual's life were
to be praised above all others, it would be youth, not childhood.
There was little if anything in the mediaeval account of the ages
that would turn a man's eyes back with longing to the beginning.
And this in spite of the story of Eden and man's primordial
happiness. That came about later when the Law of Recapitulation
was taken seriously.

The Law of Recapitulation has had two forms, one biological
and the other sociological or psychogenetic, as one sees fit to name
it. The former was frequently formulated in the simple statement
that ontogeny recapitulates phylogeny. It was maintained that
the embryological development of members of any species repeated the evolution of the genus — and even the family — to
which the species belonged. Thus human embryos went through
all the stages from fertilized egg to primate. The idea itself goes
back into the eighteenth century and was utilized by Darwin
as partial evidence of his theory of the evolution of organic
species.[1] The second formulation of the law asserted a parallelism
between the child and primitive man, a parallelism which is
obviously psychological, not physiological. The savage was
supposed to think like a child and many of the characteristics of
children's arts and beliefs about the world were held to be like
those of savages. Though no-one really knows that any group of

[1] See Jane Opperheimer, 'An Embryological Enigma in the *Origin of Species*', in
Forerunners of Darwin, 1745-1859, ed. by Bentley Glass, Owsei Temkin, and William
Straus, Jr., Baltimore, 1959, p. 293. For Darwin's use of this law, see *Origin of Species*,
chapter 14, where he quotes von Baer, whose *Ueber Entwicklungsgeschichte der Thiere*,
(1828-1837), is generally thought to be a pioneer work in this field.

human beings is literally primitive, in the sense that all men now existent evolved from a social and psychological condition similar to that of the group under consideration, the idea that some savages or pre-literates are chronologically at a standstill at just that point of development at which civilized people were at some prehistoric date, is still widely held, in spite of the efforts of some anthropologists to combat it. We speak of primitive religion, primitive science, primitive art, primitive society, as if we were talking about observable phenomena and not about conceptual constructs. The use of the adjective 'primitive' would be harmless and doubtless is harmless when made by anthropologists, but since it carries along with it a eulogistic connotation, it should be employed with the greatest care. It is easy to see that if one spoke of the beliefs of the Trobriand Islanders or Hottentots, there would be little temptation to read into them anything that we ought to emulate, but if one speaks of the primitive mind and its beliefs, the situation is quite different.

With this little sermon ended, we need only to point to a few examples of ideas which could have led to an identification of the child with primitive men. To begin with we have a celebrated outline of history in Vico which puts the emphasis upon growth and development from worse to better. Humanity, according to Vico, is as a whole moving in a definite direction. History began with the age of the gods, proceeded into the age of heroes, and ends with the age of humanity.[1] Each of these ages has certain features which are common to its customs, laws, forms of government, language and culture. The first of these, the religious, was the least rational and the most poetic. That the people of this age spoke in poetry became a commonplace later on. In it all things were supposed to be animated by spirits or gods. The notion that our primordial ancestors were not so rational as we did not make them more estimable in Vico's eyes, for he was not an anti-intellectualist, but it did at a period when the reason began to be dispraised. In the heroic age men gave up the animal

---

[1] *Scienza nuova*, Bk. IV, introduction (edition of 1744, p. 769).

life of their forebears, thought themselves to be descended from Jupiter, King of the gods, were choleric and pugnacious in the manner of Achilles, were aristocratic in government, the *aristoi* being defined as the strongest. The humane age, our own, is intelligent, modest, benign and reasonable, following the laws of conscience, reason, and duty. It believes in the equality of intelligent citizens, and governs by means of argument, not force.

The similarity between the development of a nation and that of an individual is obvious. Children had frequently been said to be best instructed by means of fables and narratives; they were usually thought of as pre-rational; and in view of the prevalence of nursery rhymes, it may be concluded that adults thought them more sensitive to the charms of verse than to those of prose. That the human race first expressed itself in poetry was no invention of Vico. It is a commonplace of Renaissance literary criticism, reiterated, with modifications, by David Hartley five years after Vico, though probably not because of Vico.[1] And all agreed that poetry and fable were characteristic of childhood. The adolescent again has often been described as rough and ready, feeling his oats, eager for a fight, aggressive, passionate, and less calm and rational than the mature man. The culmination of the individual's development in rationality is again harmonious with popular belief so that, though Vico did not emphasize the analogy between the life of an individual and that of society, it seems to lie behind his thinking as a basic metaphor. The same is true of Herder, though he is less given to attributing to all nations the same fate. Each has its own personality which lies *in potentia* in the *Volksseele*. Vico was more impressed by the parallelism in national developments; Herder, who wished above all to liberate the Germans from the prestige of Mediterranean ci-

[1] See, for example, Boccaccio, *De Genealogiis deorum*, xiv, 8-9, translated by C.G. Osgood, *Boccaccio on Poetry*, Princeton 1930, pp. 39-42, and the references given in the notes thereto. See also Osgood's introduction, pp. xxix ff., esp. xlv. For Fontenelle, see his *Digression sur les Anciens et les Modernes*, and for Hartley, his *Observations on Man*, as quoted in Basil Willey, *The Eighteenth Century Background*, London, 1940, pp. 24 and 147 respectively.

vilization, was satisfied to assign as the end of all history the actualization of *Humanität*. But just what *Humanität* consisted in he never made clear.

If one were biased in favour of the poetic and had also a primitivistic bent, it would be easy to idealize the stage of collective childhood. But if one were rationalistic and anti-primitivistic, this, of course, would not be probable. The general tendency of philosophers of history in the eighteenth and early nineteenth centuries was optimistic: the future would be better than the past. Nevertheless there were thinkers like Volney, for instance, the effect of whose work was far from optimistic, for they accentuated the death of nations as well as their growth. The influence of Gibbon's *Decline and Fall*, the popularity of Piranesi's *Ruins* and *Carceri*, the actual death of the Venetian Republic, the disastrous effect on old régimes of the French Revolution, and the Napoleonic Wars, these must never be discounted when one attempts to describe the temper of that period. Such events might well have thrown the human imagination back towards very early times and increased nostalgia on a grand scale.

It was perhaps Auguste Comte more than any other one man who did most to create the belief that the child represented the childhood of the race as a whole. He had read Vico, but his organization of historical data was somewhat different from Vico's. Yet the two agreed in one essential: the way to interpret history is on the basis of how men think. The fundamental question seemed to be, 'In what terms does a man—or a people—formulate thoughts about the world?' Comte's Law of the Three Stages is too well known to require exposition here and indeed any history of philosophy will adequately paraphrase it if it is not familiar. What concerns us is that the first stage, called by him the Theological Period, remains in modern times in the thinking of the child. In other words it is always possible that vestiges or survivals of earlier modes of thought be found in periods later than that in which they originally occurred. Both primitive man and the child, said Comte, animate the universe. All change, they

think, is produced by some agent, conceived in anthropomorphic terms, whose will suffices to produce the effects under considera- ion. Rain gods make it rain, thunder gods make it thunder, the sun and the moon gods cause the sun and the moon to shine. Everything, rivers and springs and mountains and trees, is the habitation of a god. And this is the way the child too is supposed to think. To such a cosmos scientific law has no relevance. It can be influenced not by mechanical processes but only by swaying the divine wills. Hence techniques of appeasement or of punish- ment are invoked when one desires a change. The rain god may be pleased by a rain-dance and give one rain for one's corn; or he may be frightened by withdrawal of the usual gifts and be coerced as one wishes. So the child spanks its toys, nurses its dolls, praises and denounces the local deities as occasion demands. The anthro- pology of all this is obviously wrong, but that people have fol- lowed Comte in believing it to be right is undeniable. The life of an individual, according to Comte, reproduced in miniature the life of the race, beginning with the theological stage, passing into a metaphysical stage, and maturing into a positivistic stage.

When one thinks of the child as a duplicate of primitive man, of the youth as a duplicate of mediaeval man, and of the adult as representative of positivistic man, one has taken over the Law of Recapitulation and given it a psychological turn. The social and psychological applications of this law belong on the whole to the twentieth century. As late as 1911 we find in Hastings's *En- cyclopedia of Religion and Ethics* the sentence, 'The "childhood of the race", originally a metaphor, has become an almost technical term, through the establishment of the law of recapitulation'.[1] The law is interpreted as follows: 'The theory of recapitulation, which attempts to bring under a single principle the mental

[1] Article *Childhood*, introduction, by George A. Coe. The law may have been est- ablished in 1911 but it would be hard to find anyone fifty years later who believes it to be a law. Taine in his *De l'intelligence*, 9th ed., Paris 1900, note I, p. 371, illus- trates how widespread the belief in the analogy between primitive man and childhood was. '[L'état mental d'un enfant]', he says, 'à beaucoup d'égards, c'est celui des peuples primitifs dans la période poétique et mythologique'. Taine, I suspect, but cannot prove, derived this notion from Vico via Michelet.

and the physical, the post-natal and the embryonic development, and to connect the developmental order in the individual with the evolution of the race...'. After birth, we are told, the individual follows 'in a rough schematic way' the line of mental and social evolution in the human race. This evolution involves a change from following instinct and impulse, to individual deliberation through custom. It proceeds from 'connections determined by immediate biological necessity, through stages of group loyalty like that of the clan or tribe, to some recognition of the larger humanity'. We are further informed that 'though the fact of recapitulation is unquestioned', it does not imply that the child is first a savage, then a barbarian, and finally a civilized adult. It simply tells us 'what type of interest will prevail in each period of growth'. In other words one gathers that the child has the interests of a savage but is not a savage, then the interests of a barbarian without being a barbarian, and so on. The existential diversity is obvious, the qualitative less so.   But if the child has the qualities of a savage or primitive man, and his interests, that will suffice for our present needs.

The underlying thesis was fortified by the tenets of Freud and his school. In Freud's writings the psychology of the child is frequently utilized as a substitute for that of primitive or ancient man. The adult is said to retain infantile desires and aggressions as the child manifests primitive and savage desires and aggressions. One can, for instance, extrapolate the Oedipus complex back into prehistoric societies in the form it takes in modern children. This technique appears very clearly in such a work as *Totem and Taboo*, especially in Chapter IV, 'The Infantile Recurrence of Totemism', where Freud says, 'Psychoanalysis has revealed to us that the totem animal is a substitute for the father, and this really explains the contradiction that it is usually forbidden to kill the totem animal, that the killing of it results in a holiday and that the animal is killed and yet mourned. The ambivalent emotional attitude which today still marks the father complex in our children and so often continues into adult life also extended to the father

substitute of the totem animal'.[1] Here the assumptions seem to be that human beings of any cultural period behave in the same manner and that men of the period in which totemism was a belief behaved as children do today. Therefore if one wishes to understand the beliefs and practices of 'primitive' men, one has only to study those of the child. In fact, Freud himself says in another passage of the same chapter that the castration complex is analogous to the actual castration of the priests of Cybele, who were not, of course, primitive but existed only 2500 years ago (p. 924). He also points out with his usual candour, 'It can hardly have escaped anyone that we base everything upon the assumption of a psyche of the mass in which psychic processes occur as in the psychic life of the individual'. This clearly is not the assumption of an evolutionary progression from collective infancy to collective maturity. But in the sentence which immediately follows he says, 'Moreover, we let the sense of guilt for a deed survive for thousands of years, remaining effective in generations which could not have known anything of the deed' (p. 927). This would seem to be a reformulation of the doctrine of inherited guilt as transmitted in the Christian tradition. Freud recognizes the objections that might be made to this, but maintains that it is essential to the existence of social psychology, for (p. 928) 'if psychic processes of one generation did not continue in the next, if each had to acquire its attitude towards life afresh, there would be no progress in this field and almost no development'. Unfortunately he seems to overlook the possibility that a mass of men behaves differently from any individual and that the individual in a mass behaves differently from the way he would behave in solitude. There are two distinct issues entangled here: (1) the issue of whether the date at which a child is born and at which the society into which he is born flourishes are irrelevant to studies of his behaviour, so that one can speak of child-psychology in general; and (2) the issue of how 'psychic processes' of one generation are passed

[1] See *The Basic Writings of Sigmund Freud*, translated and edited... by A.A. Brill, New York, 1938, p. 915.

on to the next. An American child born in 1850 in the state of Alabama would have, for instance, different standards of decent behaviour from those of a child born in the state of Rhode Island in 1950. Freud is fully aware of the second of these issues and expands upon it, but seems to believe that the continuity he posits is a necessary methodological assumption. In any event he utilizes it frequently.[1]

Now he does not say that adults should regress into childhood, adopt childlike ways of thinking, and play children's games. He was not a cultural primitivist. But a man who might have a nostalgia for his childhood or what he believed to be the childhood of the race would find a powerful impetus towards the cult of childhood in the supposed fact that the child survives in the adult and that he stands for the primitive mind. And the more widely accepted fact that everyone grows up would seem to such a person a misfortune. Schopenhauer, for instance, who had little respect for the intellect when it was subservient to the will, saw in the child something akin to genius. The genius, it will be recalled, to Schopenhauer was the man who could look upon himself with complete objectivity, as if he were another person. The brain of the child, he says, is fully developed at the age of seven, whereas the genital system is still embryonic.[2] 'Hence', says Schopenhauer, 'it is explicable that children, in general, are so sensible, reason-

---

[1] For instance, in explaining Leonardo's dream of the vulture in his book on that artist he referred to the significance of the vulture in Egyptian theological symbolism; in *The Interpretation of Dreams* (p. 509) he makes use of the child's crying when hungry as an example of primitive man's physical expression of his needs; in *Wit and its Relation to the Unconscious* (p. 697) he identifies individual childhood and the 'childhood of human civilization'; and again in *Totem and Taboo* (p. 875) he outlines briefly what amounts to the Comtean account of human intellectual history and concludes that 'there still lives on a fragment of this primitive belief in the omnipotence of thought' in animism. A more recent psychiatrist, the late Dr. Trigant Burrow, attributed the survival of ancient practices to instruction. The elders train the young to behave in approved ways. See his 'The Neurosis of Man' in *Science and Man's Behavior*, New York, 1953, p. 413.

[2] *The World as Will and Idea*, translated by Haldane and Kemp, 6th ed., London, 1909, vol. III, p. 161. All page references are to this edition. It would be tempting to try to support Schopenhauer's notion with the child's use of the third person when referring to himself. But I find no place where Schopenhauer does this. This may be because the child does not use the second person either. The whole world is his object.

able, desirous of information, and teachable, nay, on the whole, are more disposed and fitted for all theoretical occupation than grown-up people'. Thereupon (p. 162) follows a eulogy of childhood that is worth reprinting at length.

> They have more intellect than will, i.e., than inclination, desire, and passion. For intellect and brain are one, and so also is the genital system one with the most vehement of all desires: therefore I have called the latter the focus of the will. Just because the fearful activity of this system still slumbers, while that of the brain has already full play, childhood is the time of innocence and happiness, the paradise of life, the lost Eden on which we look longingly back through the whole remaining course of our life.

This delightful period, given over entirely to the intellect, is lost at puberty when 'the will gains the upper hand' (p. 163). 'Just because that impulse pregnant with evil is wanting in the child is its volition so adapted and subordinated to knowledge, whence arises that character of innocence, intelligence, and reasonableness which is peculiar to childhood... Every child is to a certain extent a genius, and the genius is to a certain extent a child. The relationship of the two shows itself primarily in the naiveté and sublime simplicity which is characteristic of true genius'. This appreciation of childhood is unique, for though Schopenhauer is in general an anti-intellectualist, believing that the intellect of the adult is at the service of the will, here he sees the child's mind as pure intellect, intellect which has not yet become enslaved by volition.

Freud was undoubtedly influenced by Schopenhauer whom he had read in his youth,[1] but at this point he and the philosopher

---

[1] See Ernest Jones, *The Life and Work of Sigmund Freud*, New York, 1953, vol. I, p. 375, where the influence of Kant and Schopenhauer is said to have been transmitted through Meynert; vol. II (1955), p. 226, where the theory of 'the unconscious will' is said, and rightly, to have been anticipated by Schopenhauer; and p. 415, where Freud is said to have included Schopenhauer with Goethe, Kant, Voltaire, Darwin, and Nietzsche as men whom he thought 'great'. Freud himself admits his interest in Schopenhauer in *The Interpretation of Dreams* (p. 465), in *Totem and Taboo* (p. 874),

part company. For his theory of infantile sexuality is a flat contradiction of what Schopenhauer says here and, moreover, he had none of that admiration for childhood which Schopenhauer expresses so fervently. In spite of this, some contemporary Freudians, or quasi-Freudians, do seem to set up the child as a model for the adult. This appears with special force in Norman O. Brown's *Life against Death*.[1] Accepting the thesis that 'it is in our unconscious repressed desires that we shall find the essence of our being, the clue to our neurosis...and the clue to what we might become if reality ceased to repress' (p. 23), he infers the conclusion that children are 'in some sense' unrepressed. Education is the repressive force. The adult 'in flight from repressive reality in dreams and neurosis, regresses to his own childhood because it represents a period of happier days before repression took place' (*ibid.*). Whether childhood is any happier while we are children than maturity is when we are adults and whether there is any such thing as childhood *überhaupt* need not be discussed here. If it is true that adults do look back to their childhood as a period of happiness, that is enough for our purposes. Assuming this to be the case, we can conclude, as Brown does, that childhood is regretted and that what prevents our living as children live is the repressions which family and social life insist upon. 'Children', says Brown (p. 31), 'are unable to distinguish between their souls and their bodies; in Freudian terminology, they are their own ideal'. They are also unable to distinguish between higher and lower functions and parts of the body. 'They have not acquired that sense of shame which, according to the Biblical story, expelled mankind from Paradise, and which, presumably, would be discarded if Paradise were regained' (*ibid.*). This, he continues (p. 12) is a 'scientific

and in his brief *History of the Psychoanalytic Movement* (p. 939), where he says that he thought his theory of repression to be original until Rank showed him a passage from *The World as Will and Idea* which anticipated it. He adds, 'What he states there concerning the striving against the acceptance of a painful piece of reality agrees so completely with the content of my theory of repression, that once again, I must be grateful to my not being well read, for the possibility of making a discovery'.

[1] Middletown, Connecticut, 1959.

reformulation and reaffirmation of the religious and poetical theme of the innocence of childhood'.[1] Brown grants that Freud neither advocated nor thought possible a return to childhood; he is satisfied to maintain that such a return is 'man's indestructible goal' (p. 32). But it is strange to find this goal identified with the sentiment of Jesus in Matthew XVIII, 1-3, 'Except ye become as little children...'. For if there is one thing that one may reasonably imagine the author of the Gospel not to have thought of little children, it is that they are polymorphously perverse. Nevertheless by making the identification, Brown has joined together two strands in the history of the idea which we are studying. Since he is one of the most careful expositors of Freud and since also he has interpreted the writings of poets and mystics in Freudian terms, his treatment of the theme is peculiarly significant. For it also makes clear the similarity of the life of the artist and the life of the child.

Art is a form of play, he says (Chapter V, 'Art and Eros'), which is a 'mode of instinctual liberation' (p. 64), and artistry, as the activity which eventuates in works of art, is the recovery of childhood. From this point to the assertion that the arts of the child are a model for the adult artist is but a step. Logically it would not follow that because the artist is acting like a child in so far as he is liberating his repressions, if an artist copied the paintings of a child or imitated an apparently free and undisciplined technique, he would be doing something better than if he strove to assert his maturity. Since it is impossible to find a child who has not been under the influence of some adult, if only his mother or nurse, no matter how young you catch him, his works of art

[1] The idea of the child's innocence in Roman Catholicism, whence it originated, is authoritatively defined in the *Dictionnaire de Théologie Catholique*, Paris, 1926, art. *Innocence, état d'*. It is (1) exemption from ignorance and error, (2) absence of envy, (3) immunity to pain and death along with the enjoyment of the pure joys of the Earthly Paradise. In short, this was the condition of prelapsarian Adam. It is, at a minimum, questionable that this is thought to be the condition of children in any of Freud's writings. I am not objecting to Dr. Brown's description of childhood, but rather to his belief that Freud's theory of infantile sexuality is a reformulation of the state of innocence as that state is officially defined.

will not be 'pure'. But if the theoretical freedom of the unrepressed child is our ideal, and if art is always a liberation, then there ought not to be any art produced by the theoretically pure child. For he would be living a life without repression and hence he would have nothing to recapture, nothing from which he would want liberation. This is not a very recondite observation but it should be emphasized since the only art that we have from the hands of children is that produced when the child has already been brought up in a family or a family substitute, has been at school for a year or more, or has been handed crayons and paper and told to draw. He has, to be sure, accumulated less experience than an adult has, but the intensity of the punishments which he gets for behaviour which is considered bad — by others, and of course rewards for good behaviour, cancel out the paucity of his experiences. A bad pain is not the equivalent of two mild pains.

The unconscious wisdom of the child, his similarity to the artist, his status as the adult's ideal, remind one of Schiller's essay on *Naïve and Sentimental Poetry* and lead us back to Ellen Key. For Schiller used the child as a unique example of the naïve and intuitive poet, one who was guided by the Kantian *Vernunft* rather than the *Verstand*. Appearing in 1800, Schiller's work was at heart a protest against the artifices of neo-classic poetry and a plea for a return to what he believed to be the simplicity and directness of Homer. He paid no attention to Homer's applied decoration, the elaborate similes and the ritualistic epithets, but was apparently captivated above all by his straightforward account of gestures and incidents. It is to be sure true that in Homer there is no elaboration of psychological motivation: the personages act and their acts are set down as they occurred. To express one's ideas in hexameters is not what most of us would consider to be naïve or childlike and indeed, when one has studied Schiller's essay, one finds little in it which extols the child's manners and thoughts.

The child is equated by Schiller with Nature, with the *Sitten des Landsvolks* and with the *Urwelt*. And he goes on to say:

Natur in dieser Betrachtungsart ist nichts anders, als das freiwillige Dasein, das Bestehen der Dinge durch sich selbst, die Existenz nach eignen und unabänderlichen Gesetzen. (p. 413).[1]

In short, Schiller has gone back to making a split between Nature and what has been added to Nature by human beings. He was in revolt, when he wrote this piece, against affectation and attributes affectation to what it was becoming customary in Germany to call the *Verstand*, which historically was what the *Vicaire Savoyard* called philosophy. In order to distinguish between the child's naturalness and our unnaturalness, he points out that children simply follow the demands of their inner character and make no obeisances to laws imposed by others. Oddly enough the contrast comes out as that between our freedom and their regularity (*Notwendigkeit*). But their regularity is what Kant would have called freedom, for it is self-determination. We change, he says (p. 415), and they remain.

> Wir erblicken *in uns* einen Vorzug, der ihnen fehlt, aber dessen sie entweder überhaupt niemals, wie das Vernunftlose, oder nicht anders als indem sie *unsern* Weg gehen, wie die Kindheit, teilhaftig werden können.

Their changelessness, which is also found in 'kindlichen Völkern' (p. 416), represents the Ideal, 'nicht zwar des erfüllten, aber des aufgegebenen' (*ibid.*). Hence the Child becomes a 'holy object'. We are drawn back to childhood as well as to plants, animals, and their accompaniments, by a kind of nostalgia. Children *are* (p. 414) what we *were*. But Schiller can give no explanation of our loss of holiness. If we inevitably lose it, then he might well have concluded that the loss is nothing more than growth, nothing more than the fulfilment of one natural law which adds height to our bodies. But his admiration for childhood is such that after listing the Child's noble qualities, *Schamhaftigkeit*, *Verständigkeit*, and so on, he cannot admit that these are completely gone from

---

[1] My quotations are from *Schillers Werke*, Weimar, 1962, vol. XX.

adult mankind and of course locates them in the genius: 'Naiv muss jedes wahre Genie sein, oder es ist keines' (p. 424). In all probability what Schiller did was to make the distinction which we all make between those poets or artists whom he approved of and to look in them for some quality which he could identify and thereupon called it the quality of *naïveté*, or the childlike. For he gives no examples of any infant-genius and it would have required little research amongst children to discover that the moral qualities that he praised so highly were no more present in them than in adults. Moreover, if these qualities are to be found only in geniuses and geniuses are born and not made, then he has set up no programme for a return to childhood. The case of Ellen Key is quite different.

As we have already said, she based her thoughts on the Law of Recapitulation. 'The development of the child', she says in *The Century of the Child* (p. 222), '...answers in miniature to the development of mankind as a whole. And it follows from this that children combine idealism and realism, as epic national poetry does. Great, good, heroic, supernatural traits affect them most; not only in a concrete shape sensibly perceived, with the richness of the power which comes from life, without any adaptations to our present conceptions'. That mankind as a whole in its earliest state combined idealism and realism and what such a combination would be like, she neither proves nor exemplifies. The idea itself is probably a residue from Herder, though it may be simply an echo of that bizarre theory, popular at the end of the nineteenth century, that the people as a whole were able to write ballads and epics, though how they did this was never, to the best of my knowledge, demonstrated. For that matter the *Volksseele* could also presumably compose songs and dances and invent pictorial themes.

In the second place, children are averse to abstractions. 'Children', she says (p. 225), 'do not feel drawn to abstract things... All virtues and qualities, no matter how well concealed they may be, are very quickly pronounced stupid by children'. This may

very well be true, but whether it is an admirable trait is another question. No one could operate in a world filled with concrete individual things unless he were able to group them somehow or other in his mind. The child has sooner or later to recognize milk, bread, water, to distinguish his parents, brothers, and sisters from other people, a feat which depends upon classifications. If he fails to do this sort of exercise, he fails to adjust to his vital problems. If a child should turn up who did not comprehend that each new bottle of food smelled or looked or tasted like those to which he was accustomed, he would shortly die.

Third, the child is realistic, as is shown by his conception of heaven (p. 303). In this respect he is very much like St. John with his vision of streets of gold and thrones of jasper and sardonyx, his woman clothed with the sun, the moon under her feet. This is the gift of the poetic imagination which can transform an abstraction into a concrete reality with all the sensory aspects of the concrete. But in spite of the child's revulsion against the abstract, he is credited with what is called 'invulnerable logic' (p. 286; cf. p. 301). The combination of logic and a sense for facts is not unknown amongst adults but one would imagine that the public for whom fairy tales were written and by whom they are enjoyed would not manifest the combination to the same degree as a mature man would. Happily we need not busy ourselves with the truth of Ellen Key's opinions and we must simply note them as they stand. The child's logical powers lead him, she says (p. 286), to see vividly the conflict between his elders' religious beliefs and daily life. Children are 'clear sighted in their simplicity' and cannot reconcile what they are taught with what their teachers actually do. For they are, in the fifth place, deeply interested 'in the eternal riddles of mankind' (p. 297) and this at a very early period. 'They are troubled', we are told, 'with questions of whence and whither... the honest childish nature is opposed to the Christian explanation of the world, until the child's sincerity is dulled and he either takes without question what he is taught, or in his own soul denies what his lips must repeat, or finally

allows his heart to be possessed by the only nourishment offered to his religious needs'. And finally, the child is an artist and an appreciator of art, 'in the sense that [he] desire[s] to receive an impression of its purity, not as a means to something else' (p. 222). In short the child is an aesthete in the sense in which the term was used in the nineties of the last century. The aesthete too was interested in the sensory impression in its purity and was horrified when a work of art was supposed to tell a story, to point a moral, to illustrate a text. Ellen Key's child might well have been des Esseintes.

Simple, sincere, logical, sensual, religious, the child in Ellen Key was a wonderful human being, unlike most adults. As such he becomes a sort of ideal human being whom we should all attempt to emulate. She joins hands here with Dr. Norman Brown. But what happened to the Wise Child? Why did he lapse from his ideal position, like Adam, primitive man of the Golden Age, and even the Noble Savage? The answer, as in Rousseau, Pestalozzi, and Froebel, was education. 'The desire for know-ledge', says Ellen Key (p. 203), 'the capacity for acting by oneself, the gift of observation, all qualities children bring with them to school, have, as a rule, at the close of the school period disappear-ed'. Here we have again Rousseau's diatribe against the school-master with his books, Agrippa's castigation of all the arts and sciences: the poisonous fruit of the Tree of Knowledge. The child as a child possesses all the desirable human traits, a sense of the concrete, a feeling for logicality, an artistic sense, and the result of his education is their ruination. So the beautiful life of the South Sea Islander was ruined by occidental diseases, the American Indian by rum, the Noble Savage by luxuries. The Fall of Man is repeated in each individual biography, as at each temptation the child succumbs to the Serpent. Nostalgia for childhood may account for a good deal of this idealization, for it cannot be denied that childhood has few, if any, responsibilities, and the assumption of duties towards others is as painful as it is unavoidable. Perhaps this is why Anglo-Saxon writers, even those

with a certain intelligence, Blake, Christina Rossetti, Stevenson, Walter de la Mare, will write children's jingles and songs as if writing were a liberation from the burden of maturity.

Ellen Key's point of view was paralleled by that of her friend, Rainer Maria Rilke, whose writings on the subject serve as a transition to the contemporary cult of childhood as the age of artistry.

Rilke's views on our subject are contained in his *Über Kunst*, part of which first appeared as a separate essay in the periodical *Ver Sacrum* in 1899.[1] Art to his way of thinking was a *Lebensan-schauung* in which one lost oneself completely. It involved one in 'eine weise Blindheit, die ohne Furcht einem geliebten Führer folgt'. It is something which is naïve and involuntary and resembles that period of unconsciousness 'deren bestes Merkmal ein freudiges Vertrauen ist: der Kindheit'. There follows a description of childhood which had best be left in the poet's own words:

> Die Kindheit ist das Reich der grossen Gerechtigkeit und der tiefen Liebe. Kein Ding ist wichtiger als ein anderes in den Händen des Kindes. Es spielt mit einer goldenen Brosche oder mit einer weissen Wiesenblume. Es wird in der Ermüdung beide gleich achtlos fallen lassen und vergessen, wie beide ihm gleich glänzend erschienen in dem Lichte seiner Freude. Es hat nicht die Angst des Verlustes. Die Welt ist ihm noch die schöne Schale, darin nichts verloren geht. Und es empfindet als sein Heiligtum Alles, was es einmal gesehen, gefühlt oder gehört hat. Alles, was ihm einmal begegnet ist. Es zwingt die Dinge nicht, sich anzusiedeln. Eine Schar dunkler Nomaden, wandern sie durch seine heiligen Hände, wie durch ein Triumphtor, werden eine Weile licht in seiner Liebe und verdämmern wieder dahinter; aber sie müssen alle durch diese Liebe durch. Und was einmal in der Liebe aufleuchtete, das bleibt darin im Bilde und lässt sich nie mehr verlieren. Und das Bild ist Besitz. Darum sind Kinder so reich.

[1] *Ver Sacrum*, vol. II (1899), Heft 1, pp. 10-12. I quote from Rainer Maria Rilke *Verse und Prosa, aus dem Nachlass*, Leipzig, 1929.

It would be absurd to analyse such a description, which surely has no pretension to being a contribution to scientific psychology. At the same time it would be folly to dismiss this as a statement without substance or influence. The child, as he appears in it, is the polar antithesis of Schopenhauer's sensible, rational, wise proto-genius. He is a person living in an idealized world the denizens of which are, if not created by him, at least glorified by him.

Two possibilities now confront the child, for sooner or later he will have to make a decision determinative of his future.

> Entweder es bleibt jene Fülle der Bilder unberührt hinter dem Eindringen der neuen Erkenntnisse, oder die alte Liebe versinkt wie eine sterbende Stadt in dem Aschenregen dieser unerwarteten Vulkane. Entweder das Neue wird der Wall, der ein Stück Kindsein umschirmt, oder es wird die Flut, die es rücksichtslos vernichtet, d.h. das Kind wird entweder älter und verständiger im bürgerlichen Sinn, als Keim eines brauchbaren Staatsbürgers, es tritt in den Orden seiner Zeit ein und empfängt ihre Weihen, oder es reift einfach ruhig weiter von tiefinnen, aus seinem eigensten Kindsein heraus, und das bedeutet, es wird Mensch im Geiste *aller* Zeiten: Künstler.

The contrast between the Bourgeois and the Artist was by the end of the nineteenth century fairly well established. The French aesthetes had made it quite clear that no man could be both. But it is doubtful whether any of the French critics of the Bourgeoisie had ever thought that to be an artist one had also to be a child, though Dr. Meyer Schapiro has shown how some of them identified the child's vision of the world with that of popular *imagiers*.[1] But Rilke in the article we have been quoting takes a

[1] Meyer Schapiro, 'Courbet and Popular Imagery', *Journal of the Warburg and Courtauld Institutes*, vol. IV (1940-41), pp. 164-191. See especially the references to Champfleury, pp. 178 ff., and Baudelaire, pp. 180 ff. He quotes Champfleury as saying, 'The stammering of children... offers the charm of innocence, and the charm of the modern *imagiers* comes from the fact that they have remained children...they have escaped the progress of the art of the cities' (p. 178).

different point of view. The roots of 'true' artistry 'wohnen in dieser wärmeren Erde, in der niegestörten Stille dunkler Entwicklungen, die nichts wissen von dem Mass der Zeit'. Twenty years after the first appearance of these words a group of artists in Zurich came out with a blast against all art that sprang from reason and forethought and Dada was born. The first progeny of Dada was surrealism, a movement which liberated the roots of artistry from their dark habitation and replaced schooling and tradition with spontaneity and fantasy. It then became possible, as we shall see, to praise an artist for his 'innocent eye' and childlike qualities.

V

The nineteenth century saw the rise of psychological studies of childhood, as it saw the beginnings of scientific anthropology. Amongst the studies of the child's mind were a few on the drawings made by children as evidence of their mental functions. But contemporaneously certain books and articles were written which also introduced the question of the aesthetic side of the child-artist's 'creations'. And it is sometimes found that the two aspects of his work will be treated together. A good example of this double interest is Toepffer's *Réflexions et menus-propos d'un peintre Genevois*.[1] Toepffer's book was essentially a philosophy of art, not a psychological investigation. Its main thesis was that beauty was not to be found in pictorial representations of natural objects but rather in works of art as signs of an ideal beauty. Ideal beauty when pursued to its ultimate source turned out to be God Himself, an idea which Hegel also expressed in his definition of beauty as a sensuous manifestation of the Absolute. The ability to form conceptions of beauty is innate in all human beings but some are more capable of creating signs of their conception than others are. To prove its innateness Toepffer follows the traditional

[1] 2 vols., Paris, 1848. The sub-title of this work runs *Essay sur le Beau dans les Arts*, which gives one a clearer idea of the book's subject. For the contribution of Toepffer to the growing esteem for popular imagery, see Meyer Schapiro, *op. cit.*

method of looking for it in children and in two chapters of his work (Vol. II, Book vi, Chapters xx-xxii) he discusses the drawings of children, *les petits bonshommes* which they draw on walls and in the margins of their books. But he also compares them with the *graffiti* in Pompeii and on the barracks walls of soldiers, as well as introducing a paragraph or two on the sculptures of Easter Island. We see here a fusion of the ideas of the immature, the uneducated, and the savage. But Toepffer has no extended discussion of the legitimacy of such a fusion and seems to take it for granted.

As a philosopher of art Toepffer is an anti-naturalist, in the sense that he will not grant an equivalence between beauty and correct representational drawing. He argued that if the latter were identical with the beautiful, then a wax image in colours of a human being would be more beautiful than a piece of Greek sculpture. This, he believes, is obviously false. In fact, since he is unwilling to grant that all children's drawings are equally beautiful, he maintains that some children are 'totalement dépourvus de sens artistique', just as some mature artists are. When these inartistic youngsters draw a soldier, they put in buttons, epaulettes, and other details of the uniform and think that they have succeeded in making a beautiful picture. But some drawings, on the other hand (p. 109), 'tout gauches et mal tracés qu'ils sont, reflètent vivement, à côté de l'intention imitative, l'intention de pensée, à tel point que cette dernière y est toujours, à cause même de l'ig-norance graphique du dessinateur, infiniment plus marquée et réussie que la première' (p. 108 f.). As for savages, the Pascalians make great stone images which are hideous and gross if one thinks of them as representations (p. 110). But from the point of view of the conception which their makers had, 'ce sont des êtres tout à la fois cruels, durs et supérieurs, de brutes divinités, mais divinités enfin, en qui se devine la grandeur et qui pressent la beauté: en tant donc que signes d'une conception, ces signes ont déjà la clarté à la fois et la vigueur du sens, ils vivent, ils parlent, ils proclament qu'une pensée créatrice s'y est infuse pour se manifester par leur moyen' (p. 110f.) In short, the works of art both of savages and of

THE CULT OF CHILDHOOD

some children show an instinctive sense of beauty through imitation, a 'signe d'une intention, d'un caprice, ou, pour parler plus exactement, d'un beau élémentaire, fruste, grossier, mais enfin absolument et exclusivement issu de la pensée' (*ibid.*). But when the children go to school, they learn the technique of imitation and lose the power of conceiving the beautiful (Chapter xxi).

Toepffer is thus no idolater of children. But he did call attention to their artistic possibilities and also made what became a standard comparison between their works and those of savages or 'primitive' men. His book was not a scientific psychological study but one based on personal observation and certain *a priori* metaphysical principles, amongst which was one that went back at least to Plotinus. A generation later there appeared a work by the Italian historian of art, Corrado Ricci, *L'Arte dei bambini*. Ricci, like his predecessor, began with *les petits bonshommes*, in his case drawings which he found on a wall in Bologna under an archway where he had taken shelter from a storm in 1882. The whole wall, he says (p. 3), was covered with *graffiti* and he concluded that those near the bottom had been made by children. They were *di poco valore estetico* but started a chain of thought in him which eventuated in his book. He began collecting children's drawings from various parts of Italy and was aided by his colleagues in the history of art. The result was that he was able to examine several hundred and based his conclusions therefore on what he believed to be a fair sample. He was not only interested in the subject-matter of these works of art but also in their technique. From the point of view of the former, he saw that, unlike the Creator, the child begins with Man instead of ending with him, and from that of the latter (p. 11) the child describes his subject rather than rendering it 'artistically'. A child's drawing, then, is a sign, or one might say, a hieroglyphic rather than a purely optical image: 'Fanno insomma coi segni la descrizione che nè più nè meno farebbero con la parola' (*ibid.*) He puts into his drawing what he thinks is essential, arms and legs, for instance, whether he can see them or not.

The parts which are concealed behind opaque barriers, such as the sides of a boat or the walls of a house, are drawn in as if they were visible. Anticipating Picasso, when he draws a profile, he will put in both eyes and sometimes indicate a nose frontally and a mouth as well (p. 17). Though it is not likely that Ricci had been influenced by Schopenhauer, he too appears to believe that the child is instinctively rational in the sense that he makes his visual impressions, when objectified, correspond to his knowledge.

Ricci was far from being a cultural primitivist. He made the comparison which later became ritual between the drawings of children and those of primitive peoples, in his case the American Indians. But he maintained that though both might have the same defects of technique, the drawings of the savages never had the ingenuous and amusing aspect of children's drawings (p. 24). He found that, except in a few cases, the Indian never drew a man with two arms and two eyes on the same side of the body (p. 25). Similarly (p. 27) prehistoric drawings 'sono smisuratamente superiori ai disegni dei bambini', though he believes prehistoric sculpture to have the same technical weaknesses. And in spite of his belief that Romanesque art is an art of maximum decadence, yet it is not childlike. Just as a man fallen into second childhood does foolish things, yet what he does is not the same as the foolishness of children (p. 31). So 'gli errori d'un' arte decaduta e moribonda', that is, of late mediaeval art, do not resemble those 'd'un' arte che nasce'. If this is not clear enough, he spends a page or two in attacking the legends of precocious artistic talents. No one has ever seen the marvels of the child Giotto or Masaccio and many a great painter showed no talent whatsoever as a child. Those grave and thoughtful children of whom we are always told are simply sick and, if they survive their childhood, at best end up 'magari sindaci decorati del natio luogo' but not 'gloriosi da nessuna lotta dell'animo e del corpo' (p. 74).

Ricci's little book was summarized in 1894 and partly translated by Mrs. Louise Maitland in a periodical called *The Pedagogical*

Seminary (vol. III, 1894, pp. 302 ff.).[1] It might therefore be expected to have had a certain influence upon theories of infantile development. But though it was referred to a few times, for some reason or other its calm appraisal of the child's mind and artistic products and, more probably, its author's sense of humour, seem to have prevented serious psychologists from taking its conclusions to heart. The theory of organic evolution as extended into the Law of Recapitulation seems to have crippled any objective survey of any human activity. It was essential that the child be envisioned as the father of the race and it was only when popular opinion began to praise the primitive above that which emerges out of it that the child was to be restored to the position which poets had given him. One cannot run through all the psychological literature which deals with childhood nor is it necessary to do so. I shall therefore select simply a few examples to illustrate the trend which we are tracing.

In 1895, for instance, there appeared James Sully's *Studies of Childhood*. Chapter IX of that work, which was read by all who were studying 'genetic psychology', is called 'The Child as Artist'. As might be expected, Sully without hesitation compares the art of children with that of primitive man and (p. 299) though he admits that we must not 'expect a perfect parallelism' between the two yet 'we shall find many interesting points of analogy'. To one reading these lines seventy years after they were written, the surprising thing is that anyone should begin his investigation of the child's psychology by looking for analogues in that of primitive man. But apparently the fusion of the two concepts of The

---

[1] This periodical was concerned with the psychology of children, not with art. Yet any notice appearing in it would have been seen, if not read, by hundreds of American school teachers. And most American schools included drawing in their curricula. Ricci's book itself does not seem to be in the main American libraries and, if his ideas were generally known, there is little evidence of that knowledge. Thus so careful a writer as Robert J. Goldwater in his *Primitivism in Modern Painting*, New York, 1938, p. 187, says that the first book on children's art was published in Italian in 1906, 'two years after the "discovery" of the primitive'. For first-hand information on the actual drawings of children who are quasi-primitive, see Wendell H. Oswalt 'Traditional Storyknife Tales of Yuk Girls', *Proceedings of the American Philosophical Society*, vol. CVIII, 1964, pp. 310-336.

Savage and The Child was so well established that a man so skilled in empirical investigations as Sully accepted it almost without question. He pointed out, for instance (p. 307) that the child was like the savage in having no aesthetic sentiment for nature as a whole, though 'he may feel the charm of some of her single features, a stream, a mountain, the star-spangled sky, and may even be affected by some of the awful aspects of her changing physiognomy' (p. 307). What more would be required to demonstrate the existence of the sentiment in question is not told us by Sully, and indeed it is hard to imagine what an aesthetic sentiment for nature as a whole might be. For that matter it is equally hard to imagine just what idea of nature as a whole an adult or a civilized man might entertain. Be that as it may, we are told that both the child and the savage have a vividness of imagination which leads them 'to invest a semblance with something of reality' (p. 313). But this is not said in favour of the childlike mind. On the contrary, 'we are able to control the illusory tendency and to keep it within the limits of an aesthetic semi-illusion; not so the child' (*ibid.*) Just as the savage invests his idols with the spirit of the gods that they represent, so the child invests his toys 'and even his pictures' with the same spirit. We have become accustomed nowadays to considering works of art as substitutes for reality[1] nor do we deprecate the habit. But to a man of Sully's temper, this would be something which we should abandon as maturity takes over from childhood.

For Sully is a determined evolutionist. The child's attempts at drawing, he says (p. 331), are 'pre-artistic', the term is part of the programme. In short, if there is to be artistry, there must be a stage out of which it emerges. In this stage, what is later recognized as art remains play. But nevertheless one can detect in it a suggestion of the desire to produce something which will resemble other things. And it is worth observing that Sully main-

---

[1] Or perhaps I should say, 'as forming a world of their own with its own kind of reality'. The history of this idea is not very long, going back only to the movement known as Art-for-Art's-Sake.

tains that this desire always involves other people; it awakens a social sense. That is why the child will call upon his mother to 'look at the "man", "gee-gee", or what else he fancies that he has delineated' (*ibid.*) The question of when one has succeeded in delineating something and when one only fancies that one has delineated something is one which would involve us in delicate epistemological questions. But they are themselves questions which are essential to the central issues of aesthetics. Even if one defines the end of art—in this case of drawing—as an optical illusion, someone has to be taken in. The victim may be only the artist himself, but he then becomes the artist as spectator. Such questions did not concern Sully. He believed, as most of his contemporaries believed, that anyone could tell when a picture adequately represented its subject. He was thus able to write that the 'child's drawing begins with a free aimless swinging of the pencil to and fro, which movements produce a chaos of slightly curved lines' (p. 333). A modern reader would immediately challenge the two words 'aimless' and 'chaos'. But to an evolutionist like Sully, strongly influenced by Spencer, order must come out of disorder. Hence we find him writing: 'In the first place, a child may by varying the swinging movements accidentally produce an effect which suggests an idea through a remote resemblance' (p. 333). But again, he may 'set himself to draw, and make believe that he is drawing something when he is scribbling' (*ibid.*). This would seem to involve two different things, as Sully half recognizes: (1) the emergence out of aimless scribbling of something resembling a real object; (2) the desire, unfulfilled, of a child to represent something in his drawings. The former is similar to an adult's experience when he sees animals or people in cloud-formations; the latter resembles rather theatrical acting, when one pretends to be someone other than oneself. When I say that Sully half recognizes this dual aspect of the act, it is because he wishes to emphasize mainly the way in which a form evolves from something formless. For instance when he is expounding the manner in which children draw a man, he says (p. 340, cf. p. 367), 'Here, as

in organic evolution, there is a process of specialization, the primordial indefinite form taking on more of characteristic complexity'. He is so convinced of this parallelism that he is willing to say (p. 382), 'It is clearly a movement from the vague or indefinite, a process of gradual specialization. Not only so, we may note that it begins with the representation of those rounded or ovoid contours which seem to constitute the basal forms of animal organisms, and proceeds like organic evolution by a gradual differentiation of the "homogeneous" structure through the addition of detailed parts of organs. These organs in their turn gradually assume their characteristic forms'. The outcome of all this is that the child's eye at a surprisingly early age loses its original 'innocence' and grows sophisticated. Instead of being the rational genius of Schopenhauer or the intellectualist, if the term be not too strong, of Ricci, it increases in intelligence as it matures. Does this imply that the child's drawings are of *poco valore estetico*? Just how little is not told us. But Sully was willing to say that they are 'crude, defective, self-contradictory even', but 'are not wholly destitute of artistic qualities'. For 'true art', is 'in its essential nature selective and suggestive rather than literally reproductive' (p. 396).

There is indeed little grist here for the primitivist's mill. For Sully was not an admirer of the earliest stages of man's supposed evolution. Yet by agreeing with the thesis of the fusion between childhood and savagery, he might easily strengthen the idea held by the cultural primitivist who would, of course, refuse to appraise highly the later epochs of man's history. This thesis became a commonplace, as I have suggested. In a brochure by Siegfried Otto Julius Levinstein, *Das Zeichnen der Kinder* (Leipzig, 1904), one finds it presented as an incontrovertible truth. At the very opening of his brochure Levinstein says:

Eines ist sicher. Wir dürfen dem Kinde nicht mit unseren besten Resultaten entgegentreten, sondern wir müssen es zu diesen auf derselben Leiter hinführen, die wir und

unsere Vorfahren mühselig erklommen haben. In der
Entwicklung des Kindes spiegelt sich die Entwicklung der
Rasse wieder und der daran zweifelt, den mögen die nach-
folgenden Seiten überzeugen. Sie sind geschrieben, um zu
beweisen, dass unsere Kinder im allgemeinen denselben
Gang gehen, den unsere Vorfahren einschlugen und auf
dem die Naturvölker sich noch heute bewegen. Haben wir
einmal diesen Werdegang gefunden, so wird es verhält-
nismässig leicht, die Erziehung unserer Jugend dement-
sprechend zu gestalten (p. 1).

Beginning with children's drawings of the human figure, he quotes
Ricci to prove that the first subject of interest to the child is man
and then proceeds (p. 11) to point out that just as in primitive
society clothing serves mainly as decoration, so in the child's
drawings of human beings people are 'nackt aber geschmückt'.
The decoration usually consists of buttons and headgear. Again,
in harmony with Ricci's findings, he maintains that children draw
not to make literal representations of their subject-matter nor to
express a 'rein aesthetisches Gefühl', but rather to articulate a
train of thought (p. 46). *Bildermalerei* thus is *Bilderschreiberei* and
hence children's drawings are to be read rather than taken in by
the eye alone. That we have come to think of all pictures as Levinstein thought of children's pictures is perhaps obvious, but
it may be just as well to remark, in passing, the influence of Freud
and the other depth-psychologists in directing the critic's attention
to the non-visual meanings of works of art. The fact, if it be a fact,
that children draw what they know to be present rather than
what their eyes alone show them would not prove that their
Unconscious is revealing what usually lies concealed in its depths.
Levinstein, moreover, is more interested in demonstrating the
parallelism between chidren's art and primitive art than that
between children's art and adult repressions. His chapter (Chapter
VI, pp. 61 ff.) on *Kulturhistorische und ethnologische Parallelen* which
winds up with Lamprecht's system of cultural stages (p. 68)

demonstrates that clearly enough. It is only in the latest cultural stages that we become subjective; in the earliest we are symbolical. Unfortunately we are not told what code of symbols we employ as children nor, for that matter, how our psyche expresses itself 'subjectively'.

One more psychological treatment of our theme will perhaps suffice. In 1922 G. F. Hartlaub published his *Der Genius im Kinde* (Breslau). Here we have the cult of childhood brought into the open as a *topos* with a long history, as something deep-seated within the soul of man. This cult, he says in his opening pages, is as old as Christianity. He attributes to Jesus the following thought:

> Der Begründer dieser Weltreligion [Christianity] ward nicht müde, das Geheimnis seiner Göttlichkeit als Wirkung ganz im Vater gegründeter, vom Vater ganz unabgelöster, eben darum übermenschlicher Kindschaft zu deuten und solche wahre Gotteskindschaft als ein Höchstes Gebot auch den Jüngern aufzuerlegen; im Kindschaftsstande verkündete er nicht nur eine Vergangenheit, sondern ein apokalyptisches Ziel der Menschen und der Menschheit (p. 7).

In view of the neglect of this teaching in both the early and the mediaeval writings, we may have reason to doubt Hartlaub's interpretation of Matthew XVIII. But we cannot brush off his own reading of this passage. That reading involves the belief that childhood, that is human childhood, is a symbol of the 'vollendete selige Zustand' as contrasted with 'dem erlösenden Geschehen der Passion' (p. 7). After this interesting introduction, Hartlaub proceeds to show in the usual manner that the 'sogenannten Ur- oder Naturvölker', like rural people of today, are very close to children in their way of life. Their pictures, songs, stories, and sayings are highly similar and all seem based on what he calls an anthropomorphic materialism (p. 24). But this monistic metaphysics is absolutely different from 'modern mechanism', being 'im höchsten Grade seelenvoll-religiös' (*ibid.*).

Not only does Hartlaub identify the infantile and the primitive mind, but he also maintains that the development of the child up to fourteen years, 'vielleicht bis zum 21. Jahre' (p. 25), is a recapitulation of human history 'von den frühesten Anfängen über das altsteinzeitliche Diluvium bis zur Bronze- und Eisenzeit, von der Stufe des "Jägers und Fischers" über die der "Ackerbauer und Viehzüchter" bis zu derjenigen des bewussten Kulturmenschen'. But none of this commits the author to an unlimited cult of childhood. Modern infantilism is anathema to him (pp. 61 f.). He has little confidence in Freud's theory of infantile sexuality (p. 59). He will have nothing to do with magic and demonworship. The one child in whom he is really interested is the Christchild.[1]

How much influence such writings as those we have just sampled had upon popular opinion cannot be decided without more study. That there did arise a nostalgia for childhood is indubitable and that it appeared not only in the opinions of artists and writers but also in critical essays is equally indubitable. For instance, we find Paul Klee writing in 1906:

> 'Es ist eine grosse Not und eine grosse Notwendigkeit, beim Kleinsten beginnen zu müssen. Wie neugeboren will ich sein, nichts wissen von Europa, gar nichts. Keine Dichter kennen, ganz schwunglos sein, fast Ursprung.[2]

Reading this one realizes the extent to which the rejection of maturity can go. The return to childhood may indeed be more common than the printed evidence shows, but in Klee's case we find critics praising him for having rejected the aesthetic tradition of Europe. It is instructive to read in an article on Klee by Alfred H. Barr that 'the child, the primitive man, the lunatic, the uncon-

---

[1] This may be unjust. Hartlaub's book is the typical doctoral thesis of his time, written to demonstrate industry more than critical acumen. The second edition of the work, 1930, omits a good many of the passages which I have quoted and indeed is radically altered as a whole.

[2] From Klee's Journal as given by Leopold Zahn in his *Paul Klee, Leben, Werk, Geist*, Potsdam, 1920, p. 26. A year later Klee jotted down his desire to study anatomy with the young doctors.

scious mind, all these artistic sources...offer valuable analogies to Klee's method', and this in praise of the artist.[1] In view of the findings of psychodynamics, this is understandable. For if most of our troubles are the fruits of childhood repressions, then the best remedy is to retrace our steps and recover our childhood.[2]

On the assumption of the child's already possessing the wisdom that he needs, Sylvia Ashton-Warner in her very interesting book, *Teacher* (London and New York, 1963), built a pedagogic system. Her idea is to bring into the open what the child's 'inner vision' sees. 'Act first and think last', she says (p. 150), 'is my motto'. Her purpose is to avoid cramming into her pupils her own ideas, attitudes, emotions, hopes, and crowding out their own. 'The expansion of a child's mind', she says (p. 93), 'can be a beautiful growth. And in beauty are included the qualities of equilibrium, harmony and rest'. If this is the outcome of her method, one would be stupid to find fault with it. But the question remains whether the child is thus prepared for a life in which disequilibrium, discord and movement will be the rule. That, however, is not the problem here. I am merely interested in pointing out at least one example of the positive contributions of the cult of childhood. I need not also say that as far as primary education in the United States is concerned, this theory of pedagogy is in direct opposition to the long tradition springing from New England Calvinism. For if we are all born in sin and elected by God regardless of merit, then there is no likelihood of setting up the child as an ideal. The problem of the Puritan was the rearing of children out of sin and corruption, as far as that was possible. The text-books of colonial America illustrate how this was done. Education according to that tradition would not try to bring out the inherent goodness of the child, but to prevent his inherent badness from getting the upper hand. I shrink from reading too much into popular symbols, but nevertheless point out

[1] Alfred H. Barr, *Paul Klee*, New York, 1941, p. 5. Cf. Jean Lurçat on the childlike qualities of Klee, in Will Grohmann, *Paul Klee*, n.d., p. xxv.

[2] For a somewhat caustic exposition of this, see Leslie A. Fiedler, *No! in Thunder*, Boston, 1960, p. 253.

that the little cherubs carved on old New England grave-stones were heads and wings without bodies (cf. p. 47 n. 2).

The admirers of the child-mind of course gave little thought to such gloomy views. They were more likely to share Gandhi's opinion which he wrote to the boys and girls of Sabarmati Asram, 'Children are innocent, loving and benevolent by nature. Evil comes in only when they become older'.[1] But since the individuals who introduce evil into the lives of children were perforce themselves children before they became evil, the question is still unsolved of how evil could possibly emerge out of innocent, loving, and benevolent souls. For unless human society had no beginning, it must have been started by people and these people must once have been children. And if that is so, then it is as natural to become evil as it is to be born good. And if one seeks in the child for the natural as opposed to the acquired, one is forced to grant that the argument rests mainly on question-begging epithets.

The use of such epithets is easily exemplified. Consider the following. 'People', says Jean Clay, a French writer, 'are sometimes amazed at the similarity between contemporary art and the art of children. It is not the latter who are the copyists. It is the former who in his difficult search for primordial innocence, which he seeks even in "primitive" forms of expression, Negro art, *art brut*, comes face to face with the spontaneous freshness of the child'.[2] But spontaneous freshness can result in deplorable acts as well as in those which are laudable. Moreover what is spontaneous in a child need not be spontaneous in an adult who is trying to act as a child. Similarly when Sir Herbert Read says that the paintings of Matisse resemble those of a child 'because in both you have the same pre-logical vision, the same delight of the innocent eye',[3] one wonders why pre-logicality is any better than

[1] Quoted in Louis Fischer, *The Life of Mahatma Gandhi*, New York, 1950, p. 212.
[2] Jean Clay, 'L'Académie du Jeudi', in *Réalités*, no. 189, October, 1961, p. 115. The curious syntax may be the fault of the translator.
[3] *Art Now*, New York, 1948, p. 75. It is strange that Sir Herbert should have failed to see how unchildlike Matisse's paintings are. Aside from other considera-

logicality. For not even the inventor of that term, Lévy-Bruhl, endowed it with any eulogistic connotation. And as for innocence of eye, with what crime has the adult artist been charged? But Sir Herbert here also relies on the Law of Recapitulation (p. 46) and seems to believe that one can discover more about a plant from its seed, to use his own metaphor, than from its total history. The Law of Recapitulation, he says

> has fully confirmed the general validity of the genetic method in aesthetics, and again, by drawing attention to the positive qualities of children's art, has had a direct influence on the practice of modern artists—there has been a deliberate attempt to reach back to the naivety and fresh simplicity of the childlike outlook—a retrograde step, of course, if you regard the 'march of intellect' with complacency or satisfaction.[1]

There follows a disclaimer of any appraisal of primitive and children's art which would make it superior to the art of 'civilized men', yet this in turn is followed by the opinion that the art of civilized men 'is overlaid by modes of life and manners that are not of its essence'. One would imagine that such alien elements would be detrimental to the value of those works of art in which they are found. And indeed in the last sentence of the passage under consideration Sir Herbert finds that in primitive and children's art there are 'a stirring of the pulse, a heightening of the heart's beat, a tautening of the muscles, a necessary and exigent mode of expression', all of which I take to be eulogistic.

The late Marion Richardson, who was recognized as one of the greatest teachers of art to children in her time, had no illusions about the inner vision. Her book, *Art and the Child* (London, 1948), recognizes its existence (p. 60) and preaches the need for freeing it, but to her the art of children has more of a psycho-

---

tions, every major painting of Matisse was worked out in a series of preliminary studies, beginning with careful 'realistic' drawings. There may be something pre-logical in this practice but surely nothing 'innocent'.

[1] *Op. cit.*, p. 46.

therapeutic value than an aesthetic value. I mean by this simply that its value lies more in what it will do for the child than in what it will do for the spectator. Miss Richardson objected strongly to the teaching of art through copying from casts and pre-arranged still-lifes and insisted that 'we must realize that satisfaction may be found in projecting the wish for something that real life has so far denied, the longing for grand and grown-up dresses, sentimental situations, and other deeply felt desires'. A similar attitude is expressed by Mrs. Jane Bland in her brochure, *Art of the Young Child—3 to 5 years* (New York, 1957). Mrs. Bland frequently emphasizes the individuality of small children, no two of whom are really alike. Hence generalizations about the child's artistic sense, intelligence, innocence, and all the other qualities of which we have shown some samples above, turn into loose statements which must not be taken seriously. The trend which began with Čižek in Vienna and which was furthered in England by Marion Richardson laid less emphasis upon the supposed beauty of children's art than upon its value to the child. A simple exposition of the policy was given by the late Belle Boas in her articles, 'Innocent Art' (*Baltimore Museum of Art News*, vol. XVI, 1953). Admitting that every child is an individual and that no two are precisely similar, she insisted that drawing and painting exist in the curriculum as a method for freeing the child's imagination. 'The imaginative life of children, as keen and vigorous as a young plant pushing upward through the soil, is crushed and killed if denied an outlet. This is amply proved by the investigations of the past thirty years by artists, teachers, and notably by psychologists. We are impressed by their emphasis on stimulating the imagination, on casting aside the rigid, directed lesson with its lifeless representations of the strawberry baskets or the bottle and the plate, the tulip pinned to a board, most of which we hope have passed into oblivion'. One can share this hope without also hoping that adults will try to regress into a condition which through the natural course of life they should have abandoned.

There is an appreciable distance between art as a method of

freeing a person of his inhibitions and repressions, as in 'art-therapy', and art as the making of objects which others are to look at and enjoy. The values of each type of procedure are different and it might be only by accident that what the first type produces will have the aesthetic interest of what the second produces. It is of course true that sounds and sights which are brought into being without any consideration for the human eye and ear are nevertheless found to be beautiful by us. If this were not so, the natural landscape, the songs of birds, and the odour of flowers would not be so highly appreciated as they are. Surely no one is going to suggest that the geological forces that create the landscape and the genes that create the birds have any interest whatsoever in what human beings feel about them, though Bernardin de St.-Pierre came close to suggesting it. There may be a kind of moral misdemeanour in our enjoying these things, the misdemeanour of the voyeur, if one will. And when it is a case of peering at the works of the psychically ill, the misdemeanour may be the greater. But the artist who paints for his own sake, exteriorizing his emotions, both friendly and hostile, and expects others not only to observe what he has done but also to enjoy it, is acting like the child who exhibits his drawings, his gambols, his croonings, and calls them paintings, dances, and songs.

That there exist such artists is undeniable. It is not a matter of importance that they fail to use the word 'childlike' when they issue their manifestoes. The critics are ready to use it for them and to use it in a eulogistic sense. They themselves are more likely to employ synonyms. Just as children's drawings and paintings may be said to 'come from within' and not to be copies of anything 'external', so some of our modern primitivists maintain that the origin of their works of art is subjective. Kandinsky, for instance, once argued that the objective was a projection of something internal.[1] Asserting that 'art stands above nature', he also asserts

[1] The quotations from the artists which follow in my text are all taken from *The Visual Arts Today*, edited by Gyorgy Kepes, in a special issue of *Daedalus*, the Journal of the American Academy of Arts and Sciences, Winter 1960, unless otherwise noted.

that it emerges from 'internal necessity'. Natural forms, he says (p. 80), are often impediments to the expression of this internal necessity. Rejecting such deliberate constructions as were to be found in cubism, he maintains that only hidden geometrical configurations are 'the richest in possibilities', and that they 'are meant for the soul rather than the eye'. How anything can be communicated to the soul otherwise than through some sensory organ is neither explained nor even intimated to be a problem. Apparently soul can speak to soul through sensory material — colours and shapes — and though these may be fortuitous, the very absence of any apparent connection amongst them is 'proof of its inner presence'. 'Outward loosening points to an internal merging'. This paradox loses its self-contradictory meaning as soon as one recognizes that Kandinsky was talking psychodynamics rather than aesthetics. In other words the painting looks disconnected, but from the artist's point of view it is an organized whole. What organizes it is something within the artist's psyche. We, looking at it from without, must divine the purpose concealed within it. Just as a child might expect an adult to understand his imperfect speech, as we understand a cry or a shout of joy, so the artist expects us to enter into his soul through the medium of his works of art.

Juan Gris expressed a similar idea. 'The world', he said (p. 82), 'from which I draw the elements of reality is not visual but imaginative'. So, as we have seen, the child is said to draw what he imagines rather than what he sees. The relationships amongst the visual forms which he puts on his canvases are private and subsist amongst 'the elements of an imaginary reality'. And in fact Gris said that he never knew when starting a painting what would eventuate in it. Anyone with experience of children's paintings will recall how the child will extract from his scrawls and daubs a whole story, identifying in the greatest detail the elements of the narrative. That the visual aspect of these details does not correspond to anything to which the adult is accustomed is a matter of no importance. As dreams violate all the laws of

physics, so the child's drawings violate all the laws of projective geometry. They resemble in a way those visual symbols, such as anchors, bees, lilies, which stand for hope, industry, and purity without literally resembling them.

Klee, as noted above, was frank about his desire to return to childhood. But he also reverted to Herder's old simile of the tree with roots which draw their sustenance from the soil, a trunk which feeds upon the sap drawn from the roots, and the crown 'spreading in time and space for all to see' (p. 85). The artist is the trunk in this figure of speech and his work of art the crown. But though the crown has its origin in the roots, it does not resemble them. One cannot debate with a trope and there is indeed a mysterious something in the conception of all works of art, just as there is in the formation of scientific hypotheses. But Klee's target is the thesis that the spectator has no right to demand anything whatsoever of the artist. The spectator is there to confine his attention to the source of a painting in the hidden dark roots of the work of art which, one imagines, refers to the unconscious areas of the artist's psyche. To be *ganz schwunglos, fast Ursprung, wie neugeboren*, is of course an impossibility and we all have desires that are demonstrably unattainable. But whether one call oneself a baby or someone without any experience is merely a matter of words. Klee to be sure did not try to exterminate from his psyche all experience, at least in the passage that we have quoted from his *Journal*; he would have been content to exterminate merely the experience of art and of Europe. But what makes his words of interest to us is that once again we have an artist whose programme is the rejection of the civilization in which he is living and the demand for a condition of life best exemplified in childhood.

Joan Mirò takes a similar position. The essential self, he says (p. 93), is to be found 'in that mysterious area where creation takes place and from which there flows an inexplicable radiance that finally comes to be the whole man'. This mysterious area is presumably the Unconscious where everything is 'at once distinct

and blurred'. No one aware of modern psychology would deny this. Nor would anyone deny, I suspect, that the reason why Mirò's later paintings are so often disturbing is their unique oneiric quality. There is nothing childlike in this except the child's willingness to say whatever comes into his head regardless of propriety. But when in 1925 Mirò 'wiped his canvases practically clean in order to give a final blow to any surviving fetter of reasoned form',[1] and went over to the surrealist group, it was obvious that the unchecked mind of the child had been his ideal. And though he modified the surrealist programme, he was still, like a child, drawing from his imagination rather than from his observation. In this he is very similar to his younger contemporary, Jean Dubuffet, who also paints 'from within'. For to him the reality with which one is dealing 'has nothing to do with the objects represented, rather it is distinct from them, pre-existent, surging up like an electric flow, without your ever knowing where it comes from' (p. 95). So Pierre Soulages (p. 99) insists that neither the various appearances of things nor the things themselves are reality. Reality is rather the relationship of a man 'with the objective world, his sensitivity, his myths, his ideas, the social structure they clash with'. In short, what the painter is painting is suspended between his mind and the external world. The artist does not submit himself to the external world, nor does he, like Mirò, reject it, but he lives in a realm between the two. The painting, says Soulages, is a metaphor. But since metaphors are created by human beings, a painting can be only the artist's invention and since the comparison on which a metaphor is based is one apprehended by the artist and not entirely grounded in the external world, the subjective factor predominates over the objective.[2]

[1] James Johnson Sweeney, *Joan Mirò*, New York, 1941, p. 35.
[2] One finds about the same idea in the statements by Theodore Roszak (p. 109) and Marcel Duchamp (p. 112). Matisse, according to Jan Gordon, *Modern French Painters*, New York, 1925, p. 91, once said that he tried to recapture the child's vision of the world, 'that freshness of vision which is characteristic of extreme youth, when all the world is new to it'. But no source is given for this quotation.

Though we are not concerned with the truth of what artists say about the child's mind or the child's paintings, the quotations cited do reflect what they have believed to be true. No one, except in a dream, can be purely subjective and even in dreams there may be somatic and extra-somatic influences at work, giving visual form to emotions, repressions, desires which are not themselves essentially visual. The straightforward erotic dream which is recognized as such by the dreamer reproduces after all a woman who may have been seen in daily life and not a being created out of nothing. There are, to be sure, erotic dreams which are terrifying, perverse, and humiliating. And presumably a psychoanalyst who knew enough about his patient could find the sources of such dreams in the latter's childhood. But the problem is rather that of the transmutation of something non-visual, which for the sake of brevity I shall call an emotional state, into something visual. Yet even should that be solved, there would remain the problem of how such a shape might stimulate in a person seeing it the same emotion as that which gave rise to it. It has been frequently observed that many people seeing the pictures made by the artists quoted have felt a kind of bewilderment which in some cases developed into anger. Even when they have observed what they believed to be a childlike quality in the technique, as has been true of Klee or Mirò, and apparently of Matisse as well, they have not always admired it but have felt tricked. They have known that the picture in question was not made by a child but by a man pretending to be a child. When one hears a boy of eight play a Mozart piano concerto or sees a painting done by a chimpanzee, one does not judge the result as one would were the artists adult human beings. Yet the freshness, the innocence, the lack of inhibition, the simplicity, have all been noted in the works of some of our contemporaries as if these qualities gave added value to them.

Even when painters themselves do not claim the status of children, critics sometimes claim it for them. We have seen this in Sir Herbert Read's comments on Matisse and similar remarks have been made about Chagall. Chagall, says Mr. James Johnson Swee-

ney, 'always seems to give the impression of keeping a child's innocence of eye'. But since Mr. Sweeney knows more about the difficulties of making a picture than many of his fellow-critics do, he adds, 'Yet if we consider *The Cattle Dealer* closely we will realize that what Paul Valéry said of La Fontaine also applies here. "Carelessness here is expert; laxity studied; ease the height of art... Such sustained skill and innocence, to my mind, preclude any indolence or 'simplicity' " '.[1] Appreciations of the so-called primitives of our own day, who would hardly think of themselves as anything other than professional and mature artists, strike the same note. Maximilien Gauthier, for instance, speaks of Henri Rousseau's 'making more use of heart and impulse than of mind and will, attaining "naively" the objective towards which [the Cubists] were reaching intellectually'. Of Peyronnet he says, 'In his simplicity, he believes that he is reproducing with perfect accuracy the world he sees. He does not realize that he is, instead, establishing order and equilibrium in his own soul'. So Holger Cahill speaks of a modern Primitive as externalizing 'a vision of the inner world of feeling'; of Pickett as striving for 'innocence and intensity of vision'; of Edward Hicks as showing 'innocence of vision and simplicity and freshness of expression', though he grants that he also shows knowledge; of John Kane as reacting with 'the simple-hearted affection of a child to the inspiration of his western Pennsylvania hillside towns'.[2] But when it comes to the point of praising the actual paintings of children themselves, the situation is sometimes reversed. Aldous Huxley, for instance, in his introduction to a brochure on Spanish children's drawings insists that children 'when left to themselves', 'display astonishing artistic talents... How sure is their sense of colour! I remember especially one landscape of a red-roofed house among dark trees and hills that possessed in its infantile way all the power and

---

[1] James Johnson Sweeney, *Marc Chagall*, New York, 1946, Introduction to a catalogue of an exhibition held at the Museum of Modern Art, p. 26.
[2] *Masters of Popular Painting, Modern Primitives of Europe and America*, text by Holger Cahill, Maximilien Gauthier, Jean Cassou, Dorothy Miller, *et al.*, New York, 1938, pp. 21, 36, 96, 100, and 102 respectively.

certainty of a Vlaminck... Many of these pastoral landscapes and scenes of war are composed — all unwittingly of course, and by instinct — according to the most severely elegant classical principles. Voids and masses are beautifully balanced about the central axis. Houses, trees, figures are placed exactly where the rule of the Golden Section demands that they should be placed'.[1] This is followed by a passage on the child's power of psychological and dramatic expression. No critic speaks for anyone other than himself, but here Aldous Huxley praises the child not for anything childlike but for his ability to do as well as an adult and to do it in accordance with generally accepted rules. It is true that he believed fifty per cent of children to be 'little geniuses in the field of pictorial art', whereas amongst adults the percentage goes down to one in a million. But one aspect of a genius apparently is to be precociously mature, like the infant prodigy.

Now the one adjective which appears over and over again in praise of the child is, as I hope is now clear, 'innocent'. Infantile innocence was traditionally believed to be moral, not aesthetic. The child, even when he inherited the guilt of Adam, was nevertheless innocent of any sin committed by himself as someone other than his primordial father. He was in fact prevented by his undeveloped anatomy and physiology from committing two of the more serious sins, theft and adultery. Gluttony might be attributed to him and possibly envy. But early baptism would take care of his inherited guilt and as for the rest, they could be taken care of later. Just what aesthetic innocence consists in is more difficult, if indeed possible, to define. The question boils down to whether the incapability of commiting an act should be called innocence or simply incapability. If the child does not possess the skill to reproduce in visual form the objects in the external world or knows nothing of the so-called rules of painting, he may in spite of this produce an aesthetically moving and in fact a beautiful picture. This of course is admirable. But it proves

[1] Introduction to *They Still Draw Pictures*, published by the Spanish Child Welfare Association of America, New York, 1939, pp. 3ff.

nothing more than the superfluity of rules. At best in a situation of this sort, one must revert to Rousseau and declare flatly that the child is a special variety of the species *homo sapiens*, with standards of his own to be understood in his own terms. That being so, one cannot reasonably transfer principles which are relevant to his works of art to the works made by adults. Just as one should not judge a child by the standards applicable only to his elders, so there seems to be no good reason to judge adults by the standards applicable to children.[1]

It seems reasonable to believe that if children had not been confused with men supposed to be primitive, the vogue of innocence, naiveté, freshness, and kindred qualities, would not have seemed as justified as it did. The rationalization of the confusion lay in an uncritical acceptance of the Law of Recapitulation, which can be found even in psychological treatises. If we think only of the literary documents, we have to begin with Cicero's description of the child as the *speculum naturae*, then the Christian injunction to be as little children if one would enter the Kingdom of Heaven. During the Middle Ages there was a current of anti-intellectualism that was expressed from time to time in impatient strictures on learning but there was very little cultural primitivism, if only because the great mass of people were living in sufficiently primitive conditions to be unaware of the degenerative effects of luxury. Meanwhile there was certainly a wide circulation of the apocryphal stories of the Child Jesus and by the fourteenth century mystic visions of the Child, such as those of Margaretha Ebner, became common as symbolized in statuettes later on.[2] By the sixteenth century strains of deeply-seated scepticism gathered momentum and just as there were writers willing, for paradoxical or other reasons, to estimate the beasts above mankind, so there were writers who began to think of childhood as the most blessed

[1] This comes out even in a book of children's drawings written with a definite psychological orientation, e.g., Marthe Bernson, *Du Gribouillis au Dessin*, Paris, 1957, no. 3 in the *Collection 'Techniques de l'Education Artistique'*. See especially pp. 8, 25, 41, and 79.

[2] See Hans Wentzel, 'Christkind-Bilder aus alter Zeit', in *Die Kunst und das schöne Heim*, III, 1961, pp. 93ff.

time of life. The religious and the primitivistic currents flowed along side by side, though occasionally merging, and by the nineteenth century the identification of the child with primitive man was complete. When cultural primitivism was upheld in the field of art, the moment had come to praise the art of children. And in the United States, if not elsewhere, one began to find as a sort of corollary an emphasis upon the necessity of looking young, feeling young, thinking young, acting young, that in some cases was grotesque. Thus, to take but one example, it became general to deny the reality of death as well as of old age, and all the evasions of the American funeral rites seem to have resulted from this denial. The ultimate in the cult of childhood is in all probability the opinion of Dr. Norman O. Brown which has been already cited in this essay. For in his opinion life will conquer death only when we accept the excremental vision of the child as final. To write the history of an idea is not to draw up a programme of reform, and it may well be that Dr. Brown is right. But if on the other hand men have persistently over thousands of years rejected that vision and have insisted on growing up, we can only conclude that there is after all something congenial in maturity.

This essay has stopped with the Child as Artist and the Artist as Child. Other aspects of this form of cultural primitivism have not been treated. But I have had no pretension definitively to analyse the ideas involved and have been satisfied with a broad sketch. To fill in the missing parts would demand a thorough study of pedagogical theories since *Emile*, of child-sages, of the fad for miniature objects, to say nothing of collections of dolls, toy soldiers, and electric railroads, of attempts to produce prodigies of learning and artistic skill through specialized training, and a careful examination of what adults thought of the child's nature as shown in portraits and novels and plays. The history would not be complete without at least a long chapter on literature written for children with emphasis upon what the authors thought would interest their readers. But all this would be too much for any one man to accomplish even if he began his research in his youth.

APPENDIX

TRANSLATIONS OF FOREIGN-LANGUAGE PASSAGES
IN ORDER OF APPEARANCE

P. 14, third par., lines 5 and 13, Latin: "mirrors of nature"

P. 14, third par., line 4 from bottom, Latin: "as if they were completely inanimate"

P. 14, third par., last line, cont. on p. 15, first line, Latin: "the first ... feeling of attraction to ourselves, induced in us by nature, is vague and indistinct"

P. 19, second par., line 4 from bottom, Latin: "the blameless witnesses of Christ, as yet unable to talk"

P. 19, second footnote, lines 1–2, Latin: "By what new order of things/ Does what I have not yet spoken/ Come to your ears?"

P. 20, lines 2–3, Latin: "to young girls and boys"

P. 20, first par., line 3 from bottom, Latin: "So then my grammar is Christ"

P. 21, line 9, Latin: "Briccius is not my father."

P. 22, line 3, Latin: "two-way debates"

P. 23, second par., line 8, Latin: "Indeed, the seat of truth is not in the tongue but in the heart"

P. 23, second par., lines 7–8 from bottom, Latin: "natural light"

P. 23, block quotation, Latin:
> "True happiness does not consist in the knowledge of good things, but in a good life; it is not to know, but to live with understanding; nor does even a good understanding, but a good will, join men to God, and even when applied externally a man's studies do not amount to anything else than that they bring us a certain purifying state. Though they lead to happiness somewhat, yet they do not bring understanding itself – by which our happiness may be completed – unless our present life is directed into the very nature of goodness. In fact, it has very often been remarked, as Cicero says in his defense of Archias, that nature without learning is more conducive to achieving fame and excellence than learning without understanding"

P. 24, block quotation, Latin:
> "This is that true plague which destroyed each and every race of men because of one man alone, which drove out all innocence and left us subject to so many kinds of sins and to death, which snuffed out the light of faith – casting our souls into deep darkness, which condemns truth and places errors on the highest throne."

P. 25, line 10, Latin: "praise of the ass"

P. 25, block quotation, Latin:
> "Did not Christ in the mouth of his own humble asses and rude countrymen – his own apostles and disciples – overcome and strike down all the gentiles' philosophers and the Jews' lawyers, and didn't he overthrow and destroy all human wisdom giving us to drink the water of life and eternal wisdom from the jawbone of his own asses?"

P. 26, block quotation, French:

"I find that our greatest vices are established from our earliest childhood, and that our chief education is at the hands of wet nurses."

P. 27, first block quotation, French:
"If our mind does not go a livelier pace, and if from that we do not gain a sounder judgment, I would rather have my student spend his time playing tennis. At least the body would be more agile for it."

P. 27, second block quotation, French:
"Such a person sees clearly but not correctly. Consequently he perceives goodness but does not follow it, perceives knowledge and does not use it."

P. 28, block quotation, French:
"The offspring of bears and dogs show their natural inclination. But men, throwing themselves without restraint into habits, opinions, and laws, change or disguise themselves easily, so difficult is it to overcome natural tendencies."

P. 29, lines 6–7, French: "the vilest and most abject state of human nature, with the exception of death"

P. 30, lines 16–18, French: "The first sensations of children . . . are purely affective; they experience only pleasure and pain."

P. 30, last three lines, cont. on p. 31, first line, French: "All wickedness comes from weakness. The child is naughty only because he is weak. Make him strong and he will be good. He who can do anything can never do anything evil."

P. 32, first par., lines 10–12 from bottom, French: "Free of any morality in his actions he cannot do anything which would be morally wrong and which deserves punishment or reprimand."

P. 32, second par., last two lines, cont. on p. 33, French: "lively, energetic, animated, without gnawing worries, without long and painful foresight; totally in his actual being, and rejoicing in a fullness of life which seems to want to extend outside of himself."

P. 33, lines 6–8, French: "the idea of nature in decline erases all our pleasure . . . the image of death renders everything ugly."

P. 33, first par., line 8 from bottom, French: "Farewell to joy and pleasant games"

P. 33, first par., lines 3–7 from bottom, French: "Books, what a sad furniture at his age! The poor child lets himself be led, looks regretfully at everything around him, stays quiet and leaves with eyes full of the tears he does not dare to shed, and his heart swollen with the sighs he does not dare to breathe."

P. 35, block quotation, cont. on p. 36, French:
> "My son, doing good is the happiness of virtue. There is nothing more assured and greater on earth. The planning of pleasure, rest, delights, abundance, glory, is not made for man, who is weak, wandering, and transitory. See how a step toward fortune hurled all of us from abyss to abyss. You resisted, it is true But who would not have believed that the voyage of Virginie would end with her happiness and yours? The invitation of a rich and aged relative, the advice of a wise governor, the applause of a colony, the exhortations and authority of a priest, have decided Virginie's unhappy fate. So we run to our death, misled by the very prudence of those who govern us."

P. 36, line 4 after block quotation, French: "sweet, modest, confiding"

P. 36, line 5 after block quotation, French: "A man in stature, a child in simplicity"

P. 41, first block quotation, French:
>"I am, they say, an orphan
>Thrown into the arms of God from my birth,
>And I have never known my parents."

P. 41, second block quotation, French:
>"Did God ever desert his children in need?
>To the nestling birds he gives food
>And his goodness extends over all nature."

P. 41, first footnote, lines 1–5, French: "to give him the aptitude to answer the questions asked of him. . . . I believe I did not make him say anything beyond what could be said by a child of that age who has wit and memory. . . . One must take into account that this is an extraordinary child."

P. 41, first footnote, block quotation, Latin:
>"Your god is a liar and evil;
>he is stupid, blind, deaf, and dumb."

P. 42, line 2, French: "an absolutely extraordinary child"

P. 42, first par., line 4 from bottom, French: "the natural poetry"

P. 44, second par., lines 8–9, Latin: "Limbo of children [or] little ones"

P. 45, line 1, Latin: "ability not to sin"

P. 47, second par., line 5, French: "often"

P. 47, second par., first French phrase on line 7: "almost always"

P. 47, second par., second French phrase on line 7, cont. on line 8: "in the rubrics the word 'Angeli' is often replaced by 'pueri'"

P. 53, second par., last three lines, French: "Alone and naked . . .
because the child sings even when all else is quiet."

P. 53, second par., last three words, cont. on p. 54, first block
quotation, French:
"Brother, my brother,
and standing, pink in the light
Which sanctified and warmed her,
She looked at the giant of the woods, whose eye would have forced
Typhon to recoil and Briareus to flee.
Who knows what is going on in these sacred heads?
She stood erect near the narrow bed
And threatened the monster with her little finger.
There, near the cradle of silk and lace,
The great lion placed her brother in front of her,
As a mother, lowering her arms, would have done
And he told her: here he is. Do not be angry!"

P. 54, second block quotation, French:
              "For yesterday, o gentle and strange speakers,
              You talked with stars and angels.
              In you nothing is bad.
              You bring to me, whom the clouds dishearten,
              An unknown ray of unknown dawn;
              You come from there, I go there."

P. 54, third block quotation, French:
              "Beware of this little being;
              It is very great, it contains God.
              Children are, before birth,
              Lights in the blue sky.
              God gives them to us, in his generosity;
              They come; God gives them to us;
              In their laughter, he puts his wisdom.
              In their kiss his forgiveness . . .

The affliction of innocence
Is an accusation to a vicious man.
Man holds angels in his power . . ."

P. 55, first block quotation, French:
"The babble of little children is my library;
I open each word they say, as one takes a book, and I
Discover there a great and profound meaning,
Sometimes stern."

P. 55, second block quotation, cont. on p. 56, first block quotation, French:
"In a book, everywhere, above and below, frescoes
As they are to be seen on the walls of Moorish Alhambras;
Ink spots, resembling animals
Which devour the sentence and which gnaw at words,
And, the text eaten, come and bite the margins . . .
This kid, a caprice, has mounted the verses.
The recto is the book; the verso the schoolboy.
His mischievous gaiety is mingled with the stigmata
Inflicted by the avenger who wanted to flee to Sarmatia.
The daubings are strange, deep, vigorous.
The monsters! Here they are perched, one on Codrus,
Another on Nero. Another scratches a dactyl.
A blob of ink makes its nest in the branches of the stylus . . .
Everywhere the hand of dream has traced the drawing;
And this is how, at the whim of the schoolboy, the swarm
Of scribbles, hordes hostile to literature,
Have taken wing in the midst of the dark hexameters.
Game! dream! something childish, entwining itself
Around the poem, gives it an ineffable accent,
Glosses the masterpiece, and one feels the harmony
Of naiveté complementing genius . . .
Thanks to the latter, this old text is a singular place
Where chance, boredom, jeers, erasure

Present in the background their dimly seen architecture.
His ink has brought night to the starry book.
And yet, sometimes, this black and blurred tracery
Through its branches, its porches, its pilasters,
Lets the idea pass, and lets the stars show . . ."

P. 56, second block quotation, cont. on p. 57, first block quotation, French:
"But it is innocence which is full and experience which is empty.
It is innocence which wins and experience which loses . . .
It is innocence which knows and experience which does not know . . .
There, says God, see what I make of your experience."

P. 57, second block quotation, French:
". . . I do not know anything in the world as beautiful
As this child who falls asleep while praying
(As this little being who falls asleep trustingly)
And who mixes up "Our Father" with "Hail Mary,"
Nothing is as beautiful, and this is even a point
On which the Holy Virgin agrees with me.
On that point.
And I can certainly say that this is the only point upon which we agree.
For generally we disagree.
Because she is for mercy.
And I, I have to be for justice."

P. 57, second footnote, last line, Italian: "A little girl, playing, in laughter and in tears"

P. 65, first footnote, lines 2–3 from bottom, French: "[The mental state of a child] . . . in many respects is that of primitive peoples in the poetic and mythological period."

P. 73, first block quotation, German:
"Nature seen in this light is simply spontaneous being, the

existence of things through and by themselves according to the inherent and unalterable laws which govern them."

P. 73, second block quotation, German:
> "We perceive in ourselves an advantage which they lack, an advantage which they, like all unreasoning beings, cannot share except if they advance along the path we have gone, as children must do."

P. 73, first line after second block quotation, German: "childlike peoples"

P. 73, lines 2–3 after second block quotation, German: "not, indeed, of fulfillment, but rather of renunciation"

P. 73, line 2 from bottom, German: "Modesty, Perceptiveness"

P. 74, lines 1–2, German: "Naiveté must be the characteristic of every genius, for there can be no genius who lacks it."

P. 77, third par., lines 3–4, German: "vision of life"

P. 77, third par., lines 5–6, German: "a blind wisdom, that fearlessly follows a beloved leader"

P. 77, third par., lines 2–3 up from block quotation, German: "of which the best feature is joyous confidence: childhood"

P. 77, block quotation, German:
> "Childhood is the realm of supreme Justice and profound Love. In the hands of a child one thing is never more important than another. The child may be playing with a golden brooch or a wild flower: if he grows tired he will drop and forget either without paying the least attention, though either of them had seemed equally splendid in the light of joy. He

quite lacks the fear of loss. To him the world is still the beautiful shell where nothing can ever get lost. Everything he has
ever seen, sensed or heard, everything he has ever encountered
is sacred to him. He does not compel things to settle within his
reach. A host of dusky nomads pass through his holy hands as
through a triumphal arch; his love lights them up for a while
and then they fade away, yet pass through his love they must.
For whatever had once been lit up by his love remains in it
as an image and can never be lost again. The image is possession, and this is the reason why children are so rich."

P. 78, block quotation, German:
"Either the fullness of images remains untouched after the in-
rush of new knowledge, or the old love is obliterated like a dy-
ing city under a rain of ashes from these unexpected eruptions.
Either the new becomes a dike that protects a piece of childhood, or it is turned into the flood that ruthlessly annihilates it.
In the first instance, the child may become older and more sensible in the everyday sense of the word, in which case he
becomes a budding citizen who will join the Order of his historical epoch and be ordained as a member. Or, again, he may
ripen quietly and simply from the depths of his being, nour-
ished by his own existence as a child, in which case he will
belong to the spirit of all epochs – he will be an artist."

P. 79, lines 1–3, German: "live in this warm earth, in the never-
disturbed silence of dark unfoldings, which knows nothing of the
measure of time."

P. 80, second par., lines 8–9, French: "totally deprived of artistic
sense"

P. 80, second par., lines 10–14 from bottom, French: "awkward and
badly drawn as they are, they reflect vividly, besides their imitative intention, the intention of thought – so much so that the latter is

always there but, because of the artist's incapacity in execution, in a way that is infinitely more marked and effective than the former."

P. 80, second par., last 7 lines, French: "these beings are at the same time cruel, harsh, superior, rough divinities, but they are still divinities, in whom greatness is intimated and beauty foreshadowed; insofar as they are signs of a concept, these signs already possess the clarity and the vigor of meaning at the same time. They live, they speak, they proclaim that a creative thought has infused itself there and is to be manifested through them."

P. 81, lines 2–4, French: "the token of an intention, of an imaginativeness or, to speak more precisely, of a beauty elemental, beautiful, raw and rude, but finally absolutely and exclusively having its origin in thought."

P. 81, second par., lines 13–14, Italian: "of little aesthetic value"

P. 81, second par., lines 2–4 from bottom, Italian: "In short, they describe in signs what they would by no means describe in words"

P. 82, second par., lines 9–10, Italian: "are immeasurably superior to the designs of children"

P. 82, second par., lines 9–10 from bottom, Italian: "the errors of a decadent and moribund art"

P. 82, second par., line 8 from bottom, Italian: "of a new-born art"

P. 82, second par., first Italian usage on line 2 from bottom: "Perhaps beribboned mayors of their home towns"

P. 82, second par., last two lines, Italian: "glorious in any struggle between soul and body"

P. 86, block quotation, cont. on p. 87, German:
> "One thing is certain. We must not confront the child with our highest achievements. On the contrary we must guide him up the same ladder which we and our forefathers have climbed with difficulty. Reflected in the development of the child is the development of the race, and whoever doubts this will, we hope, be convinced by the following pages. They are written in order to prove that our children travel over the same trail that our ancestors blazed, and on which the primitive races still move today. Once we have discovered this evolutionary process, it becomes relatively easy to fashion the education of our young accordingly."

P. 87, line 5 after block quotation, German: "naked but adorned"

P. 87, line 9 after block quotation, German: "purely aesthetic feeling"

P. 87, first German usage on line 10: "image-painting"

P. 87, second German usage on line 10: "image-writing"

P. 88, block quotation, German:
> "The founder of that world-religion [Christianity] never tired of explaining the mystery of his divinity as grounded on his relationship to the Father, as derived from him and yet quite divorced from him in a superhuman state of childhood. It was this awareness that we are all god's children which he demanded of his disciples as their supreme law. This state of childhood represented for him not merely a phase of the past but also the apocalyptic goal of the individual and of mankind."

P. 88, lines 5–6 after block quotation, German: "perfected blessed condition"

P. 88, lines 6–7 after block quotation, German: "the redeeming event of the Passion"

P. 88, lines 6–7 from bottom, German: "so-called primitive or un-civilized people"

P. 88, last line, German: "soulful/religious in the highest degree"

P. 89, line 3, German: "perhaps until the 21st year"

P. 89, lines 4–8, German: "from the earliest beginnings through the old stone age until the bronze and iron ages, from the stage of the 'hunter and fisher' through that of the 'planter and cattle-herder' until the stage of conscious, civilized humans."

P. 89, block quotation, German:
> "It is a great affliction and a great necessity that we must begin with the smallest. I want to be like a newborn, know nothing of Europe, absolutely nothing. Know no poets, be entirely passive, almost primeval."

P. 96, second par., lines 11–12 from bottom, German: "passive, almost primeval, like the newly born"

German and French translations by Stephen C. Simmer, Latin by Eric Purchase, Italian by Jay Livernois. All translations edited by J. B. Trapp and Professor E. H. Gombrich of The Warburg Institute.

INDEX

Aaron 22

Abraham 22

Adam 11, 18f., 24, 29, 35, 36, 42ff., 76, 100

Adeodatus 41n.

Agrippa, Henricus Cornelius 10, 22, 23ff., 76

Alcott, Louisa M. 60

Aldrich, Thomas Bailey 60

Alexamenos 25n.

Andersen, Hans 59

Andrea dal Castagno 17n.

Anselm, St. 20n.

Anthony of Padua, St. 21

Apollo 13, 41n.

Aristotle 12f., 20, 30

Aristotle, ps.- 13

Arnold, Thomas 29n.

Ascanius 13

Ashton-Warner, Sylvia 90

Astyanax 13

Augustine of Hippo, St. 21, 23, 44f., 60f.

Ausonius 19

Baïf, J. A. de 22

Balaam 25

Baldovinetti, A. 17n.

Balzac, H. de 59

Barop, J. A. 40

Barr, Alfred H. 89f.

Baudelaire, C. 78n.

Baudry of Bourgueil 20

Bayle, P. 22

Bellegambe, Jean 46

Bergson, H. 40n.

Bernardin de Saint-Pierre, J. H. 29, 34ff., 94

Bernson, Marthe 101n.

Bérulle, Pierre de 29

Birgitta, St. 17

Blake, W. 49, 77

Bland, Jane 93

Boas, Belle 93

Boccaccio, G. 63n.

Borgognone, Il 46

Bramante 17n.

Brice, St. 21

Brown, Norman O. 70f., 76, 102

Brués, Guy de 22

Brunelleschi, F. 17n.

Burrow, Trigant 68n.

Cahill, Holger 99

Calvinism 44f.

Catherine of Siena, St. 17

Cebes 12

Chagall, Marc 98

Champfleury (J. F. F. Husson) 78n.

Charron, P. 22

Chatterton, T. 51

Chaucer, G. 46, 50

Cherubs 46ff.

Christopher, St. 21

Cicero 14f., 17, 23, 40, 101

Čižek, Franz 93

Clay, Jean 91

Clement of Alexandria 18f.

Coe, George A. 65n.

Coleridge, H. 51

Comte, Auguste 64f., 68n.

IN THE SPRING FAMILY . . .

INSCAPES OF THE CHILD'S WORLD   John Allan
The fruit of over twenty years of clinical work with children – "normal" ones as well as those abused, neglected, and terminally ill. The book describes different ways to use art: drawing, guided imagery, and active imagination. Rooted in Jung's theory of the regenerative ability of psyche, Allan's approach is pragmatic and sensitive. The drawings, paintings, and writings by children open a profound dimension of their suffering and strength. (235 pp.)

ROSEGARDEN AND LABYRINTH   Seonaid M. Robertson
With care and precision and aided by years of experience, Seonaid Robertson explores the relationship between art and psyche. Focusing on the drawings of children and adolescents, she views these first products of the imagination against the background of artistic and cultural history. Illustrations, index. (216 pp.)

BROODMALES   Nor Hall, Warren R. Dawson
In the folk customs of couvade – "brooding" or "hatching" – a man takes on the events of a woman's body – pregnancy, labor, nursing – so that her experience becomes his. Introducing Dawson's 1929 "Custom of Couvade," Hall supplies biographical data about social anthropologists at the turn of the century and then looks inward at the physical ground of the customs they labeled bizarre. Dawson's ethnological classic collects material from accounts of explorers and informants from the 1700s to the 1900s. (173 pp.)

FATHERS AND MOTHERS   Jung, Neumann, Hillman et al.
A psychology book *imagined* by writers, this second, revised edition of a Spring classic addresses its topic archetypally, personally, and in concrete clinical detail reaching toward the poetic and the mythic. Robert Bly, James Hillman, and Augusto Vitale explore fathers and sons wrestling through the agonies of that relationship, resulting in a sober vision of the dark father. The imagination of mother is treated with novelty and beauty by Marion Woodman (changes in the mother image during analytical work), Patricia Berry (the depth of matter in mother), Mary Watkins (the mothering of desire in actual child and inner child), Jackie Schectman (the step-mother), Ursula K. Le Guin (woman between Persephone and mother), and James Hillman (differentiating the puer, hero, and Great Mother's son). Foundational essays by Erich Neumann (on moon consciousness) and C. G. Jung (the original 1938 version of "The Psychological Aspects of the Mother Archetype") conclude this rich and useful volume.

SPRING PUBLICATIONS, INC.   P.O. BOX 222069   DALLAS, TEXAS   75222